Probability at Saint-Flour

For further volumes:
http://www.springer.com/series/10212

Saint-Flour Probability Summer School

Founded in 1971, the Saint-Flour Probability Summer School is organised every year by the mathematics department of the Université Blaise Pascal at Clermont-Ferrand, France, and held in the pleasant surroundings of an 18th century seminary building in the city of Saint-Flour, located in the French Massif Central, at an altitude of 900 m.

It attracts a mixed audience of up to 70 PhD students, instructors and researchers interested in probability theory, statistics, and their applications, and lasts 2 weeks. Each summer it provides, in three high-level courses presented by international specialists, a comprehensive study of some subfields in probability theory or statistics. The participants thus have the opportunity to interact with these specialists and also to present their own research work in short lectures.

The lecture courses are written up by their authors for publication in the LNM series.

The Saint-Flour Probability Summer School is supported by:

- Université Blaise Pascal
- Centre National de la Recherche Scientifique (C.N.R.S.)
- Ministère délégué à l'Enseignement supérieur et à la Recherche

For more information, see back pages of the book and
http://math.univ-bpclermont.fr/stflour/

Jean Picard
Summer School Chairman
Laboratoire de Mathématiques
Université Blaise Pascal
63177 Aubière Cedex
France

Sergio Albeverio • Hans Föllmer • Leonard Gross
Edward Nelson

Mathematical Physics at Saint-Flour

Springer

Sergio Albeverio
Institute of Applied Mathematics
University of Bonn
Bonn, Germany

Hans Föllmer
Institut für Mathematik
Humboldt University
Berlin, Germany

Leonard Gross
Department of Mathematics
Cornell University
Ithaca, NY, USA

Edward Nelson
Department of Mathematics
Princeton University
Princeton, NJ, USA

Reprint of lectures originally published in the Lecture Notes in Mathematics volumes 929 (1982), 1362 (1988) and 1816 (2003).

ISBN 978-3-642-25955-5
Springer Heidelberg Dordrecht London New York

Library of Congress Control Number: 2011945778

Mathematics Subject Classification (2010): 60J45; 60K35; 60G60; 81P2; 82B05; 60J99; 60F10

Printed on acid-free paper

Springer is part of Springer Science+Business Media (www.springer.com)

Preface

The *École d'Été de Saint-Flour*, founded in 1971 is organised every year by the *Laboratoire de Mathématiques* of the *Université Blaise Pascal* (Clermont-Ferrand II) and the *CNRS*. It is intended for PhD students, teachers and researchers who are interested in probability theory, statistics, and in applications of stochastic techniques. The summer school has been so successful in its 40 years of existence that it has long since become one of the institutions of probability as a field of scholarship.

The school has always had three main simultaneous goals:
1. to provide, in three high-level courses, a comprehensive study of 3 fields of probability theory or statistics;
2. to facilitate exchange and interaction between junior and senior participants;
3. to enable the participants to explain their own work in lectures.

The lecturers and topics of each year are chosen by the Scientific Board of the school. Further information may be found at http://math.univ-bpclermont.fr/stflour/

The published courses of Saint-Flour have, since the school's beginnings, been published in the *Lecture Notes in Mathematics* series, originally and for many years in a single annual volume, collecting 3 courses. More recently, as lecturers chose to write up their courses at greater length, they were published as individual, single-author volumes. See www.springer.com/series/7098. These books have become standard references in many subjects and are cited frequently in the literature.
As probability and statistics evolve over time, and as generations of mathematicians succeed each other, some important subtopics have been revisited more than once at Saint-Flour, at intervals of 10 years or so .

On the occasion of the 40th anniversary of the *École d'Été de Saint-Flour,* a small ad hoc committee was formed to create selections of some courses on related topics from different decades of the school's existence that would seem interesting viewed and read together. As a result Springer is releasing a number of such theme volumes under the collective name "Probability at Saint-Flour".

Jean Bertoin, Erwin Bolthausen and K. David Elworthy

Jean Picard, Pierre Bernard, Paul-Louis Hennequin
(current and past Directors of the *École d'Été de Saint-Flour*)

September 2011

Table of Contents

THERMODYNAMICS, STATISTICAL MECHANICS AND RANDOM FIELDS

PAR LEONARD GROSS

Originally published in: *Ecole d'Eté de Probabilités de Saint-Flour X – 1980*, Lecture Notes in Mathematics, Vol. **929**, 101–204, DOI: 10.1007/BFb0095619, © Springer-Verlag Berlin Heidelberg 1982, Reprint by Springer-Verlag Berlin Heidelberg 2012

THERMODYNAMICS, STATISTICAL MECHANICS, AND RANDOM FIELDS

Leonard Gross
Department of Mathematics
Cornell University
Ithaca, New York 14853/USA

1. Introduction. Our objective is to give an introduction to random fields with emphasis on those aspects which are pertinent to statistical mechanics. Indeed, statistical mechanics has been a rich source of problems in random fields. We give therefore an introduction to statistical mechanics also. Since these notes are intended for mathematicians we try to profit from the mathematical sophistication of the reader in order to make the treatment of the physics succinct. Nevertheless, even in those sections which are superficially self contained, these notes are not intended as a substitute for a text on statistical mechanics.

The immediate connection of statistical mechanics with everyday experience is via thermodynamics, whose laws statistical mechanics attempts to explain. In order to understand this goal of statistical mechanics we give also an introduction to the foundations of equilibrium thermodynamics.

There are numerous expositions of thermodynamics and statistical mechanics as topics in physics. There are also several axiomatic treatments of portions of these subjects as topics in mathematics. It is hoped that these notes will fill what I perceive as a gap between these two types of exposition. We shall attempt to show how the physics of heat leads, via partly physical and partly mathematical argument, to the formal mathematical structures of thermodynamics, statistical mechanics and random fields. Because these notes attempt to survey a very broad field they have of necessity a certain superficiality. I believe that a reader who wishes to learn the subject, rather than merely what the subject is about, will have to read several other more specialized works in conjunction with these notes. I shall rely on some of the existing mathematical expositions extensively.

In the section on thermodynamics our objective is to describe the arguments which lead from our physical experience with heat to its mathematical formulation via a single convex function on R^n. Of necessity, much of this type of exposition proceeds by example rather than by mathematical definitions, theorems, proofs, etc. Thus in classical mechanics the initial mathematical structure associated to a mechanical system is its configuration space--a manifold whose points are in one to one correspondence with the possible "configurations" of the system. But what is a "configuration"? This can be explained only by giving examples: the configuration space of a free particle moving in space is R^3. The configuration space of a particle attached by a rigid rod of length L to a fixed point is a sphere of radius L. The configuration space of a rigid body with one point fixed in space is the rotation group $SO(3)$. (Fix an orthonormal frame in space and an orthonormal frame in the body. The orientation of the body is then determined by the rotation which takes one frame to the other.) The configuration space of a freely moving rigid body is $R^3 \times SO(3)$. Thus the configuration space of a classical mechanical system is some differentiable manifold, $\mathbf{C}$, which captures some aspects of the physical system.

Recall that in Newtonian mechanics the instantaneous state of a mechanical system is determined by the configuration of the system together with a "velocity", i.e., a tangent vector to $\mathbf{C}$. That is, the state is a point of the tangent bundle, $T_*(\mathbf{C})$. The basic problem of Newtonian mechanics is this. Given the state of a mechanical system, and all the forces acting on the system, find the state for all future times. As is well known, the forces may be specified mathematically by giving a possibly time dependent vector field on $T_*(\mathbf{C})$, and the Newtonian evolution is then the flow determined by this vector field. Thus the process of converting a classical mechanical problem to a mathematical one and then solving it may be summarized as follows.

A. (Mathematical description of system.) Find the configuration space, $\mathbf{C}$ (some differentiable manifold) corresponding to the mechanical system.

B. (Mathematical description of forces.) Specify a (perhaps time dependent) vector field on the tangent bundle $T_*(\mathbf{C})$ corresponding to the forces acting on the system.

C. (Time evolution.) Solve the (first order) differential equation (Newtons equations) determined by the vector field in part B, with the desired initial condition in $T_*(\mathbf{C})$.

What are the analogous problems, mathematical structures and solution methods for equilibrium thermodynamics?

In Chapter I we shall give an account of Gibb's form of the answer to this question. The basic mathematical object is a convex function (the negative of the entropy) on a convex cone in R^n. (See Section 9: The ABC of equilibrium thermodynamics). In Section 11 we describe the basic facts about the Legendre transform--a useful tool in studying convex functions and in thermodynamics. In Section 12 we describe its physical significance. Although this chapter is more self contained than the next two, I think it very useful to read in conjunction with it the carefully written physics books [C] and [Z1].

In Chapter II we sketch how classical statistical mechanics yields the entropy function of an equilibrium thermodynamic system from a knowledge of the forces between molecules. We rely heavily on the excellent exposition of A. Martin-Löf [ML].

In Chapter III we outline some of the general theory of random fields associated with the (classical) statistical mechanics of crystals. In this part we rely for many details on the recent books by D.Ruelle [R2] and R. Israel [Is 2]. We conclude with the uniqueness theorem of Dobrushin.

It is a pleasure to acknowledge many illuminating discussions with Michael Fisher, Giovanni Gallavotti and Benjamin Widom.

Chapter I: Equilibrium Thermodynamics

2. Thermometers, Calorimeters and a Bit of History.

The earliest attempt to make the intuitive notions of "hotter" or "colder" quantitative was probably that of Galileo. In about 1600 he invented a device intended to assign a number to the degree of hotness of a body--what we would refer to now as temperature. Based on the fact that air expands when heated, the device failed to distinguish between expansion due to temperature change and expansion due to change in the surrounding atmospheric pressure. But the 17th century and early 18th century saw the development of this device into a scientifically reliable instrument. A thermometer based on expansion of water (and later alcohol) in a thin capillary tube was invented in 1631. The change in volume could be read off from a linear scale marked along the capillary. Typically, two points on the scale were chosen to coincide with some describable temperature. (E.g. that of the "coldest winter day" or "hottest summer day" were used in the 17th century. Later the temperature of melting ice or boiling water at one atmosphere pressure was adopted, yielding a more reproducible temperature scale. In between the two fixed points the thermometer maker used a linear scale with the number of division points up to him. In 1732 Fahrenheit, a renowned instrument maker, introduced the mercury thermometer. For a given liquid the various temperature scales are then affinely related ($x \to ax + b$), and, for the precision of the seventeenth and eighteenth centuries, different liquids gave rise to temperature scales which were (approximately) affinely related in a reasonable temperature range.

Underlying the meaningfulness and usefulness of the concept of temperature are the following facts. Consider two substances, say liquids, separated by a thin metal wall. If one is hotter than the other then the hotter one will get cooler and the cooler one will get hotter till no further change takes place. They are then said to be in thermal equilibrium. Thus two substances are in thermal equilibrium if, when brought into contact with only a thin wall separating them, no changes (in their temperatures) occur. Now if substance A is in thermal equilibrium with substance B and substance B is in thermal equilibrium with substance C (as determined by bringing B subsequently into contact with C) then A and C are in thermal equilibrium, as experience shows. A thermometer works by bringing it into contact with the substance to be measured and waiting for them to come to thermal equilibrium. One reads the temperature from the thermometer after thermal equilibrium is reached. Thus it follows from the preceding transitive law that two substances which are at the same temperature are in thermal equilibrium while two substances which are in thermal equilibrium are at the same temperature. In an axiomatic development of thermodynamics one could therefore postulate the existence of an equivalence relation called "thermal equilibrium" on the set of systems of interest and define temperature as an element of the quotient space. The postulate of the existence of such an equivalence relation is sometimes called the "zeroth law of thermodynamics" (Vintage 1927, R. H. Fowler). We shall follow a more concrete, instrument oriented approach.

By 1723 another type of instrument, allowing another type of measurement, began to occupy center stage in the story of thermodynamics. Imagine a closed container full of hot water immersed in a bath of cold water. We know from experience that the hot water will cool and the cold water temperature will rise. But the rate at which this occurs depends on the material of which the container is made. If it is made of thin metal or glass the temperature changes will be quick. If it is made of styrofoam the temperature change will be very slow. Better yet, if the walls of the container are made of two separated layers of glass, if the region between the two layers of glass is evacuated, and if the two glass surfaces adjacent to the evacuated region are covered with a mirror-like layer of silver, (the reader may recognize this description of a Dewar flask) then the temperature of the hot water will change extremely slowly. Ideally, one can conceive of a container which prevents any change of temperature at all. The walls of such an ideal container are called adiabatic. A container whose walls are approximately adiabatic, when used for experiments, is called a calorimeter. It is a problem for the experimenter to devise a container which approximates an ideal calorimeter with sufficient precision for the purpose at hand.

Here is a typical experiment using a calorimeter and a thermometer. The first experiments of this sort were apparently done by Brook Taylor about 1723. We shall describe the experimental facts as they were known to Joseph Black in 1760. Black, a physician and chemist, whose students included James Watt, developer of the steam engine, advanced the art of these experiments considerably and provided an "explanation" for the results.

Suppose n grams of water at t° centigrade are mixed with m grams of water at s° centigrade in a calorimeter. The resulting mixture will, after a little stirring, come to equilibrium at a temperature of $(nt + ms)/(n + m)$ degrees centigrade. The same formula holds if one mixes mercury with mercury. On the other hand, if one mixes n grams of water at t° centigrade with m grams of mercury at s° centigrade it will be found, after a while, that the mercury (which falls to the bottom and has the same temperature as the water) is at a temperature of $T = (nt + cms)/(n + cm)$ where $c = .033$.

Here is a theory which "explains" these facts and which was widely accepted for nearly a century. Let us suppose that every substance contains an invisible, indestructible, fluid, which we shall call caloric. Suppose further that the more caloric the substance contains, the higher is the temperature of the substance. Assume also that the fluid is capable of flowing from one body to another if the two bodies are in physical contact, but that such flow always takes place from the hotter body to the cooler one. Thus if one adds some caloric to some water the temperature will go up. Now to make the theory quantitative define one unit of caloric to be the amount which must be added to one gram of water to raise its temperature one degree centigrade. We shall call this unit a calorie. Of course caloric distributes itself uniformly throughout a homogeneous substance, so that, for example, addition of 6 calories of caloric to 3 grams of water will raise the temperature by 2° centigrade.

Now we are ready to derive the above mixture formulas. We may suppose $t > s$. When the two water samples are mixed (it suffices actually to bring their containers into contact with only a thin metal wall separating the two water samples) caloric flows from the hotter one to the cooler one, thereby lowering the temperature of the hotter water and raising the temperature of the cooler water until the temperatures are equal. Assume further that no caloric can flow through adiabatic walls. If the final temperature is T° centigrade then the n grams of hot water must have given up $n(t - T)$ calories of caloric and the cooler water must have absorbed $m(T - s)$ calories. Since caloric is indestructible we must have $n(t - T) = m(T - s)$. Solving for T gives $T = (nt + ms)/(n + m)$, in agreement with experiment.

Now Black proposed that a substance different from water, say mercury, requires not one calorie of caloric to raise the temperature of one gram one degree centigrade but, say, c calories. In this case n grams of water at t° centigrade will give up $n(t - T)$ calories in coming to equilibrium with m grams of mercury at temperature s (assume $t > s$), while the mercury absorbs $cm(T - s)$ calories in its temperature rise. Once again the indestructibility hypothesis yields $n(t - T) = cm(T - s)$, yielding an equilibrium temperature of

$T = (nt + cms)/(n + cm)$, in agreement with experiment if one takes for c (known as the specific heat of mercury) the value $c = .033$.

Black discovered also that a lot of caloric must be added to ice at 0° centigrade to convert it to water at 0° centigrade. This fact requires a modification of the assumption that adding caloric raises temperature. But Black also showed experimentally the important fact that upon refreezing the same amount of caloric is recovered. Thus there continues to be no creation or destruction of caloric but merely a storing of it in a latent form (called latent heat of fusion). It should be mentioned that the caloric theory was also successful, at least qualitatively, in explaining other phenomena, such as the decrease of pressure of a gas upon expansion. But the only feature of the theory which persists now is the very important idea that in calorimeter experiments some kind of "heat quantity" is conserved. In the period from 1760 to 1840 the conserved quantity was widely believed to be the indestructible fluid caloric. In addition to Black it had such influential supporters as Lavoisier. Although the idea of an uncreatable and indestructible fluid was frequently under attack, most notably by Count Rumford (circa 1800) who claimed he could produce caloric by friction, (as in boring a cannon, which was part of his job) nevertheless clever rebuttals to his evidence and arguments were adduced by the theory's defenders. Throughout history wrong explanations of natural phenomena have maintained adherents simply because there was no available answer to the question "What else can it be?" In the period 1840 to 1850 an answer to this question was forthcoming.

Robert Joule began a series of experiments in 1840 in which he converted mechanical, electrical and chemical forms of energy into each other and into heat. The simplest to understand of these experiments is the famous paddlewheel experiment. Imagine a paddlewheel immersed in water contained in a calorimeter. A weight is attached to the axle of the paddlewheel by a rope in such a way that the falling weight causes the paddlewheel to turn. The turning of the paddlewheel is opposed by the resistance of the water. The water temperature increases. When the weight comes to rest at the bottom of its range it has done some work, W (= weight times distance). The final water temperature can be measured. The water and paddlewheel appear to have gained Q calories of caloric, where Q = temperature increment × [mass of water + (specific heat of paddlewheel) × (mass of paddlewheel)]. The ratio $J = W/Q$ proves to be independent of the many parameters of the experiment.

Joules interpretation of the experiment was this: the work done by the falling weight was converted by friction into internal energy of the water. Thus energy is conserved. The mechanical energy of the raised weight has merely been converted into internal energy of the water. It is the increased internal energy of the water which accounts for its higher temperature, not the presence of more of the invisible fluid caloric.

The idea that heat is a form of energy (e.g. the kinetic energy of the particles making up the substance) had been around for a long time. Already in the early part of the seventeenth century Francis Bacon had proposed that "heat" was a form of energy, as had Leibniz and Newton later. But a quantitative theory was apparently too difficult to come by in the seventeenth century and conclusive experimental evidence was not available till the very experiments of Joule that we are describing. Joule believed that transfer of "caloric" from one body to another was really a transfer of internal energy--energy which resided perhaps in the kinetic energy of motion of the molecules of which the substances were made.

Thus the highly successful caloric theory--based on conservation of caloric--could be replaced by conservation of energy, thereby enlarging the caloric conservation laws to include processes involving work, as well as the calorimetric experiments of Black described previously.

Converting heat (i.e. internal energy) back into work, however, is another matter. Carnot pointed out in 1824 that cyclical engines do not simply absorb some heat from a source at temperature T, say, and convert it into work, but in fact must reject some of absorbed heat to a heat reservoir at a lower temperature. There is, therefore, some constraint imposed by real processes on the interconvertibility of internal energy and mechanical energy. Only after Clausius and Kelvin showed, in 1850, the comensurability of conservation of energy with these constraints on their interconversion did classical thermodynamics take its present form--to which we now turn. For further discussion of this period see [Ro] or [T].

3. Equilibrium States

We are concerned only with systems in thermodynamic equilibrium and their transformations. The relevant physical notion of thermodynamic equilibrium can only be understood by examples. Consider a liquid in a closed container. If the fluid is not rotating or translating, does not appear to be moving in any other way, if there is no net evaporation (which in fact implies that the vapor in the space, if any, above the liquid is not changing its density), if the temperature of the liquid is constant in time (and therefore the same throughout the liquid) and if the volume of the container is not changing with time, then the system is probably in thermodynamic equilibrium. If any of these conditions fail then the system is not in thermodynamic equilibrium. A system is in thermodynamic equilibrium if neither its mechanical, chemical or thermal attributes are changing with time. By mechanical attributes we exclude consideration of random molecular motion and include only macroscopic properties. There should be no macroscopic motion. Often a good deal of judgement is involved in determining whether a system is an equilibrium system. A salt crystal sitting on a table is in equilibrium. But a piece of glass sitting on a table is not in equilibrium because it is actually in the process of flowing--very slowly. We are interested only in equilibrium thermodynamic systems. These are systems in which temperature plays a significant role. Thus two massive particles suspended by a rope that goes over a pulley will be in equilibrium if they are balanced, but typical classical mechanical questions about such a system do not involve temperature.

For the most part we shall focus on the simplest thermodynamic system--a fixed mass of gas contained in a cylinder fitted with a piston which allows its volume, V, to be controlled.

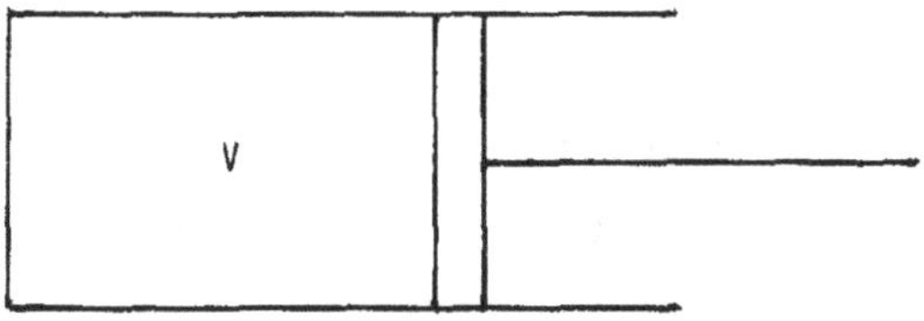

Figure 3.1

4. The First Law of Thermodynamics

Consider two equilibrium states i and f of an adiabatically enclosed thermodynamic system. Suppose that it is possible to change the state of the system from i to f by doing work on the system. For example if the system is a gas in an adiabatically enclosed cylinder fitted with a piston and if the piston is allowed to be pushed out by the gas then, since the force on the piston is pressure x area (= PA), the work done by the gas is $\int_{x_0}^{x_1} PA\, dx = \int_{V_0}^{V_1} P\, dV$, and therefore the work done on the gas is $-\int_{V_0}^{V_1} P\, dV$. Suppose moreover that a small resistor is located in the interior of the cylinder and a current is supplied by a generator which is operated by a falling weight. As the weight falls the work it does is dissipated as heat inside the adiabatically enclosed gas and the gas heats up. By alternating compressions, expansions and resistance heatings one can imagine many ways to change the state from an initial state i to a final state f, as indeed there are. Joules' experiments suggested the following general law.

First Law: If it is possible to change the state of an adiabatically enclosed system from an equilibrium state i to an equilibrium state f by doing work on the system then the work done is the same for all such adiabatic processes.

Note that we have allowed the possibility that there may be no way to change from state i to state f by doing work alone. But Joules' experiments also suggested that, given two states A and B of a thermodynamic system of fixed mass, it is always possible by doing work on the system to change it either from state A to state B or from state B to state A (sometimes both directions).

It follows from the first law together with the preceding accessibility assumption that there exists a function U of the state of a system of fixed mass such that

$$-W_{i\to f} = U_f - U_i \tag{4.1}$$

where $-W_{i\to f}$ is the work done on the system in bringing it from state i to state f. (The minus sign is to adhere to conventions.) The function U is determined only up to an additive constant (for a system of fixed mass). U is called the internal energy function of the system.

Now if the system is not adiabatically enclosed then a process which changes the system from i to f may yield a non zero value for $U_f - U_i - (-W_{i\to f})$ because heat energy has been absorbed from the surroundings. We define

$$Q = U_f - U_i + W_{i\to f} \tag{4.2}$$

to be the heat energy transferred in the process to the system from its surroundings. The sign conventions used here are the traditional ones. Q represents heat transferred to to the system, while $W_{i\to f}$ represents work done by the system.

5. The State Space

So far we have relied on the reader's intuitive understanding of the meaning of the phrase "equilibrium state". At some point in the mathematical development it is necessary to identify the set of equilibrium states of a thermodynamic system with some well defined set (e.g. some topological space). Just as in the definition of classical mechanical configuration space discussed in the introduction we shall look at some examples.

According to the classical gas laws the pressure, volume and absolute temperature, (which will be defined later) are related by $PV = NRT$ when N is the number of moles of gas and R is a constant. (This actually holds for real gases only approximately.) Recall that one mole of a chemically pure compound consists of m grams of the substance, where m is its molecular weight. A relation such as $PV = NRT$ is called an equation of state. Our intuition about a gas in a cylinder suggests that everything we might want to know about it thermodynamically is determined by its pressure, volume, temperature and N. But because of the equation of state only three of these parameters suffice to determine the state. Thus the state space for a gas is a three dimensional manifold. The internal energy function is then a function on this manifold. For example if we choose P, V, and N as state coordinates then for fixed N a state is given by a point in the PV plane. However real gases condense, and we wish to include the condensed systems in the mathematical formalism. Thus while P, V, N are good coordinates for water vapor the following well known difficulty occurs for water near 4° C.

At 4° centigrade water passes through a state of maximum density at atmospheric pressure. Thus there is a temperature T_1 slightly below $4^\circ C$ and a temperature T_2 slightly above $4^\circ C$ at which $P_1 = P_2 = 1$ atmosphere while $V_1 = V_2$ and $N_1 = N_2$. Hence P, V, N are not valid global coordinates on the thermodynamic state space for the gas-liquid-solid system known as H_2O because there are two distinct states with the same P, V, and N. They cannot therefore be used to define the state space mathematically. T, P and N are also not valid global coordinates. For example two ice-water mixtures at $0^\circ C$ (e.g. 25% ice in one case, 50% ice in the other) have the same P, T and N. A similar difficulty arises with T, V, N.

In Gibbs' first paper (1873) he proposed using U, V, N for global coordinates. This is, in fact, free of the previous difficulties, and, as he showed, allows all of the thermodynamic information of the system to be expressed in a single function of U, V, N -- the entropy, to which we shall soon turn. We define then, the state space of a system consisting of a chemically pure gas together with its various phases as the convex cone in R^3 determined by $V > 0$, $N > 0$, $U > \alpha N$ where α is a constant--the lower bound of the energy of one mole of the substance. So $-\infty \leq \alpha < \infty$.

Similarly, for a system consisting of r chemically pure non interacting substances the state space is parametrized by the total energy U, the total volume V and the numbers $N_1, \ldots, N_r$ of moles of each compound. Thus the state space for this system is a convex cone in R^{r+2}.

This discussion of the state space is incomplete, however, because the measurement process for internal energy (e.g. by using paddlewheels) does not give a meaningful comparison between internal energies of different quantities of a substance. A useful notion of internal energy requires the intuitive concept of extensivity.

<u>Extensivity</u>: Consider two moles of H_2O. Suppose that we partition it by a wall with one mole on the left and one mole on the right. If the total internal energy of the two moles is U, is the internal energy of the one mole on the left $U/2$? Not necessarily. First of all the arbitrary additive constants that appear in the definition of U for one mole and for two moles may not have been chosen commensurably. Secondly, there may have been an ice cube in the left half and none in the right half inspite of the fact that the whole was in equilibrium. Thirdly the two separated moles may have a larger surface area than the two combined moles and surface energy effects may contribute to a lack of additivity of the energy.

We shall consider only so called "bulk thermodynamics" in which one allows only reasonably shaped walls, and substances for which the shape of the container has "negligible" effect on the total energy. Accordingly, if each of the two moles of H_2O has the same percentage of ice as the other, the same percentage of water, and the same percentage of water vapor, then it is reasonable to define the total energy of the two moles to be twice the energy of one of the moles. Thus in order to avoid unpleasantness associated with the ambiguity in the additive constant we can define the internal energy of N moles of a substance as follows.

Given N moles of a pure substance consider one mole of the substance which would be in equilibrium with it (thermal and mechanical) if placed in contact thus:

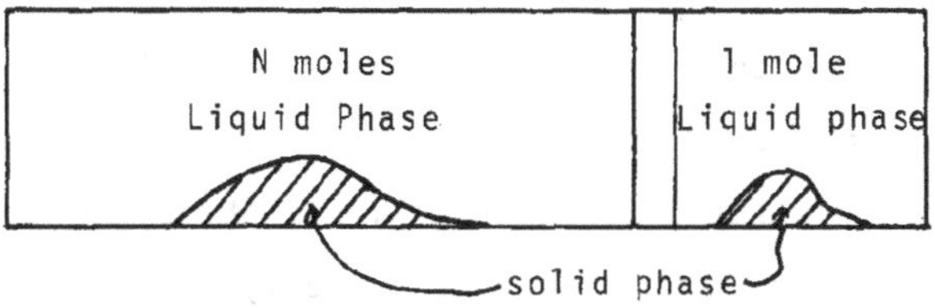

Figure 5.1

Assume moreover that if several phases of the substance are present then each occurs in the same percentage (by mole number, say) in the N mole sample as in the one mole sample. If the internal energy of the one mole is u then we define the internal energy of the N moles to be Nu. Of course the internal energy, u, per mole has an additive arbitrary constant in it.

The internal energy U defined this way is called extensive. Roughly speaking this means that if you have N times as much of the "same kind of stuff" then its internal energy is N times as large.

The volume is "clearly" extensive. So is N. Thus the global coordinates for the state space proposed by Gibbs are all extensive.

By way of contrast, note that the temperature and pressure in the preceding picture are the same for the one mole as for the N moles. Temperature and pressure are called intensive variables. Roughly speaking, a function on state space is called intensive if it takes the same value on one sample as on any other sample of the "same kind of stuff." It is, in fact, homogeneous in U, V, N of degree zero. Later we shall see that this dichotomy between extensive and intensive state functions is (with some qualification) exactly the dichotomy between coordinates in a vector space and coordinates in the dual space.

For a system consisting of r chemically pure non interacting substances we can define the total energy to be the sum of the energies of its pure components. (See Section 10.)

6. Processes

Consider again gas in a cylinder fitted with a piston. Assume for the moment that the walls of the cylinder are not adiabatic. One says in this case that the walls are diathermal. If the cylinder is immersed in a large bath of water at 80° centigrade, say, then the temperature of the gas inside the cylinder will gradually change until its temperature coincides with that of the water. If the mass of the water bath is extremely large in comparison to that of the cylinder and gas then the final equilibrium temperature will be approximately 80°C. The water bath gives up (or receives) some of its internal energy to (or from) the gas without changing its temperature substantially. By a heat reservoir at temperature T is meant an idealized system which can increase or decrease its internal energy without changing its temperature. A prototype would be an infinite water bath at temperature T.

In the preceding example the gas went through intermediate states which are not equilibrium states. Not only was its temperature changing, but it was heating up first near the walls of the cylinder and convection currents were set up. By a process we mean a change from one equilibrium state to another, of which the preceding is an example. Here are some more examples.

Example 2. Assume the walls of the cylinder are adiabatic. If we rapidly pull out the piston, shock waves will be set up in the gas and will gradually disappear, converting by internal friction to heat. Eventually an equilibrium state will be reached again.

Example 3. Suppose that in the preceding example we pull out the piston very slowly. The effect of the shock waves will be very small--negligible in the limit of "infinite slowness". Each intermediate position of the piston defines an equilibrium state and the change from the initial to the final state is accompanied by a "continuous succession" of intermediate equilibrium states. This is called a quasistatic process. We wish to emphasize that time evolution is not really involved here in an ordinary way. No real process is quasistatic. It is easy to be somewhat more precise with this definition than heretofore: a quasistatic process is a piecewise continuously differentiable curve in state space. The differentiability is convenient for later technical purposes.

Note that in the last example if we push the piston back in very slowly we can retrace the trajectory in the state space, ending at the initial state. Moreover the work done by the piston on the way out is equal to the work done by us on the way in. Thus if the piston is made to raise a weight on the way out then the work done by the falling weight on the piston as the weight falls will be just enough to push the piston back to its original position. Both the gas in the cylinder as well as the external universe (i.e., the weight) are in exactly the same state as at the beginning. A quasistatic process which can be retraced in such a way as to leave the entire universe in the original state is called <u>reversible</u>. By way of contrast consider a process in which a very slowly falling weight runs a generator which heats a resistor located in the gas. The gas heats up and in the "limit" of infinite slowness the gas temperature rises uniformly, the system goes through a succession of equilibrium states, and the process is quasi-static. But it is a fact of experience that the mechanical energy expended by the falling weight, and which has been dissipated in heat, cannot be completely recovered. Thus this quasistatic process is not reversible. It is <u>irreversible</u>. Any piecewise differentiable curve in U, V space can be carried out reversibly with the help of external heat reservoirs (perhaps continuum many) and work. It may also be carried out irreversibly. We shall soon formalize the impossibility of converting heat energy completely into work in the second law. But first I wish to consider one more type of process.

A process in which the system begins and ends in the same state is called <u>cyclic</u>. The rest of the universe need not be in the same state at the beginning and end of the process. The most important cyclic process for theoretical purposes is the <u>Carnot</u> cycle. This is the reversible process consisting of four parts, which we describe for a gas.

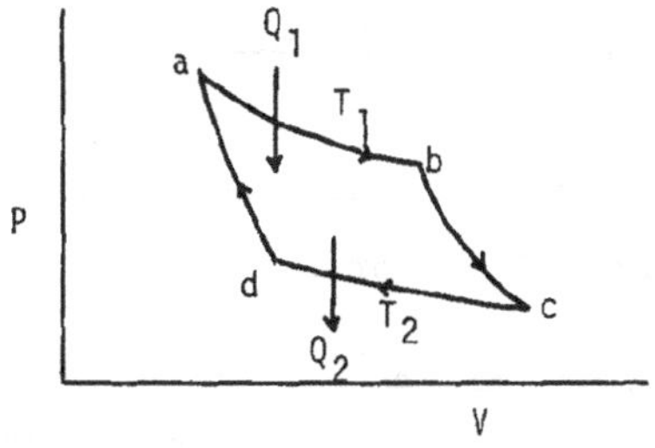

Figure 6.1

a → b. The system expands isothermally at temperature T_1, absorbing heat reversibly from a heat reservoir at temperature T_1.

b → c. Adiabatic expansion. Temperature decreases to T_2.

c → d. Isothermal compression. The system compresses at constant temperature T_2 giving up heat energy reversibly to a second heat reservoir at temperature T_2.

d → a. The system compresses adiabatically. The temperature increases to T_1.

Observe that when a Carnot cycle has been completed the system is in its original state. But some heat energy, Q_1, say, has been transferred out of the high temperature reservoir while some heat energy Q_2 has been absorbed by the low temperature reservoir. Moreover the system has done some work W

$$W = \int_{\text{cycle}} PdV$$

$$= \text{area enclosed by curve in P-V diagram.}$$

By the first law

$$W = Q_1 - Q_2 .$$

7. The Second Law

The second law of thermodynamics is a sharp formulation of the familiar fact that heat flows naturally from hot bodies to cooler bodies and not the other way around.

Kelvin Planck's statement of the second law: It is impossible to construct an engine which, operating in a cycle, has no effect other than to convert internal energy from a heat reservoir at some fixed temperature entirely into mechanical work.

Clausius' statement of the second law: It is impossible to construct an engine which, operating in a cycle, has no effect other than to convert internal energy of a heat reservoir at temperature T_2 to internal energy of a heat reservoir at temperature $T_1 > T_2$.

Remark: The very terminology "second law" as well as our use in this section of the words "theorem", "proof", etc. may suggest to the reader that there is some mathematics in progress here. But the phrase "to construct an engine" has not been given a precise mathematical meaning by us and the "proofs" below frankly rely on one's physical understanding rather than on the conventional forms of mathematical reasoning. This is, on the one hand, unavoidable, since we have not yet defined suitable mathematical structures in which to make these intuitive notions precise, while on the other hand this type of proof is desirable at this stage since the arguments illustrate the physical ideas behind the eventual mathematical format. For a completely axiomatic treatment of some of this material see [Gi] or [Fa].

Proposition. The Kelvin-Planck and Clausius forms of the second law are equivalent.

Proof. If Clausius' statement is false then an engine violating Kelvins' statement can be constructed as follows. Let E_1 be any cyclic engine (i.e., cyclic process) which absorbs heat from a reservior at temperature $T_1 > T_2$ and discharges heat, say Q_2, to a reservior at temperature T_2 in each cycle, doing work $W > 0$. If Clausius' statement is false we may (by scaling the size) construct a cyclic engine E_2 which has no effect other than to absorb internal energy Q_2 from the reservoir at temperature T_2 and convert it into internal energy of the reservoir at temperature T_1. The two engines combined constitute a cyclic engine violating Kelvin's statement.

Conversely, assume Kelvin's statement is false. Then there exists a cyclic engine E which absorbs heat $Q_1' > 0$ from a reservoir at temperature T_1 and converts it completely into work $W = Q_1'$. Let $T_2 < T_1$. Use the work W to run a refrigerator R, e.g. a Carnot engine run backwards, which operates between reservoirs at temperatures T_1 and T_2. If the refrigerator absorbs internal energy $Q_2 > 0$ from the reservoir at temperature T_2 then the combined cyclic engine E and R has no effect other than to "raise" the internal energy Q_2 from the reservoir at temperature T_2 to that at temperature T_1, violating Clausius' statement. q.e.d.

Henceforth we shall be more careful with our use of the word temperature because we now want to show that there is a "natural" temperature scale which is independent of which materials one happens to construct a thermometer out of. We shall therefore refer to temperatures read from some particular thermometer as an empirical temperature scale. We insist only that hotter bodies have higher empirical temperature than cooler bodies.

Definition. If an engine (i.e., a cyclic process) absorbs internal energy Q_1 from a heat reservoir at empirical temperature θ_1 and transmits energy Q_2 to a reservoir at empirical temperature θ_2, and does work W in each cycle then the efficiency of the engine is

$$\eta = \frac{W}{Q_1} .$$

Since $W = Q_1 - Q_2$ we have

$$\eta = 1 - \frac{Q_2}{Q_1} .$$

So $0 \leq \eta \leq 1$. But by the Kelvin-Planck form of the second law $Q_2 > 0$. So $\eta < 1$.

We assume henceforth the validity of the second law and we describe its consequences.

Theorem (Carnot). No cyclic process operating between reservoirs at empirical temperatures θ_1 and θ_2 is more efficient than a reversible Carnot cycle operating between these reservoirs. Any two Carnot cycles operating between the same two reservoirs have the same efficiency.

Proof. Let $\theta_1 > \theta_2$. Suppose that there is a cyclic engine E which absorbs internal energy $Q_1 > 0$ from a heat reservoir at temperature θ_1, discharges heat energy $Q_2 > 0$ to a heat reservoir at temperature θ_2, does work W (which equals $Q_1 - Q_2$ by the first law), and whose efficiency $\eta \equiv W/Q_1$ is strictly greater than the efficiency η' of some reversible Carnot cycle C operating between the same two reservoirs. By scaling the size of the Carnot cycle we may suppose that C also produces work W in each cycle. Let Q_1' be the heat absorbed by C at temperature θ_1 and Q_2' the heat discharged at temperature θ_2. Now run C backwards, using the work output W of engine E to operate C.

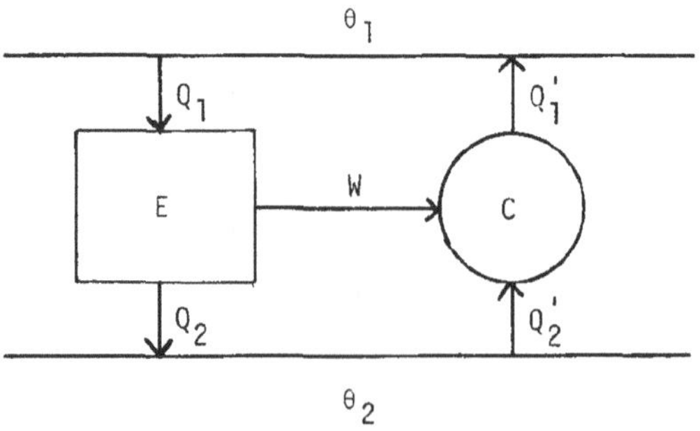

Figure 7.1

The net effect of the two engines is to transfer heat $Q_2' - Q_2$ (which may conceivably be negative) from the low temperature reservoir to the high temperature reservoir. But in fact $Q_2' - Q_2 > 0$. For $W/Q_1 = \eta > \eta' = W/Q_1'$. Hence $Q_1' > Q_1$. However $Q_1' - Q_2' = W = Q_1 - Q_2$. So

$Q_2' - Q_2 = Q_1' - Q_1 > 0$. Thus Clausius' form of the second law has been violated. Hence $\eta \leq \eta'$.

The last statement in the theorem now follows from the first.

Remark. The idea that an engine can be scaled in size without changing its efficiency, which we have used in the preceding two proofs, has a direct intuitive appeal, but could be replaced by a more careful argument in which one starts with an arbitrary reversible Carnot cycle C_0 that produces work W_0 different from W. If the efficiency of C_0 is η' and $\eta' < \eta$ then one produces a violation of the second law by running m copies of C_0 backward, using the work output of n copies of E, where n and m are positive integers such that $1 \leq nW/mW_0 \leq 1+\epsilon$ and $0 < \epsilon < (\eta-\eta')/\eta$. The extra work $nW - mW_0$ can be dissipated as heat at temperature θ_1. In fact, putting $\delta = \eta - \eta'$, $\eta' = W_0/Q_1'$, $Q_2' = Q_1' - W_0$ we have $mW_0/mQ_1' = \eta' = \eta - \delta = nW/nQ_1 - \delta$ which, upon multiplying by $mQ_1'nQ_1/mW_0$ gives

$$mQ_1' - nQ_1 \geq \delta nQ_1Q_1'/W_0 - \epsilon mQ_1' .$$

But $n(Q_1 - Q_2) = nW \geq mW_0 = m(Q_1' - Q_2')$. Hence

$$\begin{aligned} mQ_2' - nQ_2 &\geq mQ_1' - nQ_1 \\ &\geq (\delta nQ_1/W_0 - \epsilon m)Q_1' \\ &\geq (\delta (m/W)Q_1 - \epsilon m)Q_1' \\ &= (\delta/\eta - \epsilon)mQ_1' \\ &> 0 \end{aligned}$$

Thus the cyclic engine composed of n copies of E operating m copies of C_0 backward produces no effect other than to raise a positive amount of heat energy, $mQ_2' - nQ_2$, from a low temperature reservoir to a high temperature reservoir--contradicting the second law.

8. Consequences of the Second Law: Absolute Temperature and Entropy

Carnot's theorem has the following four fundamental corollaries.

Corollary 1. (Existence of absolute temperature) There exists a strictly positive function $T(\cdot)$ on the state space for a substance of fixed composition such that

a) for any empirical temperature scale θ, T is a strictly increasing function of θ.

b) If the substance undergoes a reversible Carnot cycle between reservoirs at temperatures θ_1 and θ_2 with $\theta_1 > \theta_2$, then the efficiency is

$$\eta = 1 - \frac{T(\theta_2)}{T(\theta_1)} . \tag{8.1}$$

Moreover any function satisfying a) and b) is unique up to a constant scalar multiple.

Proof. For $\theta_1 > \theta_2$ let $f(\theta_1,\theta_2) = 1 - \eta$ where η is the efficiency of any reversible Carnot cycle operating between reservoirs at empirical temperatures θ_1 and θ_2. Suppose that $\theta_1 > \theta_2 > \theta_3$. Consider two reversible Carnot cycles, one operating between temperatures θ_1 and θ_2, absorbing heat energy Q_1 at temperature θ_1 and discharging heat energy Q_2 at temperature θ_2, while the other operates between θ_2 and θ_3, absorbs heat Q_2 and discharges heat Q_3. Then $f(\theta_1,\theta_2) = Q_2/Q_1$, $f(\theta_2,\theta_3) = Q_3/Q_2$ and, since the combined system is also a reversible Carnot cycle, $f(\theta_1,\theta_3) = Q_3/Q_1$. Thus

$$f(\theta_1,\theta_3) = f(\theta_1,\theta_2)f(\theta_2,\theta_3) \quad \theta_1 > \theta_2 > \theta_3 . \tag{8.2}$$

Fix some empirical temperature θ_0 and some number $T_0 > 0$ and define

$$T(\theta) = \begin{cases} T_0 f(\theta_0,\theta) & \text{if } \theta \leq \theta_0 \\ T_0/f(\theta,\theta_0) & \text{if } \theta > \theta_0 . \end{cases}$$

Now if, say, $\theta_1 > \theta_2 > \theta_0$ then $T(\theta_2)/T(\theta_1) = f(\theta_1,\theta_0)/f(\theta_2,\theta_0)$. But by (8.2) with θ_3 replaced by θ_0 this is $f(\theta_1,\theta_2)$. Thus

$$\frac{T(\theta_2)}{T(\theta_1)} = f(\theta_1,\theta_2) \qquad \theta_1 > \theta_2 \tag{8.3}$$

The reader can verify similarly that (8.3) holds in the other two cases $\theta_1 > \theta_0 > \theta_2$ and $\theta_0 > \theta_1 > \theta_2$. Hence $\eta = 1 - T(\theta_2)/T(\theta_1)$ in all cases.

The essential uniqueness of $T(\cdot)$ follows from the fact that for any other function $T'(\cdot)$ satisfying (8.1) there holds $T'(\theta_2)/T(\theta_2) = T'(\theta_1)/T(\theta_1)$. Finally, since $Q_1 > Q_2 > Q_3$ in the preceding notation, it follows that $f(\theta_1,\theta_2)$ is strictly increasing in θ_2 and strictly decreasing in θ_1. Hence $T(\theta)$ is a strictly increasing function of θ.

Remark 8.1. In the formulation of the preceding corollary it should be understood that the states referred to are thermodynamic equilibrium states so that for any empirical temperature scale, θ, the temperature may be regarded as a function on state space.

Definition. The absolute temperature scale is the function $T(\cdot)$ on a state space determined as in Corollary 1 and which takes the value 273° (Kelvin) when the system is in thermal equilibrium with an ice-water mixture at one atmosphere pressure.

Henceforth T will always refer to the absolute temperature.

Now consider a fixed quantity of a substance whose thermodynamic states can be parametrized by a half space or quadrant $U > U_0$ in the U, V plane. For a Carnot cycle C as in Figure 6.1 we have $Q_2/Q_1 = T_2/T_1$ by Corollary 1, and so $Q_2/T_2 = Q_1/T_1$ where Q_1 is the heat absorbed at temperature T_1 and Q_2 is the heat discharged at temperature T_2. Thus

$$\int_C \frac{đQ}{T} = 0 \tag{8.4}$$

where $đQ$ is now the heat transferred <u>to</u> the system along the corresponding part of the cycle. Thus $đQ$ is negative in the first compression portion of the cycle ($c \to d$ in Figure 6.1) and of course is zero on the adiabatic legs, and positive on leg $a \to b$. Here we are following the standard convention of putting a bar through the d in dQ to emphasize that this is not the differential of some function Q.

Corollary 2. (Existence of entropy) Consider a fixed quantity of a substance whose thermodynamic states are parametrized by U and V. There exists a function $S(U,V)$ whose differential is

$$dS = \frac{dU + PdV}{T} \tag{8.5}$$

Proof. The heat transferred to the system in a quasistatic process is by the first law, the integral of the one form $đQ \equiv dU + PdV$ (cf. Equation 4.2). Consider a reversible process made up of the envelope of a finite number of Carnot cycles thus:

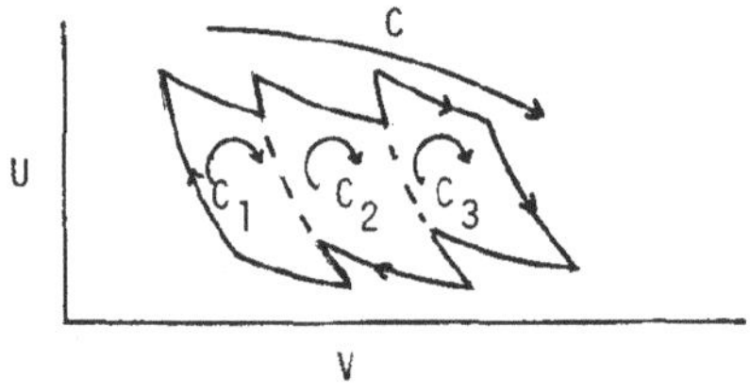

Figure 8.1

But $\int_{C_j} \frac{đQ}{T} = 0$ by (8.4). Moreover the integrals over the dotted lines cancel, leaving

$$\int_C \frac{đQ}{T} = 0 \tag{8.6}$$

where C is the envelope. The corollary would therefore follow immediately if we could prove this for any smooth curve C or, just as well, if we knew that the isotherms, T = constant, and adiabatic curves, $dU + PdV = 0$, really look the way I have been drawing them. Such a proof would require further technical information about the two functions $T(U,V)$ and $P(U,V)$. (One expects that T is at least continuous in U, V and non decreasing in U, while P is Lip 1 but may not be differentiable at points of phase coexistence.) I wish to avoid such technicalities because the main ideas behind the argument leading to the existence of the function S is already apparent and sufficient for this premathematical exposition. I will therefore not complete the proof. A reader interested in the state of the art of approximating smooth curves by Carnot cycle envelopes should see [Z1, Sec.10.1].

The important role played by envelopes of Carnot cycles should be noted. Their role is similar to that played by simple functions in Lebesgue integration theory.

The function S is called the <u>entropy</u>. It is defined only up to an additive constant for a fixed quantity of substance. If we fix a reference state (U_0,V_0) we may define

$$S(U,V) = \int_{U_0,V_0}^{U,V} \frac{dU + PdV}{T} \tag{8.7}$$

along any path.

We wish to consider now how the entropy is to be defined for a substance whose state depends not only on U and V but also on other parameters. To begin with, we consider in this section a chemically pure substance. Its state is determined, as in Section 5, by the extensive parameters U, V, and N. For each fixed N the entropy is determined up to an additive constant, by Corollary 2. The additive constant may, of course, depend on N. In order to make the dependence on N reasonable we use the following heuristics. Consider two moles of a pure substance divided by an imaginary wall into two identical parts, and a reversible process in which the two moles are at each stage identical (compare the disucssion of extensivity in Section 5). Then at each stage the contribution to the heat absorbed, dQ, will be identical, and the entropy difference,

$$S(B) - S(A) = \int_A^B \frac{\bar{d}Q}{T} ,$$

between the initial and final states A and B of the 2 mole system will be twice the entropy difference for each mole. Because of the extensivity of internal energy and volume the state, A, of the two mole system will have energy and volume $2U_0$ and $2V_0$ respectively, where U_0 and V_0 are the parameters for the "left half" of the two mole system when the two mole system is in state A. Therefore if the reference state for the two mole system is taken to be $(2U_0, 2V_0)$ when the reference state for the one mole system is (U_0, V_0) then the entropy will double for twice as much of the "same stuff". That is, we shall have extensivity. With this as motivation we then <u>define</u>

$$S(U,V,N) = \int_{NU_0, NV_0}^{U,V} \frac{dU' + PdV'}{T} \tag{8.8}$$

along any path in the U, V cone, where U_0, V_0 is a chosen reference state for one mole. Here P and T are the pressure and temperature at the point U', V', N.

If we make the change of variables $U' = NU''$, $V' = NV''$ and use the fact that P and T are intensive: $P(NU'', NV'', N) = P(U'', V'', 1)$, then (8.8) gives

$$S(U,V,N) = NS(U/N, V/N, 1). \tag{8.9}$$

Thus $S(U,V,N)$ is determined up to an additive constant multiple of N.

Note that S is homogeneous of degree one:

$$S(\lambda U, \lambda V, \lambda N) = \lambda S(U,V,N) \quad \lambda > 0 \tag{8.10}$$

as follows from the previous equation. Hence S is indeed <u>extensive</u>: If you have λ times as much of the "same kind of stuff" its entropy is λ times as large.

We now define the chemical potential $\mu(U,V,N)$ by

$$\mu = -T\, \partial S/\partial N \, . \tag{8.11}$$

In view of (8.8), we have

$$\frac{\partial S}{\partial U} = \frac{1}{T} \qquad \qquad N \text{ fixed}$$
$$\frac{\partial S}{\partial V} = \frac{P}{T}$$

so that the exterior derivative (differential) of S is the one form given by

$$TdS = dU + PdV - \mu dN. \tag{8.12}$$

Of course,

$$đQ = TdS \quad (N \text{ fixed})$$

is the basic defining relation for the entropy by Corollary 2. We emphasize that in a process in which N changes it is meaningless to say that TdS represents the (infinitesimal) heat energy transferred to the system. See Section 4, on the First Law, for the definition of heat transfer.

Remark 8.2. As we noted before, the one form $đQ = dU + PdV$ on state space (fixed N) is not exact, which is the reason for the (conventional) use of the bar through the d in $đQ$. There is no function $Q(U,V)$. Corollary 2 says, however, that $1/T$ is an integrating factor for $đQ$.

Corollary 3. (Principle of increase of entropy) Consider two thermodynamic systems of fixed composition each of whose thermodynamic states can be parametrized by U, V. Let $S_1(U,V)$ and $S_2(U,V)$ be their entropy functions. Suppose that the two systems are enclosed in an adiabatic cylinder and separated by an adiabatic immovable wall.

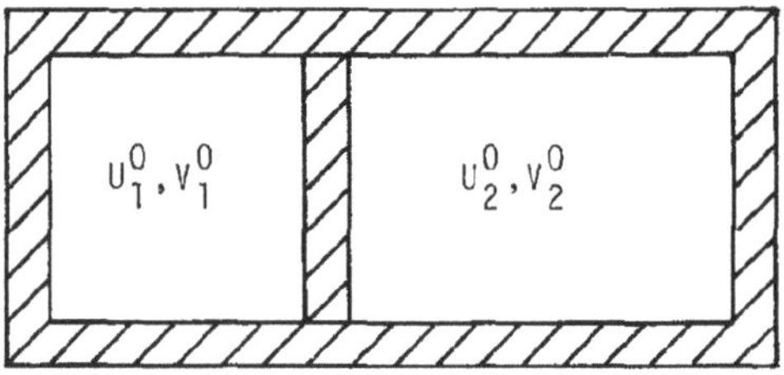

Figure 8.2

Let U_j^0 and V_j^0 be the parameters of system j, $j = 1,2$. Suppose that one or both of the two constraints on the wall are removed. (E.g. we may make the wall diathermal, or movable, or both.) When the system returns to equilibrium there holds

$$S_1(U_1,V_1) + S_2(U_2,V_2) \geq S_1(U_1^0,V_1^0) + S_2(U_2^0,V_2^0)$$

where U_j, V_j are the new equilibrium values.

Remark 8.3. The process that occurs after removing the constraint need not be quasistatic nor, a fortiori, reversible.

Proof. After equilibrium returns isolate the systems from each other. Fix a temperature T. By a reversible adiabatic process (e.g. expansion or compression) bring each of the systems to the common temperature T.

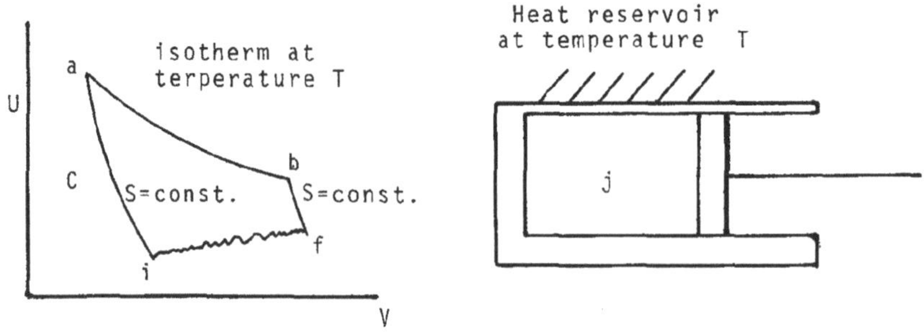

Figure 8.3

Let b_j be the resulting state for system j, $j = 1$ or 2. Let C_j be the adiabatic curve for system j through the initial state, i, of system j and let a_j be the state at the intersection of C_j with the isothermal curve of temperature T.

Consider the cyclic process of the two systems together consisting of $i \to f \to b \to a \to i$ for each system. If Q_j is the heat energy absorbed from the reservoir at temperature T by system j in going from b_j to a_j then the total heat energy absorbed in the cyclic process is

$$Q = Q_1 + Q_2$$

because the total heat transferred during $i \to f$ is zero, the two systems having been thermally isolated from the exterior during this part of the cycle. By the first law the work done by the composite system during the cycle is $W = Q$. But since Q was absorbed entirely from a single reservoir at a fixed temperature the second law implies $Q \leq 0$. But

$$S_{a_j} - S_{b_j} = Q_j/T.$$

Hence $S(i_j) - S(f_j) = Q_j/T$. So

$$\Sigma_{j=1}^{2}(S(i_j) - S(f_j)) = (\Sigma_{j=1}^{2} Q_j)/T$$

$$\leq 0.$$

Thus $\Sigma_{j=1}^{2} S(f_j) \geq \Sigma_{j=1}^{2} S(i_j)$, completing the proof of Corollary 3.

We shall need the following Lemma.

Lemma. (Restricted additivity of entropy) If, for a pure substance, the two states (U_1, V_1, N_1) and (U_2, V_2, N_2) have the same temperature and pressure then

$$S(U_1 + U_2, V_1 + V_2, N_1 + N_2) = S(U_1, V_1, N_1) + S(U_2, V_2, N_2).$$

Proof. Put the two systems in the given states in a cylinder separated by a diathermal movable wall.

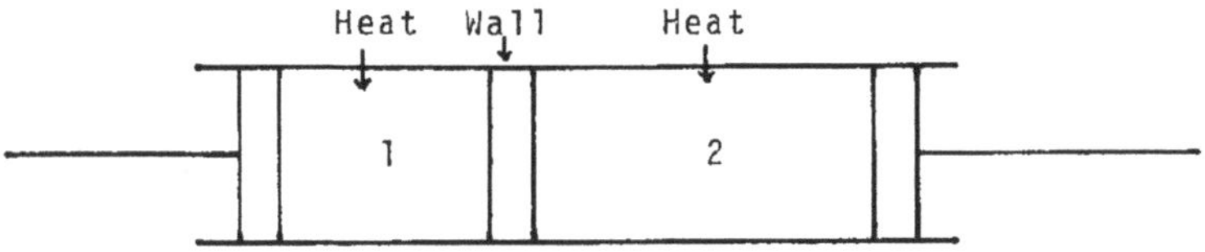

Figure 8.4

The system is then in equilibrium because $P_1 = P_2$ and $T_1 = T_2$. In the reference states (N_1U_0, N_1V_0) and (N_2U_0, N_2V_0) the two systems are also in equilibrium, (see discussion of extensivity in §5) and the joint system is an equilibrium state with coordinates $((N_1+N_2)U_0, (N_1+N_2)V_0)$ which is the reference state for the combined systems. Take any path of the joint system joining the reference state to the state $U_1 + U_2$, $V_1 + V_2$ and beginning with subsystem j in state (N_jU_0, N_jV_0). The wall does not affect the work done on or heat absorbed by the joint system. As we proceed along the path by a reversible process some phases may disappear and some new phases may arise (e.g. ice may form). Arrange the distribution of heat absorbed so that any <u>new</u> phase that arises is always distributed between the two subsystems in proportion to its relative amount in the final states (U_1, V_1, N_1), (U_2, V_2, N_2) (if there is any of this phase in the final states). This procedure determines a path in each subsystem along which the pressure is the same in the two subsystems and so is the temperature. This path for subsystem j begins at (N_jU_0, N_jV_0, N_j) and ends at (U_j, V_j, N_j). Thus if $U' = U_1' + U_2'$ and $V' = V_1' + V_2'$ where U_j', V_j' traverses the path of system j, then U', V' traverses the path of the joint system and

$$S(U_1 + U_2, V_1 + V_2, N_1 + N_2) = \int \frac{dU_1' + dU_2' + P(dV_1' + dV_2')}{T}$$

$$= S(U_1, V_1, N_1) + S(U_2, V_2, N_2) \; .$$

Corollary 4. (Concavity of entropy) The entropy of a pure substance is a concave function of U, V, N. (That is, $-S$ is convex.)

Proof. Given states (U_1,V_1,N_1) and (U_2,V_2,N_2) and α in $(0,1)$ consider the system shown in Figure 8.5.

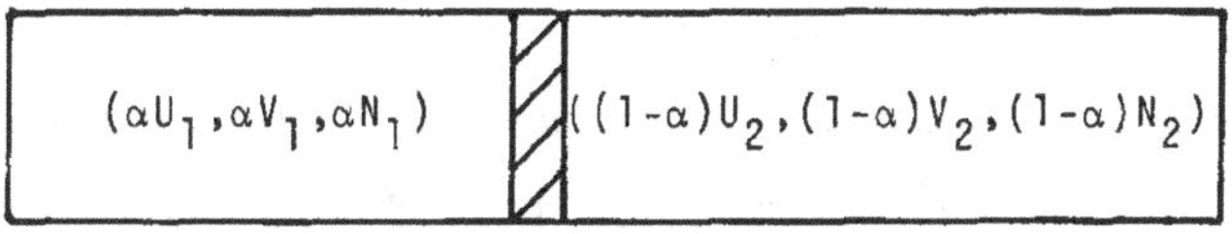

Figure 8.5

With adiabatic, rigid walls initially, replace the separating wall by a movable, diathermal wall. After equilibrium returns we have, by Corollary 3.

$$S(U_1',V_1',\alpha N_1) + S(U_2',V_2',(1-\alpha)N_2)$$

$$\geq S(\alpha U_1,\alpha V_1,\alpha N_1) + S((1-\alpha)U_1,(1-\alpha)V_1,(1-\alpha)N_1)$$

where the primed quantities are equilibrium values. But by the preceding lemma the left side is $S(U_1'+U_2',V_1'+V_2',\alpha N_1+(1-\alpha)N_2)$. Moreover since the composite system is closed, $U_1'+U_2' = \alpha U_1 + (1-\alpha)U_2$ and $V_1'+V_2' = \alpha V_1 + (1-\alpha)V_2$. Thus

$$S(\alpha U_1 + (1-\alpha)U_2, \alpha V_1 + (1-\alpha)V_2, \alpha N_1 + (1-\alpha)N_2)$$

$$\geq \alpha S(U_1,V_1,N_1) + (1-\alpha)S(U_2,V_2,N_2).$$

On the right we have used the homogeneity of S. But this inequality is exactly the definition of concavity.

9. The ABC of Equilibrium Thermodynamics

The basic problem of thermodynamics can be stated with the help of the apparatus of Figure 8.2. Given two thermodynamic systems which are thermally and mechanically isolated as in the figure, what is the new equilibrium state after one or both of the internal constraints is removed?

Corollary 3 gives an answer to the question. For definiteness lets suppose that the original wall is replaced by a diathermal, but still immovable wall. Then, after equilibrium returns, we still have $V_1 = V_1^0$, and $V_2 = V_2^0$ (in the notation of Corollary 3). Only the internal energies have changed. Of course, since the system is isolated we have $U_1 + U_2 = U_1^0 + U_2^0$. Thus Corollary 3 says that the equilibrium values of U_1 and U_2, V_1 and V_2 are those that make the total entropy of the two subsystems together (i.e., the sum of their entropies) a maximum, consistent with the restraints $[U_1 + U_2 = \text{const.}, V_1 = \text{const.} = V_1^0, V_2 = \text{const.} = V_2^0]$. Hence if this maximization problem has a unique solution for U_1, U_2 then Corollary 3 tells us how to find the new equilibrium state parameters U, V. So knowledge of the entropy function solves the basic problem of equilibrium thermodynamics! (Non uniqueness of the solution is necessarily associated with coexistence of phases.) Similar maximization problems are obtained, if the wall is made adiabatic but movable, or diathermal and movable.

Now let $K = U_1^0 + U_2^0$. Then since the equilibrium values of the internal energy also satisfy $U_1 + U_2 = K$, the total entropy will be a maximum only if the derivative of $U_1 \to S_1(U_1, V_1^0) + S_2(K-U_1, V_2^0)$ is zero. Thus at equilibrium

$$\frac{\partial S_1}{\partial U}(U_1, V_1^0) - \frac{\partial S_2}{\partial U}(K-U_1, V_2^0) = 0.$$

But $\frac{\partial S_j}{\partial U} = \frac{1}{T_j}$. Hence equilibrium occurs only if

$$\frac{1}{T_1(U_1, V_1^0)} = \frac{1}{T_2(U_2, V_2^0)}.$$

In other words the two systems are in equilibrium when separated by a diathermal wall only if they are at the same temperature--(which of course we knew all along).

We can now state the A, B, C of equilibrium thermodynamics. The reader should compare this with the ABC for mechanics given in the introduction. We state B and C in the case of a one component system. For the generalization to a multicomponent system see Callen [C].

A. The state space for an r component thermodynamic system is a convex cone $\mathbf{C}$ in R^{r+2} (coordinates $U, V, N_1, \ldots, N_r$).

B. The "dynamics" is determined (for $r = 1$) by a function $S(U,V,N)$ on $\mathbf{C}$ which is

a) once continuously differentiable with $\partial S/\partial U > 0$.

b) homogeneous of order one:

$$S(\lambda U, \lambda V, \lambda N) = \lambda S(U,V,N) \quad \text{for all} \quad \lambda > 0.$$

c) concave

S is called the entropy function of the system.

C. When an internal constraint is removed (as in Corollary 3) the new equilibrium state is that which maximizes the total entropy consistent with the new constraints.

Remarks 9.1. The "axioms" in A, B, C are phrased without reference to temperature or pressure--important thermodynamic functions. But the temperature and pressure may be recovered from S by

$$1/T = \frac{\partial S}{\partial U}(U,V,N) \quad \text{and} \quad P = T\,\partial S/\partial V.$$

As we have seen, the conditions for equilibrium involve P and T in just the way one expects. [But see discussion in Callen [C] for mechanical equilibrium in absence of thermal equilibrium.]

9.2. There is a very pretty and important connection between geometrical properties of the graph of $U,V \to S(U,V,1)$ and the coexistence of phases. Consider the phase diagram for water

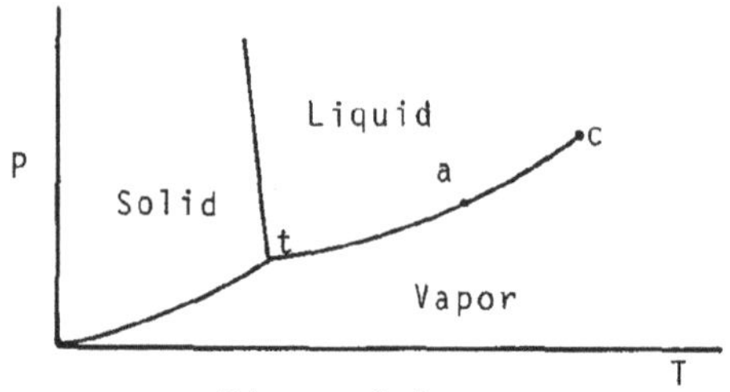

Figure 9.1

At a point a on the curve t-c the liquid and gas phase coexist: At this particular temperature and pressure one can have one mole of water-water vapor mixture in all proportions (0% water to 100% water). In the U, V plane the set of these states will constitute a straight line segment, L. Above this line segment the graph of S also contains a line segment. Thus S fails to be strictly convex in each neighborhood of any point on the line segment L. (See [Z2, §11.9] for the reasons for this, or see Wightman's introduction to [Is 3].) The point t (triple point) in the phase diagram is a point where all three phases can coexist. In the U, V plane the set of points for which $(T(U,V),P(U,V)) = t$ turns out to be a (closed) triangle. Lying above this triangle, in the graph of the entropy function, is also a triangle. In general, the "flat places" in the graph of S correspond to coexisting phases. Moreover any support plane for S touches S in a simplex. If the simplex is a point then the corresponding U, V point below it is a pure phase. If the simplex is a line segment then its projection in the U, V plane is a set of coexisting phase points (except the endpoints). If the simplex is a triangle then the interior of its projection consists of states of three mixed phases. Similarly for higher component systems. The fact that support planes touch the graph of S in points, segments or triangles, etc. is not a consequence of the second law. We refer the reader to Wightman's introduction to [Is 2] for an illuminating discussion of this.

10. Multicomponent Systems

For a pure substance we saw in Corollary 4 that the entropy is a concave function of U, V and N. We wish to extend this now to a mixture of r non interacting pure substances. The state space is coordinatized by $U, V, N_1, \ldots, N_r$ as we have noted before. For a substance of fixed composition the equation (8.7) determines the entropy. But the reference state U_o, V_0 may depend on the composition in any way. In order to achieve an entropy function $S(U, V, N_1, \ldots, N_r)$ which is both homogeneous and concave we proceed in a different way, based on the heuristics associated with a Gibbs separator--a device which separates a compound substance into its pure subcomponents.

Consider a mixture of two chemically pure gases A and B. A semipermeable membrane is a wall which allows just one of the gases, say A, to pass through it but not the other. Semi-permeable membranes are known to exist for some gases (see [Z1, Z2]) and are therefore regarded as acceptable devices for theoretical discussion of arbitrary gases and fluids. A Gibbs separator consists of two cylinders, as shown in Figure 10.1. One cylinder fits snugly inside the other with the semipermeable membranes interlaced as shown.

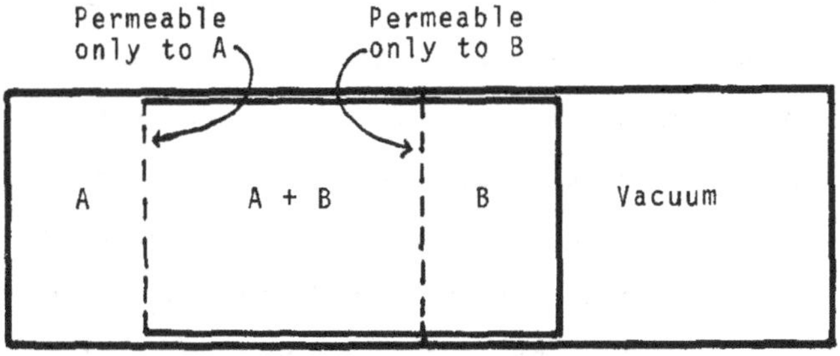

Figure 10.1

(The right hand membrane can be attached to the outer cylinder at three points with the help of three slits in the inner cylinder.) One cylinder can move inside the other without friction. With the inner cylinder all the way to the left the two gases are completely mixed and have a volume, V, say. After the inner cylinder is moved ("very slowly") to the far right, with the membranes just touching, gases A and B are separated with gas A occupying volume V in the left half cylinder and gas B occupying volume V in the right cylinder. We assume that the walls are thermally insulated so that no heat is absorbed in the process but that A, B and A + B are always in thermal contact. Now the assumption that the two gases do not interact with each other means for us that the force exerted on a section of wall by the molecules of the gas mixture is the sum of the forces exerted by each gas. That is to say the total pressure is the sum of the so-called partial pressures--the pressure that each gas would exert if it occupied the same volume by itself. Under this assumption the force on the right hand wall of the inner cylinder, which is due to the pressure of gas B, is equal and opposite to the net force on the left hand wall of the inner cylinder since the pressure of gas A on the two sides of the left wall cancels. Thus there is no net force on the inner cylinder. Consequently no work is done in separating the two gases. (To be operationally meaningful, the statement "would exert if ... by itself" four sentences back can be taken to mean that there is no net force on the inner cylinder.) Suppose that U_1 and U_2 are the internal energy coordinates of the two separated gases, the arbitrary additive constant having been chosen for each gas. With the aim of determining an internal energy coordinate for the mixed gas, imagine doing some work W adiabatically on the mixed gas and then once more separating it as above into its two components, this time with internal energies U_1' and U_2'. Since the separation process involves no work it follows that $W = (U_1' - U_1) + (U_2' - U_2)$. By definition the internal energy change of the mixed gas equals W. It is therefore consistent to define the internal energy coordinate of the mixed gas to be $U = U_1 + U_2$ in the initial state. This amounts to making a choice of the additive constant for the internal energy of the mixed gas. It is easy to see that U is an extensive coordinate.

Now we must make a second assumption to carry out our construction of a concave entropy function. We assume that the temperature T of the system before separation is equal to the common temperature of the two gases after separation. This assumption is consistent with statistical mechanical models of non interacting real gases and can actually be shown for "ideal" gases (see [Z1, Ch.17] or [Z2, Ch.16]).

With these assumptions we then define

$$S(U,V,N_1,N_2) = S_1(U_1,V,N_1) + S_2(U_2,V,N_2) \tag{10.1}$$

where S_1 and S_2 are the entropy functions of A and B respectively. Then,

$$\begin{aligned}\frac{\partial S}{\partial U} &= \frac{\partial S_1}{\partial U_1}\frac{\partial U_1}{\partial U} + \frac{\partial S_2}{\partial U_2}\frac{\partial U_2}{\partial U} \\ &= \frac{1}{T}\left(\frac{\partial U_1}{\partial U} + \frac{\partial U_2}{\partial U}\right) \\ &= \frac{1}{T}\frac{\partial U}{\partial U} \\ &= \frac{1}{T}\,.\end{aligned}$$

We have used in this informal derivation the technical assumption that U_1 and U_2 are differentiable functions of U. Similarly, keeping in mind that U_1 and U_2 may depend also on V (but always $U_1 + U_2 = U$), we have

$$\begin{aligned}\frac{\partial S}{\partial V} &= (\partial S_1/\partial U_1)(\partial U_1/\partial V) + (\partial S_2/\partial U_2)(\partial U_2/\partial V) \\ &\quad + \partial S_1/\partial V + \partial S_2/\partial V \\ &= (1/T)\partial(U_1 + U_2)/\partial V + (P_1 + P_2)/T \\ &= 0 + P/T.\end{aligned}$$

Hence $S(\cdot,\cdot,N_1,N_2)$ is an entropy function for the mixed system. It is easy to see that S is positive homogeneous of degree one in its four dimensional argument.

Before proving the concavity of S let us note that the preceding discussion can be carried out without use of the Gibbs separator although some of the physics is lost. Thus for fixed V the function $T^{-1} \equiv \partial S_1(U_1,V,N_1)/\partial U_1$ is decreasing (not strictly) in U_1 by the concavity of S_1 while $U_1 \to \partial S_2(U-U_1,V,N_1)/\partial U_1$ is increasing. Both functions have range $(0,\infty)$ under the mild assumption that the temperature of each system can vary over $(0,\infty)$. It follows that for fixed U,V,N_1,N_2 there is at least one internal energy U_1 such that the preceding two derivatives are equal. We may then define $S(U,V,N_1,N_2)$ by (10.1) as before, where $U_2 = U - U_1$. Even if U_1 is not unique, which can happen only if the temperatures of the two systems are internal energy independent on some interval of U_1, S is well defined because $S_1(U_1,\ldots)$ and $S_2(U-U_1,\ldots)$, respectively, increase and decrease linearly on this interval at the same rate. The resulting function $S(U,V,N_1,N_2)$ is then the entropy function of some system whose pressure function is the sum of the pressures of the two subsystems. This approach raises the conceptual difficulty that if non uniqueness of U_1 actually occurs (which implies a common boiling temperature) then U,V,N_1,N_2 does not characterize the equilibrium state of the combined system uniquely. We ignore this unresolved possibility by excluding such systems from our consideration.

Proposition. The homogeneous entropy function given by (10.1) is concave.

Proof. Since S is homogeneous of degree one it suffices to prove that it is superadditive. That is,

(10.2) $$S(U,V,N_1,N_2) + S(U',V',N_1',N_2') \leq S(U+U',V+V',N_1+N_1',N_2+N_2').$$

Writing $U = U_1 + U_2$ as in the preceding discussion and similarly $U' = U_1' + U_2'$ the left side of the last inequality equals $S_1(U_1, V, N_1) + S_2(U_2, V, N_2) + S_1(U_1', V', N_1') + S_2(U_2', V', N_2')$ which is less or equal to

$$S_1(U_1 + U_1', V + V', N_1 + N_1') + S_2(U_2 + U_2', V + V', N_2 + N_2')$$

by superadditivity of S_1 and S_2. Now the two systems parametrized by $(U_1 + U_1', V + V', N_1 + N_1')$ and by $(U_2 + U_2', V + V', N_2 + N_2')$ respectively are not necessarily at the same temperature. By means of a calorimeter such as used in Corollary 3, allow the two systems to come to mutual thermal equilibrium without changing their volumes. (Use a diathermal immovable wall in Figure 8.2.) If W_1 and W_2 are the new equilibrium internal energies then by Corollary 3 (the principle of increase of entropy) we have

$$S_1(U_1 + U_1', V + V', N_1 + N_1') + S_2(U_2 + U_2', V + V', N_2 + N_2')$$
$$\leq S_1(W_1, V + V', N_1 + N_1') + S_2(W_2, V + V', N_2 + N_2').$$

Since the two systems are now at the same temperature and volume we may use (10.1) to conclude that the right side of the last inequality is exactly $S(W_1 + W_2, V + V', N_1 + N_1', N_2 + N_2')$. But $W_1 + W_2 = (U_1 + U_1') + (U_2 + U_2') = U + U'$. This proves (10.2).

Remark 10.1. For multicomponent systems not satisfying these two stringent non-interaction conditions I have not been able to find a phenomenological definition of the entropy which yields concavity. Statistical mechanics, however, does yield concave entropy functions as we shall discuss.

Nevertheless it is tempting to derive as much as possible of classi cal thermodynamics (and at least concavity of S) in a purely phenomenological manner, without appeal to the molecular structure of matter.

For recent work on the foundations of thermodynamics we refer the reader to [Se 1,2,3] and its bibliography.

11. The Legendre Transform

A basic operation on convex functions, the Legendre transform plays a central role in equilibrium thermodynamics. In this section we define the Legendre transform of a convex function on a Banach space and in the next section we give its physical interpretation in the important case in which the Banach space is finite dimensional. Cf. [Rk].

Consider, to begin with, the one dimensional case. Let f be a twice continuously differentiable real valued function on an interval I (finite or infinite, closed or open or neither). Suppose that $f'' > 0$ on I. Then f is convex and f' is strictly increasing. The range, J, of f' is an interval. For any number α in J define $f^*(\alpha)$ as follows: let u_0 be the unique point in I such that $f'(u_0) = \alpha$. Let $-b$ be the height at which the tangent line to the graph of f at u_0 crosses the y axis. Define $f^*(\alpha) = b$.

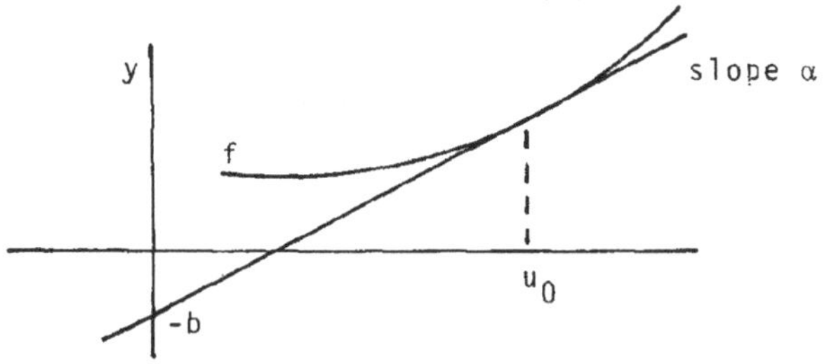

Figure 11.1

It is clear from Figure 11.1 that

$$-b = \sup\{s : \alpha u + s \leq f(u) \text{ for all } u \in I\} = \inf\{f(u) - \alpha u : u \in I\}.$$

Hence

$$f^*(\alpha) = \sup\{\alpha u - f(u) : u \in I\}. \tag{11.1}$$

Since the tangent line is given by $y - f(u_0) = \alpha(u - u_0)$ we have also $-b = f(u_0) - \alpha u_0$ and so

(11.2) $$f^*(\alpha) = \alpha u_0 - f(u_0)$$

where u_0 is the unique solution to $f'(u_0) = \alpha$.

We shall need to dispense with the strict convexity, $f'' > 0$, and even with differentiability. Note first that if f is merely convex and once continuously differentiable (and so may have a flat spot), and even if the equation $f'(u_0) = \alpha$ has more than one solution u_0, (all will lie below the flat spot) then (11.2) defines a unique number, $f^*(\alpha)$, because f is linear with slope α on the interval of non uniqueness. Moreover equation (11.1) clearly continues to represent $f^*(\alpha)$ correctly in this case. Now if f is merely convex and not differentiable then (11.1) continues to be well defined although (11.2) may not be. We therefore take (11.1) as the definition below.

Let B be a real Banach space. Recall that a real valued function f defined on a subset $\mathcal{D}_f$ of B is <u>convex</u> if $\mathcal{D}_f$ is a convex set and $f(\beta x + (1-\beta)y) \leq \beta f(x) + (1-\beta)f(y)$ for all x and y in $\mathcal{D}_f$ and all β in $[0,1]$. The <u>epigraph</u> of f is the subset of $\mathcal{D}_f \times R$ given by

$$\text{epi} f = \{(u,y) \in \mathcal{D}_f \times R : y \geq f(u)\}$$

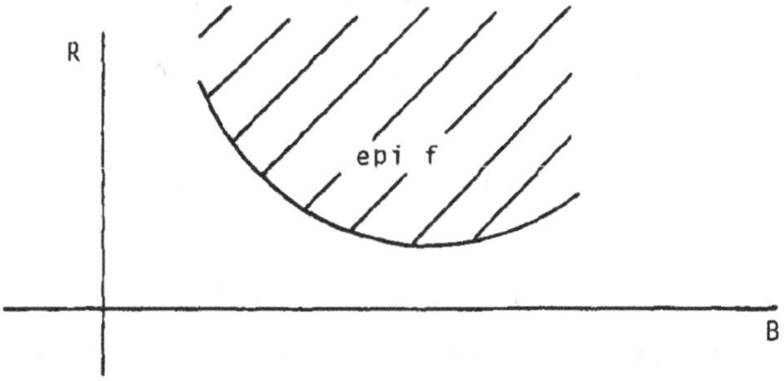

Figure 11.2

A function is convex if and only if its epigraph is a convex set.

We call a function f <u>closed</u> if its epigraph is closed. For example the function which is zero on a bounded open interval and one at the right hand endpoint is convex but not closed because of trouble at both endpoints. But the function which is $1/x$ on $(0,1]$ is convex and closed. Denote by B^* the dual space of B. The Legendre transform of a convex function f is the function f^* whose domain is

$$\mathcal{D}_{f^*} = \{\alpha \in B^*: \sup_{u\in\mathcal{D}_f}[\alpha(u) - f(u)] < \infty\}.$$

For α in $\mathcal{D}_{f^*}$ we define

$$f^*(\alpha) = \sup\{\alpha(u) - f(u): u \in \mathcal{D}_f\}.$$

The basic theorem showing that f can be recovered from its Legendre transform is the following.

Theorem. a) If $\mathcal{D}_{f^*}$ is not empty then f^* is a closed convex function.

b) If f is a closed convex function then $\mathcal{D}_{f^*}$ is not empty and

$$f^{**}|B = f.$$

Proof. a) The epigraph of f^* is

$$\begin{aligned} \text{Epi } f^* &= \{(\alpha,y): y \geq f^*(\alpha), \alpha \in \mathcal{D}_{f^*}\} \\ &= \{(\alpha,y) \in B^* \times R: y \geq \alpha(u) - f(u) \ \forall u \in \mathcal{D}_f\} \\ &= \bigcap_{u\in\mathcal{D}_f} \{(\alpha,y) \in B^* \times R: y \geq \alpha(u) - f(u)\}. \end{aligned}$$

The right side is an intersection of closed half spaces in the Banach space $B^* \times R$ (sum norm, say). Hence it is a closed convex set. This proves a).

To prove b) assume f is closed. A picture suggests that if (x_0,y_0) is not in epi f then there is a hyperplane passing through (x_0,y_0) and lying "below" epi f. More precisely, there is a linear functional λ in B^* such that

(11.3) $$y_0 + \lambda(z-x_0) \leq f(z) \quad \text{for all } z \text{ in } \mathcal{D}_f.$$

To see this note that the closed convex set epi f can be separated from (x_0, y_0) by a continuous linear functional on $B \times R$. That is, there exists α in B^* and β in R such that for some real number c

(11.4) $$\alpha(x_0) + \beta y_0 < c < \alpha(z) + \beta y \text{ for all } z \text{ in } \mathcal{D}_f \text{ and all } y \geq f(z).$$

Suppose first that x_0 is in $\mathcal{D}_f$. Then, putting $z = x_0$ in (11.4) we see that $\beta(y-y_0) > 0$ if $y \geq f(x_0)$, which implies $\beta > 0$. In this case we may put $\lambda = -\beta^{-1}\alpha$ in (11.4) to obtain (11.3). If x_0 is not in $\mathcal{D}_f$ then y_0 is arbitrary and (11.4) implies $\beta \geq 0$. If $\beta > 0$ we proceed as before to obtain λ. If $\beta = 0$, however, we have $\alpha(z-x_0) > c - \alpha(x_0) > 0$ for all z in $\mathcal{D}_f$ and, letting λ_0 be some linear functional (whose existence we have already established using some x_0 in $\mathcal{D}_f$) satisfying, for some real number a, $a + \lambda_0(z) \leq f(z)$ for all z in $\mathcal{D}_f$, we put $\lambda = \lambda_0 - s\alpha$ for some large positive real number s. Then for all z in $\mathcal{D}_f$

$$\begin{aligned} y_0 + \lambda(z-x_0) &= y_0 + \lambda_0(z) - \lambda_0(x_0) - s\alpha(z-x_0) \\ &\leq f(z) + y_0 - a - \lambda_0(x_0) - s(c - \alpha(x_0)) \\ &\leq f(z) \end{aligned}$$

if s is sufficiently large. This establishes (11.3) in all cases.

Now for any x_0, y_0 and λ as in (11.3) we have $\lambda(z) - f(z) \leq \lambda(x_0) - y_0$ for all z in $\mathcal{D}_f$. So λ is in $\mathcal{D}_{f^*}$ and $f^*(\lambda) \leq \lambda(x_0) - y_0$. Thus $y_0 \leq \lambda(x_0) - f^*(\lambda)$, and $y_0 \leq \sup\{\lambda(x_0) - f^*(\lambda) : \lambda \in \mathcal{D}_{f^*}\}$. If x_0 is not in $\mathcal{D}_f$ then y_0 is unrestricted and it follows that the right side of the last inequality is $+\infty$, so that x_0 is not in $\mathcal{D}_{f^{**}}$. Hence $\mathcal{D}_{f^{**}} \subset \mathcal{D}_f$. On the other hand if x_0 is in $\mathcal{D}_f$ then the last inequality holds whenever $y_0 < f(x_0)$. It follows that $f^{**}(x_0) \geq f(x_0)$ whenever $f^{**}(x_0)$ exists. But $f^*(\alpha) \geq \alpha(x_0) - f(x_0)$

for all α in $\mathcal{D}_{f^*}$ so that $f(x_0) \geq \alpha(x_0) - f^*(\alpha)$ for all such α. Hence x_0 is in $\mathcal{D}_{f^{**}}$ and $f(x_0) \geq f^{**}(x_0)$. This concludes the proof.

12. Application of the Legendre Transform

Consider a gas of N moles in a cylinder with a piston. How much work will be done by the gas when it expands from volume V_1 to V_2? If the gas is thermally isolated then the work done will be $U_1 - U_2$, by the definition of internal energy. But suppose that the cylinder is kept in thermal contact with a heat reservoir at temperature T and that the process is carried out reversibly.

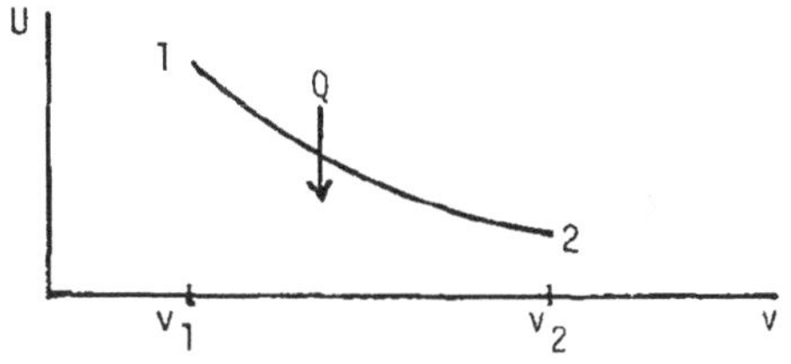

Figure 12.1

The work done by the gas is

(12.1) $$\begin{aligned} W &= U_1 - U_2 + Q \\ &= U_1 - U_2 + \int_1^2 T dS \\ &= U_1 - U_2 + T(S_2 - S_1) \\ &= (U_1 - TS_1) - (U_2 - TS_2). \end{aligned}$$

Now since the entropy $S = S(U,V,N)$ is a strictly increasing function of U with derivative $\partial S/\partial U = 1/T$, we may invert the function $U \to S(U,V,N)$ to get U as a strictly increasing function of S for each V,N: $U = U(S,V,N)$, and of course, $\partial U/\partial S = T$. Moreover one sees easily from a picture that the concavity of $U \to S(U,V,N)$ implies the convexity of $S \to U(S,V,N)$. In fact U is a convex function of S, V, N, as we shall show in the next lemma. But keeping V and N fixed we may take the Legendre transform of $S \to U(S,V,N)$. Denoting it by $-F$ we have $-F(T,V,N) = TS - U(S,V,N)$ by equation (11.2) where S is chosen to satisfy $\partial U/\partial S = T$. That is, $F = U - TS$. Thus by (12.1)

$$W = F(T,V_1,N) - F(T,V_2,N).$$

$F(T,V,N)$ is called the <u>free energy</u>. Because of the invertibility of the Legendre transform $U(\cdot,V,N)$ can be recovered from $F(\cdot,V,N)$. So F determines the thermodynamics of the system as well as $S(\cdot)$ or $U(\cdot)$. We note however that F is not a function on state space. Since a Legendre transform is clearly to be regarded, by virtue of the theorem of Section 11, as a function on a dual space, the temperature T is to be regarded as a coordinate dual to that of S. Here we regard the state space as parametrized by S, V and N instead of U, V and N. In some treatments the surface which is the graph of $S = S(U,V,N)$ is regarded as the state space. The state space thus becomes some C^1 manifold with (U,V,N) or (S,V,N) as two global coordinate systems. Generalizations of this viewpoint have been investigated. See e.g. [Mis].

In the preceding discussion we have seen that the work available in an isothermal reversible process is determined by the Legendre transform of $U(\cdot,V,N)$. Similarly the heat energy available from the gas in an isobaric process (constant pressure) is determined by the Legendre transform of $U(S,V,N)$ with respect to V. The resulting function, $H(S,P,N) = U - PV$ is called the enthalpy. Enthalpy is important because processes which take place when exposed to the atmosphere (e.g. a chemical reaction in an open test tube) are isobaric.

Before proceeding further let us prove the convexity of $U(\cdot,\cdot,\cdot)$.

<u>Lemma</u>. Let f be a concave function defined on a convex set A in R^{n+1}. Assume that for each point x in R^n for which $\{u \in R; (u,x) \in A\}$ is non empty, the function $u \to f(u,x)$ is strictly increasing. Let $g(\cdot,x)$ be the inverse function for each such x.

Then a) $g(s,x)$ is strictly increasing in s for each such x and b) g is a convex function on any convex subset of its domain in R^{n+1}.

Proof. a) is clear. Suppose B is a convex subset of the domain of g. Let (s_1,x_1) and (s_2,x_2) be two points of B and let $0 < \alpha < 1$. Put $u_j = g(s_j,x_j)$ for $j = 1,2$. Then $s_j = f(u_j,x_j)$ for $j = 1,2$, and

$$s_0 \equiv f(\alpha u_1 + (1-\alpha)u_2, \alpha x_1 + (1-\alpha)x_2)$$

$$\geq \alpha s_1 + (1-\alpha)s_2$$

by the concavity of f. Hence

$$\begin{aligned} \alpha g(s_1,x_1) + (1-\alpha)g(s_2,x_2) &= \alpha u_1 + (1-\alpha)u_2 \\ &= g(s_0, \alpha x_1 + (1-\alpha)x_2) \\ &\geq g(\alpha s_1 + (1-\alpha)s_2, \alpha x_1 + (1-\alpha)x_2) \end{aligned}$$

by the definition of s_0 and the monotonicity of g. This proves b).

The lemma shows that $U(\cdot,\cdot,\cdot)$ is a convex function on any convex subset of its domain. But the convexity of its entire domain does not appear to follow from the properties of $S(U,V,N)$ already established and we hereby make the technical assumption that its entire domain is convex. The homogeneity of S shows that the domain of $U(\cdot,\cdot,\cdot)$ is also a cone in R^3.

We conclude this section by computing the Legendre transform of U with respect to S and N.

The derivative $\mu = \partial U(S,V,N)/\partial N$ (cf. Equs. (8.11) and (8.12)), which is now, of course, a function of S, V, and N, is called the chemical potential of the system. It measures the rate of increase of internal energy per added mole of substance when the entropy and volume are kept constant. In order to get a good intuitive understanding of its physical significance it is essential to study its role in determining the equilibrium concentration of chemically interacting systems. (See e.g. [W].) This is beyond the scope of these notes. However the reader can easily appreciate that in a chemical interaction (in which mole numbers can change) equilibrium concentrations may be determined by some "force for change of mole numbers" analogous to the "force for change of volume" which we call pressure. Indeed the (negative) pressure measures the rate of increase of internal energy per unit increase of volume when the entropy and mole number remain constant. That is $-P = \partial U(S,V,N)/\partial V$.

Because $\partial U/\partial S = T$ and $\partial U/\partial N = \mu$ (chemical potential) it is appropriate and customary to denote the coordinates in the two dimensional vector space dual to the planes $V = \text{const.}$ (in S, V, N space) by T and μ, analogously to our first example in this section. With V fixed we must locate (in accordance with the obvious two dimensional generalization of (11.2)) any point S, N such that $\partial U/\partial S = T$ and $\partial U/\partial N = \mu$ where T and μ are given. Then the Legendre transform $(\mathcal{L}U)(T,\mu)$ is

$$-(\mathcal{L}U)(T,\mu) = U(S,V,N) - TS - \mu N.$$

Now the definition of $U(\cdot)$ and the homogeneity of $S(\cdot)$ yields easily the homogeneity of $U(\cdot)$: $U(\lambda S,\lambda V,\lambda N) = \lambda U(S,V,N)$. Upon differentiating this with respect to λ at $\lambda = 1$ we get the Euler relation

$$U = TS - PV + \mu N$$

from which it follows that

$$[\mathcal{L}U(\cdot,V,\cdot)](T,\mu) = PV. \tag{12.2}$$

It is not hard to see that the left side is actually linear in V for fixed T and μ. Hence (12.2) shows that the pressure is a function of temperature and chemical potential.

Finally we note that since the derivative of a function and the derivative of its inverse are simply related, the first and last of the preceding three Legendre transforms discussed can be expressed directly in terms of the entropy function. We leave it as an exercise for the reader to show that

$$P = \frac{T}{V}[\mathcal{L}(-S(\cdot,V,\cdot)](-\frac{1}{T},\frac{\mu}{T}) \; . \tag{12.3}$$

We shall see this equation arising later in statistical mechanical models.

Chapter II: Equilibrium Statistical Mechanics

13. The Single Particle Distribution: Pressure.

By 1851 the experiments of Joule and the formulation of the second law by Kelvin and Clausius provided a consistent and convincing (to many) view that heat is a form of energy. The next question: How is this energy stored in matter? Is it stored in the vibrational energy of a continuum out of which matter is made? Or, if matter, say a gas, is made of discrete particles (molecules), are the particles more or less stationary, with their energy stored as potential energy, or are the particles of a gas moving quickly, with their energy stored to a significant extent as kinetic energy? Deductions of the classical gas law, $PV = NRT$, from the assumption that the gas consists of large numbers of rapidly moving molecules (kinetic theory of gases) were published by Krönig (1856), Clausius (1857), Maxwell (1859) and Boltzmann (1868) during the next 20 years after the establishment of the first and second laws of thermodynamics.

The point of view that emerged from the work of these four (primarily the last three) was this. Suppose that in a region $\Lambda \subset R^3$ there are N molecules. For an ordinary gas N will be very large--on the order of 10^{23}. Moreover the motions will be rapid and complicated. Assume that when the gas is in macroscopic equilibrium the frequency distribution of position and momentum of the particles is given by a measure ν on $\Lambda \times R^3$. Thus if $A \subset \Lambda \times R^3$, then $\nu(A)$ is the "expected number" of particles with position and momentum, $(\vec{x},\vec{p})$, in the set A. In particular $\nu(\Lambda \times R^3) = N$. Maxwell, for example, argued that in the absence of external forces, such as gravity, the measure ν should have the form $d\nu(\vec{x},\vec{p})$ = constant times $e^{-\alpha|\vec{p}|^2}dx\,dp$ for some constant α. (He later rejected his original arguments while keeping the conclusion.)

Let us see how the pressure of the gas can be deduced from such a probabilistic, microscopic description. Consider a small flat section, A, of the wall as shown in Figure 13.1. We assume that a molecule hitting the wall bounces off with its angle of incidence equal to its angle of reflection (specular reflection) and that all of the molecules have mass m.

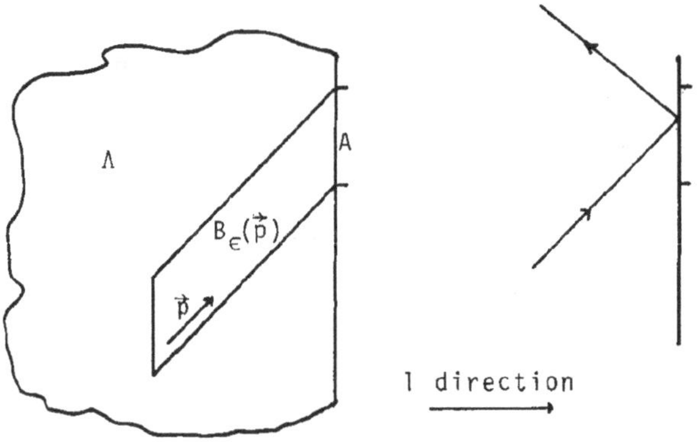

Figure 13.1

The force necessary to change the momentum of a particle of mass m is determined by Newton's equation: $\vec{F} = m\vec{a} = d\vec{p}/dt$ where $\vec{p} = m\vec{v}$, as usual. Thus in a time interval (a,b) the average force on the particle is

$$\frac{1}{b-a}\int_a^b \vec{F}\,dt = \frac{1}{b-a}\int_a^b (d\vec{p}/dt)dt$$
$$= (\vec{p}(b)-\vec{p}(a))/(b-a) \ .$$

Since the pressure on the wall at A is, by definition, the force on A exerted by the gas divided by the area of A, we shall consider first the average force exerted by the particles in Λ on the surface A in the time interval $(0,\epsilon)$, and then we shall let ϵ go to zero. Let $B_\epsilon(\vec{p})$ be the set of initial points in Λ from which a particle starting at time zero with momentum $\vec{p}$ will collide with A at some time in $(0,\epsilon)$:

$$B_\epsilon(p) = \{\vec{x} \in \Lambda: \vec{x} + (\vec{p}/m)\, s \in A \text{ for some } s \text{ in } (0,\epsilon)\}.$$

We are ignoring here possible collisions with other molecules and other parts of the wall. This is heuristically justifiable on the grounds that we will let ϵ go to zero in the end and this is in fact the reason for taking this limit. Now a particle that collides with A changes only its first component of momentum and the change is $-2p_1$. Thus the average force on A in time ϵ exerted by this particle is $2p_1/\epsilon$. For the total average force in time ϵ for the N particles we integrate $2p_1/\epsilon$ over the set, C_ϵ, of all possible positions and momenta that can cause a collision with A during $(0,\epsilon)$. Ignoring intermolecular collisions we have

$$C_\epsilon = \{(\vec{x},\vec{p}) \in \Lambda \times R^3: \vec{x} \in B_\epsilon(\vec{p})\}.$$

Hence the pressure at A is

$$(13.1) \qquad P = (\text{area } A)^{-1} \lim_{\epsilon \downarrow 0} \epsilon^{-1} \int_{C_\epsilon} 2p_1 \, d\nu(\vec{x},\vec{p}).$$

Examples. Suppose, as Maxwell argued for a gas of hard spheres, that ν is Gaussian in the momentum variables and uniform in the spatial variables. Maxwell, in fact, deduced this simply from the assumption of independence of orthogonal momentum variables and spherical symmetry in these variables. Then we may write

$$d\nu(\vec{x},\vec{p}) = (N/V)(D \exp[-\beta|\vec{p}|^2/2m])dx\, dp$$

where $D = (2\pi m/\beta)^{-3/2}$ is the nomalization constant for the momentum factor and V is the Lebesgue measure of Λ. Thus $\nu(\Lambda \times R^3) = N$. Referring to Figure 13.1 we see that $B_\epsilon(p)$ is empty if $p_1 < 0$, while for $p_1 \geq 0$ we have, writing $|C|$ for the volume of a set C,

$|B_\varepsilon(p)| = \varepsilon(p_1/m)$area A, if the parallelopiped shown is actually contained in Λ. In any case, we have

$$\frac{|B_\varepsilon(\vec{p})|}{\varepsilon \text{ area } A} \leq p_1/m, \quad p_1 \geq 0$$

with equality in the limit as $\varepsilon \downarrow 0$. Hence, using Fubini's theorem and then the dominated convergence theorem, we have, from (13.1)

$$\begin{aligned} P &= \lim_{\varepsilon \downarrow 0} \frac{N}{V} D \int_{R^3} 2p_1 e^{-\beta|p|^2/2m} \Big(\int_{B_\varepsilon(\vec{p})} \frac{1}{\varepsilon \text{ area } A} dx\Big) dp \\ &= \lim_{\varepsilon \downarrow 0} \frac{N}{V} D \int_{\{\vec{p}: p_1 \geq 0\}} 2p_1 e^{-\beta|p|^2/2m} \Big(\frac{|B_\varepsilon(\vec{p})|}{\varepsilon \text{ area } A}\Big) dp \\ &= \frac{N}{V} (D/m) \int_{\{\vec{p}: p_1 \geq 0\}} 2p_1^2 \, e^{-\beta|\vec{p}|^2/2m} \, dp \\ &= N/(V\beta). \end{aligned}$$

That is, $PV = N/\beta$. Now the classical gas law of Boyle and Gay-Lussac, $PV = NkT$, [where N is now number of particles instead of number of moles and k is a constant--Boltzmann's constant] suggests that the parameter β in the statistical distribution should have the thermodynamic interpretation of a constant times inverse temperature. Indeed we shall see later that this is consistent with the currently accepted derivation of thermodynamics from mechanics.

Boltzmann [1868] proposed a generalization of the preceding equilibrium distribution of Maxwell to a gas whose particles interact with an external field (e.g. a gravitational field). He proposed $d\nu = \text{const.} \exp[-\beta E(\vec{x},\vec{p})] dx\, dp$ with $\beta = (kT)^{-1}$ and $E(\vec{x},\vec{p})$ the energy of a particle at $\vec{x}$ of momentum $\vec{p}$. For example in a gravitational field the energy of a particle of mass m is $E(\vec{x},\vec{p}) = |\vec{p}|^2/(2m) + mgz$ where $\vec{x} = (x,y,z)$. If one takes for Λ a vertical infinite cylinder with horizontal base B located at $z = 0$ then one can deduce from (13.1) that the pressure at height z and temperature T is in this case

$$P = Nmg\, e^{-mgz/kT}/(\text{area of base } B). \tag{13.2}$$

The derivation is similar to the preceding derivation of $PV = NkT$. One must take the section of wall A to be at height z and very small. I.e., one should take a limit in (13.1) as A decreases to a point. We leave this as an exercise for the reader. (13.2) is called the barometric formula.

Much of the kinetic theory of gases from the 1850's to the present is concerned with the question of the time evolution of the frequency distribution ν when the initial distribution is not an equilibrium distribution. The time evolution of ν is important to understand because it pertains to the following non-equilibrium questions. Why and how does a system not in thermal equilibirum approach equilibrium? How does the kinetic theory explain quantitatively the conduction of heat through a gas, the diffusion of one gas through another, or the viscosity of a gas? These questions, especially the first one, are intimately connected with equilibirum statistical mechanics. But we are exclusively concerned in these notes with the equilibrium case. For a historical account of these non-equilibrium questions, we refer the reader to the books [Br, E]. We shall here simply describe Gibb's formulation, (1902), of equilibirum statistical mechanics--a culmination of the previous 40 years of development of these ideas by Boltzmann and others.

14. Phase Space and Liouville Measure.

Consider N point particles each of mass m in a bounded region $\Lambda \subset R^3$. We assume Λ is the closure of its interior. The phase space, $\mathcal{P}$, for this system is $(\Lambda \times R^3)^N$. We write $q = (\vec{x}_1, \vec{x}_2, \ldots, \vec{x}_N) = (q_1, \ldots, q_{3N}) \in \Lambda^N$ and $p = (\vec{p}_1, \ldots, \vec{p}_N) \in R^{3N}$ for the configuration and momentum, respectively, of the system. Write $H_\Lambda(q,p)$ for the energy of the system in state q,p. Thus

$$(14.1) \qquad H_\Lambda(q,p) = \Sigma_{j=1}^N |\vec{p}_j|^2/(2m) + U_\Lambda(q)$$

where $U_\Lambda(q)$ is the potential energy of the configuration q. If the particles interact via a potential $V(\vec{x}_i-\vec{x}_j)$ (so force of particle j on particle i is $-\text{grad}_i\, V(\vec{x}_i-\vec{x}_j)$) then for $q \in \Lambda^N$

$$U_\Lambda(q) = \frac{1}{2} \sum_{i\neq j} V(\vec{x}_i-\vec{x}_j) + \Sigma_{i=1}^N \varphi(\vec{x}_i)$$

where φ is the potential of external forces. We assume here that $V(-\vec{x}) = V(\vec{x})$. We recall that Newton's equations of motion can be written

$$d\vec{x}_j/dt = \partial H/\partial\vec{p}_j, \quad d\vec{p}_j/dt = -\partial H/\partial\vec{x}_j.$$

(These derivatives should be interpreted, of course, componentwise.) Now under reasonable conditions on V and φ, and under the assumption that a particle that collides with the wall (which we assume smooth) is reflected specularly, the initial value problem has a unique left continuous solution which exists for all time. We assume that this is the case and write $(q(t),p(t)) = T_t(q^0,p^0)$ for the solution with initial condition (q^0,p^0). It is a basic fact of classical mechanics that the transformation $T_t: \mathcal{P} \to \mathcal{P}$ leaves invariant the Liouville measure $dqdp \equiv dq_1 \ldots dq_{3N} dp_1 \ldots dp_{3N}$. We refer the reader to [ML, pages 3-5] or [K] for a proof.

Let E be a real number. Assume that the set $\mathbf{C}_E \equiv \{(q,p) \in \mathcal{P}: H(q,p) = E\}$ is compact. This will be the case, for example, if V and φ are smooth and bounded below on Λ, which we assume. Since the gradient of H is nowhere zero, except perhaps on $(\Lambda \times \{0\})^N$, which is a set of measure zero, the measure on the real line induced by H from Liouville measure is absolutely continuous with respect to Lebesgue measure and therefore has a density $Z(E)$. That is,

$$Z(E) = (d/dE)|\{u \in \mathcal{P}: H(u) \leq E\}|$$

where $|A|$ denotes the Liouville measure of a set A.

Moreover for each real number E for which $Z(E) > 0$ there exists a unique probability measure α_E on $\mathcal{C}_E$ such that

$$(14.2) \qquad \int_{-\infty}^{\infty} \alpha_E(A \cap \mathcal{C}_E) Z(E)\, dE = |A|$$

for every Borel set $A \subset \mathcal{P}$. The measures α_E may be described in the following ways.

i. For a purely measure theoretic description we may define α_E as a conditional expectation with respect to Liouville measure:

$$\alpha_E(A \cap \mathcal{C}_E) = E(\chi_A | H = E)$$

where χ_A is the indicator function of the set A.

ii. If $B \subset \mathcal{C}_E$ then

$$Z(E)\alpha_E(B) = \lim_{\epsilon \downarrow 0} |B_\epsilon|/\epsilon$$

where $B_\epsilon = \{u+s \text{ grad } H(u) \colon u \in B,\ 0 \le s \le \epsilon\}$

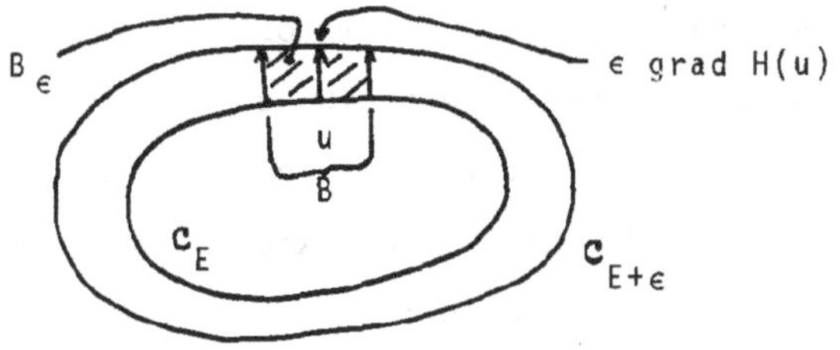

Figure 14.1

The set B_ϵ is approximately the region "between" the surfaces $\mathcal{C}_E$ and $\mathcal{C}_{E+\epsilon}$ obtained by extending the surface element B normal to $\mathcal{C}_E$ to fill up the shaded region shown in Figure 14.1.

iii. If σ denotes Lebesgue surface area on $\mathcal{C}_E$, computed using the metric $(\Sigma_{j=1}^{3N}(q_j^2 + p_j^2))^{1/2}$, then we see from Figure 14.1 that $Z(E)d\alpha_E = d\sigma/|\text{grad } H|$. This is the most customary description of α_E. (See e.g. [K].) We note that the Euclidean metric used to compute surface measure σ is a sum of dimensionally inhomogeneous terms, and is therefore physically meaningless. Moreover, the same objection applies to $|\text{grad } H|$. The quotient $d\sigma/|\text{grad } H|$ is, however, meaningful, being definable as in i or ii and without reference to any metric.

iv. For a reader who enjoys exterior differential calculus, I want to point out that if the exterior derivative dH is not zero at some point u of C_E then there is neighborhood of u in $\mathcal{P}$ and a $6N-1$ form ω on that neighborhood such that $dH \wedge \omega = dq_1 \wedge \ldots \wedge dp_{3N}$. The existence of ω can be proved using a local coordinate system with H as the first coordinate. ω is not unique. But its restriction to C_E is unique and gives, locally, exactly the measure $Z(E)\alpha_E$.

Now the energy is constant along the orbits of the Newtonian flow T_t. Consequently T_t maps each surface C_E into itself. Moreover, since the volume of B_ϵ is unchanged by T_t (up to terms of order ϵ^2, the way B_ϵ is defined). T_t leaves the measure α_E invariant. We shall make use of this fact soon.

Suppose now that a gas of N point particles is confined to the region Λ in R^3. Assume that the boundary of Λ is insulated so that the energy of the gas is fixed, with value E, say. The phase space of the gas is $(\Lambda \times R^3)^N$. If the gas molecules interact with each other via a potential V and with an external field via the potential φ as before, then we may use the preceding measure theoretic structure to determine the frequency measure ν on $\Lambda \times R^3$ via the following arguments.

Let $A \subset \Lambda \times R^3$. Suppose that the instantaneous state of the system at time $t = 0$ is given by a point γ in C_E. Then the number of molecules whose phase point lies in A is

$$n(\gamma) = \Sigma_{k=1}^{N} \chi_A(\Pi_k \gamma) \tag{14.3}$$

where Π_k: $(\Lambda \times R^3)^N \to \Lambda \times R^3$ is the kth coordinate projection and χ_A is the function which is one on A and zero elsewhere. We have already seen what kind of sets, A, are of interest for predicting the pressure. Now any measurement of the system takes a long time compared to the rapid movements of the molecules and during this time the point γ moves according to the flow T_t. A measurement of the gas therefore is really a measurement of the average

$$\frac{1}{T}\int_0^T n(T_t\gamma)dt$$

where T is "large". It is reasonable and customary to approximate this by the limit as $T \to \infty$. Thus we obtain

$$(14.4) \qquad \nu(A) = \lim_{T\to\infty} T^{-1}\int_0^T n(T_t\gamma)dt$$

for the expected number of molecules with phase point in the set $A \subset \Lambda \times R^3$.

Next, recall that if the flow T_t is ergodic with respect to the measure α_E (that is, if the only sets in $\mathbf{C}_E$ which are invariant under T_t for all t are of α_E measure 0 or 1), then by the ergodic theorem the limit in (14.4) may be replaced, at least for almost all initial points γ, by the expectation with respect to α_E:

$$(14.5) \qquad \nu(A) = \int_{\mathbf{C}_E} n(\gamma)\alpha_E(d\gamma).$$

Now ergodicity has never been proven for realistic systems except in the case of colliding billiard balls (Sinai, 1966). Nevertheless the equation (14.5) and its extensions to other problems have had such far reaching agreement with experiment that it is now generally accepted as the correct starting point for equilibirum statistical mechanics even though possible lack of ergodicity represents a gap in the argument which led to it. From now on we too will ignore time evolution and take the corresponding phase space average as the basic ingredient of the theory.

It is clear that (14.5) defines a measure ν on $\Lambda \times R^3$ with $\nu(\Lambda \times R^3) = N$ as it should. We have thus shown how the equilibirum frequency measure ν is to be determined--in principle--from classical mechanics. Since ν depends on the energy E, the equation (13.1) will give the pressure as a function of E and Λ. But to recover thermodynamics from mechanics one must recover entropy and/or temperature. We have so far mentioned temperature only in the context of the two special frequency distribution laws of the preceding examples. A means for defining entropy for a general mechanical system of particles began with the paper of Boltzmann [B]. By 1902 the basic structure of the equilibrium theory of statistical mechanics took the form which is today regarded (with modifications from quantum mechanics) as correct. We shall turn now to this formulation, which was first published in Gibb's book "Statistical Mechanics" (1902). In so doing we are skipping the years of controversy between Boltzmann on the one hand and Loschmidt and Zermelo and others on the other, concerning the meaning and mechanism of the approach to equilibrium--a controversy not yet settled. For a history of this period see [Br] and [E].

15. Three Ways to Recover Thermodynamics from Mechanics.

I. The Micro Canonical Ensemble.

Let h be a positive real number--which we take to have the dimension length times momentum. Let τ_N be the dimensionless measure on $(\Lambda \times R^3)^N$ given by

$$(15.1) \qquad d\tau_N = h^{-3N}(N!)^{-1}dq_1 \ldots dq_{3N}dp_1 \ldots dp_{3N}.$$

As in the preceding section Λ is a bounded region in R^3. Fix $\epsilon > 0$, permanently, and let

$$Z_\Lambda^m(E,V,N) = \tau_N(\{(q,p) \in (\Lambda \times R^3)^N\colon E-\epsilon \leq H_\Lambda(q,p) \leq E\}) \quad .$$

where $H_\Lambda(q,p)$ is the total energy function of the system described in the preceding section and V is the volume of Λ. [Since V depends on Λ the notation is admittedly redundant.]

We recall that the classical gas law of Boyle and Gay-Lussac for one mole of a gas is $PV = RT$ where R is a constant independent of the particular gas. Boltzmann's constant, k, is defined by $k = R/N_a$ where N_a is Avogadro's number--the number of molecules making up one mole ($N_a = 6.022 \times 10^{23}$ approximately.).

Define

$$(15.2) \qquad S_\Lambda(E,V,N) = -k \log[Z^m_\Lambda(E,V,N)].$$

Claim #1. S_Λ is the entropy of the system in Λ.

Remark. We shall see later (Section 16) that in a certain sense $S_\Lambda(E,V,N)$ depends on Λ only through its volume V. This is the reason for making explicit the dependence of the preceding functions on V. Granting then, for the moment, that $S(E,V,N)$ is the desired entropy function of the system, one should note that the temperature and pressure of the system can be recovered (cf. Section 9) by the usual thermodynamic equations $T^{-1} = \partial S/\partial E$ and $P = T\, \partial S/\partial V$. Thus the definition (15.2) yields temperature and pressure from mechanics as well as entropy. Of course one must address the question of extensivity and concavity of the entropy in addition to its independence of the shape of Λ before we can even assert that we have an internally consistent theory, let alone a correct one. We shall discuss these questions in Section 16.

II. The Canonical Ensemble.

Define for $T > 0$

$$(15.3) \qquad Z^c_\Lambda(T,V,N) = \int_{(\Lambda\times R^3)^N} \exp[-(kT)^{-1}H_\Lambda(q,p)]d\tau_N(q,p).$$

Put

$$(15.4) \qquad F_\Lambda(T,V,N) = -kT \log(Z^c_\Lambda).$$

Let

$$(15.5) \qquad \rho_N(q,p) = (Z_\Lambda^c)^{-1}\exp[-H_\Lambda(q,p)/kT].$$

Then $\int \rho_N \, d\tau_N = 1$. For any function f on phase space $\mathcal{P}$ we write its expectation with respect to the density ρ_N as

$$(15.6) \qquad \langle f\rangle_{T,\Lambda,N} = \int_{\mathcal{P}} f(\gamma)\rho_N(\gamma)d\tau_N(\gamma)$$

when the integral exists. Define

$$(15.7) \qquad U = \langle H_\Lambda\rangle_{T,\Lambda,N}$$

and

$$(15.8) \qquad S = -k\langle \log \rho_N\rangle_{T,\Lambda,N}.$$

Claim #2. $F_\Lambda(T,V,N)$ is the Helmholtz free energy function of the system at temperature T. U is its internal energy, and S is its entropy.

Internal consistency. As in the microcanonical ensemble, independence of the shape of Λ must be shown along with extensivity. We leave further discussion of this for the next section. But in the present ensemble there is in addition another, and easier, aspect of internal consistency to be shown. Since the free energy, internal energy, and entropy are all directly given in this ensemble, we must show that they are related in the appropriate way, considering that in thermodynamics (cf. Section 12), the free energy is the negative of the Legendre transform of U with respect to S. Thus we must verify that $\partial F/\partial T = -S$ and $F = U-TS$. But from (15.8) and (15.5)

$$\begin{aligned} S &= -k\langle -H_\Lambda/kT - \log Z_\Lambda^c\rangle_{T,\Lambda,N} \\ &= T^{-1}\langle H_\Lambda\rangle + k \log Z_\Lambda^c \\ &= U/T - F_\Lambda/T. \end{aligned}$$

So $F_\Lambda = U - TS$. Moreover, from the definitions (15.4) and (15.3) we have

$$\partial F_\Lambda/\partial T = -k \log Z_\Lambda^c - kT\langle (kT^2)^{-1}H_\Lambda\rangle_{T,\Lambda,N}$$
$$= -k \log Z_\Lambda^c - T^{-1}\langle H_\Lambda\rangle_{T,\Lambda,N}$$
$$= -S.$$

Since the pressure, P, and chemical potential, μ, are not explicitly given in this ensemble, the thermodynamic identities $-P = \partial F/\partial V$ and $\mu = \partial F/\partial N$ need not be verified, but must be taken as definitions of P and μ.

III. The Grand Canonical Ensemble.

Recall that the Hamiltonian (14.1) depends on N. But we will not indicate this dependence explicitly. Let

(15.9) $$Z_\Lambda^{g.c.}(T,\mu,V) = \Sigma_{N=0}^{\infty}\int_{(\Lambda\times R^3)^N} \exp[-(kT)^{-1}(H_\Lambda-\mu N)]d\tau_N.$$

The $N = 0$ term is to be interpreted as one. Define

(15.10) $$P_\Lambda(T,\mu) = (kT/V)\log Z_\Lambda^{g.c.}(T,\mu,V).$$

Write $\Gamma_\Lambda = \cup_{N=0}^{\infty}(\Lambda\times R^3)^N$, with the $N = 0$ term defined as a single point, $\emptyset$. Define ρ on Γ_Λ by

(15.11) $$\rho(q_1,\ldots,q_N,p_1,\ldots,p_N) = (Z_\Lambda^{q.c.})^{-1}\exp[-(kT)^{-1}(H_\Lambda-\mu N)]$$

and write τ for the measure on Γ_Λ whose restriction to $(\Lambda\times R^3)^N$ is τ_N, with $\tau(\{\emptyset\}) = 1$. Then $\int_{\Gamma_\Lambda}\rho\, d\tau = 1$. For any reasonable function f on Γ_Λ we define

$$\langle f\rangle_{T,\mu,\Lambda} = \int_{\Gamma_\Lambda} f\,\rho\, d\tau.$$

Finally, we define

(15.12) $$U = \langle H_\Lambda \rangle_{T,\mu,\Lambda}$$

(15.13) $$\overline{N} = \langle N \rangle_{T,\mu,\Lambda}$$

and

(15.14) $$S = -k\langle \log \rho \rangle_{T,\mu,\Lambda}.$$

Claim #3. $P_\Lambda(T,\mu)$ is the pressure of the system at temperature T and chemical potential μ. (Cf. Equs. (8.11) and (8.12) and Sec. 12.) U is the internal energy, S is the entropy and $\overline{N}$ is the number of particles.

Internal Consistency. As in the preceding two ensembles, the presumably extensive quantity $VP_\Lambda(T,\mu)$ must be shown to be independent of the shape of Λ and actually extensive. As before, we leave this for the next section. However, in thermodynamics, VP is the Legendre transform of U with respect to S and N because of the Euler relation (cf. Sec. 12). Since all of the thermodynamic variables U, S, N, P, T, and μ appear explicitly in this ensemble, it behooves us to show that the Euler relation, a) $TS = U + PV - \mu N$, and the correct derivative relations, b) $VdP = SdT + Nd\mu$, hold for the quantities defined in (15.10), (15.12), (15.13) and (15.14). The derivative relations b) must be verified because in the presence of a) they are equivalent to (8.12). The latter cannot be verified directly in this ensemble because S is not given explicitly as a function of U, V and N.

To prove a) we use the definitions for this ensemble to get

$$\begin{aligned} TS &= -kT\langle \log \rho \rangle \\ &= -kT\langle -(kT)^{-1}(H_\Lambda - \mu N) - \log Z_\Lambda^{g.c.} \rangle \\ &= \langle H_\Lambda \rangle - \mu\langle N \rangle + kT \log Z_\Lambda^{g.c.} \\ &= U - \mu\overline{N} + PV \end{aligned}$$

as required. Of course, we are now measuring quantity of matter by number of particles rather than the (proportional) number of moles. For the derivative relations we have

$$V\,\partial P_\Lambda/\partial T = k \log Z_\Lambda^{g.c.} + kT(Z_\Lambda^{g.c.})^{-1}(kT^2)^{-1}\int_{\Gamma_\Lambda}(H_\Lambda - \mu N)e^{-(kT)^{-1}(H_\Lambda-\mu N)}d\tau$$

$$= k \log Z_\Lambda^{g.c.} + T^{-1}\langle H_\Lambda\rangle - (\mu/T)\langle N\rangle$$

$$= S$$

by the third line of the preceding TS equalities. Moreover,

$$V(\partial P_\Lambda/\partial\mu) = kT(Z_\Lambda^{g.c.})^{-1}\int_{\Gamma_\Lambda}(kT)^{-1}Ne^{-(kT)^{-1}(H_\Lambda-\mu N)}d\tau$$

$$= \overline{N}.$$

Remarks. Each of the three preceding formalisms yields some of the thermodynamic variables as functions of others. We have postponed the discussion in all three cases of whether the respective thermodynamic function S_Λ, F_Λ or P_Λ is actually independent of the shape of the region Λ--as it must be in order to recover thermodynamics. Now the idea that heat, pressure, and temperature are reflections of the laws of mechanics when there are "large numbers" of particles present entails the assumption that Λ is very large on a molecular scale and that N is of course also very large. It is only for such "large" regions Λ and large N that one should expect (approximate) shape independence. Mathematically this requirement takes the form that the entropy per unit volume, S_Λ/V should converge to a limit as Λ increases to R^3 through regions of a reasonably arbitrary shape--provided of course that E and N also increase so as to yield finite energy density, $u = \lim E/V$, and particle density $d = \lim N/V$ as Λ increases to R^3. This limiting process is called "taking the thermodynamic limit", and applies also to F_Λ/V and P_Λ as we shall see in the next section.

Of course, the presence of the factor $(N!)^{-1}$ in the definition of the measure τ_N affects the existence of the thermodynamic limit of S_Λ/V, F_Λ/V and P_Λ. Its presence is necessary (or at least some factor with the same assymptotic behavior is necessary) in order to permit the existence of these limits, as it turns out. Some treatments argue for its presence on the ground that the correct volume in phase space should take account of the fact that the particles are identical. But in classical mechanics identical particles are in fact distinguishable. In quantum mechanics, however, identical particles are indistinguishable and the factor $(N!)^{-1}$ appears automatically in the quantum mechanical formalism. We, however, shall have to rely on the fact of validity of the existence theorem of the next section for justification of the appearance of the factor $(N!)^{-1}$ in the measure τ_N.

We remark also that in the thermodynamic limit the dependence of S_Λ/V on the arbitrarily chosen ε, which has been fixed in the microcanonical ensemble, disappears.

We note finally that any one of the ensembles gives rise to a frequency measure ν on $\Lambda \times R^3$ in a manner similar to (14.3) and (14.5). For example in the grand canonical emsemble one puts $n(\gamma) = \Sigma_{N=0}^{\infty}\Sigma_{k=1}^{N}\chi_A(\pi_k\gamma)$ for γ in Γ_Λ and $A \subset \Lambda \times R^3$ and then $\nu(A) = \langle n(\cdot)\rangle_{T,\mu,\Lambda}$. In this case $\nu(\Lambda \times R^3) = \langle N\rangle_{T,\mu,\Lambda} = \bar{N}$.

16. The Thermodynamic Limit and Equivalence of Ensembles.

Consider now

$$H_\Lambda(q,p) = \Sigma_{j=1}^{N}|\vec{p}_j|^2/2m + U_\Lambda(q)$$

where $U_\Lambda(q) = (1/2)\Sigma_{i\neq j=1}^{N} V(\vec{x}_i - x_j)$ and $V(x)$ is a symmetric potential of interaction between two molecules in the region Λ. Under reasonable conditions on the potential V the following theorem may be proved. For precise statements and proofs we refer the reader to the original work [L] or [ML, Sec. 3.4.1]. The limit as Λ approaches

R^3 referred to in the theorem allows Λ to increase to R^3 through any sequence of sets whose "surface area" is not disproportionately large compared to its volume. For a precise description of this kind of convergence (Van Hove convergence), see [ML, page 89].

Theorem. (Lanford 1972)

1) $$s(u,d) = \lim_{\Lambda\to R^3} \frac{1}{V} S_\Lambda(E,V,N)$$

exists if E and N are chosen for each Λ so as to go to ∞ with the volume V in such a way that E/V converges to u and N/V converges to d.

2) $$f(T,d) \equiv \lim_{\Lambda\to R^3} \frac{1}{V} F_\Lambda(T,V,N)$$

exists if N is chosen to go to ∞ with the volume V in such a way that N/V converges to the density d.

3) $$P(T,\mu) \equiv \lim_{\Lambda\to R^3} P_\Lambda(T,\mu)$$

exists.

Moreover the functions $S(U,V,N) \equiv Vs(U/V,N/V)$, $F(T,V,N) \equiv Vf(T,N/V)$ and $P(T,\mu)$ have the correct monotonicity, convexity and extensivity properties for the entropy, free energy, and pressure functions of a thermodynamic system. Furthermore they define the same thermodynamics. That is, they are related to each other via the Legendre transform as in Section 12.

Remarks 16.1. The last statement of the theorem is called the equivalence of ensembles because it asserts that, in the thermodynamic limit, the three ensembles give the same thermodynamics. It tends to be easier, in practice, to use the canonical ensemble than the micro-canonical ensemble and easier to use the grand canonical ensemble than the canonical ensemble, for computations of explicit formulas.

16.2. In regard to phase transitions, we saw in our discussion of thermodynamics that coexistence of phases is reflected by flat spots on the entropy function and therefore by non differentiability of the pressure, $P(T,\mu)$ and free energy, $F(T,N)$. But the expression $P_\Lambda(T,\mu)$ in Section 15 is easily seen to be continuously differentiable (in fact real analytic) in T and μ if the interaction is reasonable. Thus the pressure function P_Λ does not allow phase transitions in the sense we have discussed. One must take the thermodynamic limit--where a loss of smoothness of P_Λ and F_Λ is possible and actually occurs--in order to see phase transitions. Of course real systems are finite, though large $(N \sim 10^{23})$ and one expects that phase transitions are reflected by "approximate" flat spots in S_Λ for large Λ, and corresponding "nearly non differentiable" points in P_Λ and F_Λ. But it may be difficult to quantify these approximate notions.

Chapter III: Random Fields

17. Beyond the Thermodynamic Functions: The Configurational Ensemble.

In addition to the thermodynamic functions, Section 15 also described three measure spaces. There was, first, the energy shell, $E-\varepsilon < H \leq E$, with Liouville measure. Second, there was phase space, $(\Lambda \times R^3)^N$, with the probability measure $\rho_N d\tau_N$. Third, we considered $\Gamma_\Lambda \equiv \cup_{N=0}^{\infty}(\Lambda \times R^3)^N$, with the probability measure $\rho d\tau$. These three measures contain useful information about the physical system. For example the moments of the random variable $n(\cdot)$ defined in (14.3), which determines the frequency measure ν as in (14.5), are particularly useful in studying the behavior of a gas near its critical point (See the point c in Figure 9.1). Critical phenomena are currently under intense investigation in the physics literature. See e.g. [Fi], [Fr], [Do] and their bibliographies. It is the behavior of these moments and the correlations between such random variables for <u>very large Λ</u> which are of interest. Now it is possible to take the thermodynamic limit of these measure spaces and obtain a probability space in the limit. One expects to obtain the same measure spaces whichever ensemble one starts with, [DT], [Th], [A]. This extends the notion of equivalence of ensembles discussed in Section 16. We shall discuss only the grand canonical ensemble here.

The integral defining the partition function $Z_\Lambda^{g.c.}(T,\mu)$ can be partly carried out explicitly because the momentum integrals are Gaussian. Thus, since $H_\Lambda = \Sigma_{j=1}^{3N} p_j^2/(2m) + U_\Lambda(q)$ and $\int_{-\infty}^{\infty}\exp[-\beta p_j^2/(2m)]dp_j = (2\pi m\beta^{-1})^{1/2}$, we have from (15.9)

$$Z_\Lambda^{g.c.}(T,\mu) = \Sigma_{N=0}^{\infty} \int_{\Lambda^N} \left(\frac{2\pi mkT}{h^2}\right)^{\frac{3N}{2}} e^{-(kT)^{-1}(U_\Lambda-\mu N)} d\omega_N(q)$$

where

$$(17.1) \qquad d\omega_N(q) = (N!)^{-1} dq_1 \cdots dq_{3N}.$$

It is customary to put

$$z = \left(\frac{(2\pi mkT)^{1/2}}{h}\right)^3 e^{\mu/(kT)}$$

so that the partition function appears as a power series in z:

$$(17.2) \qquad Z_\Lambda^{g.c.}(T,\mu) = \Sigma_{N=0}^{\infty} z^N \int_{\Lambda^N} e^{-\beta U_\Lambda(q)} d\omega_N(q).$$

We have put, as usual, $\beta = (kT)^{-1}$. z is called the activity.

As we have seen, the partition function in each ensemble determines all the thermodynamic functions. At the same time, in each ensemble, it is the normalization constant for some finite measure. Written in the form (17.2), it is the normalization constant for the following measure space. Let

$$Q_\Lambda = \cup_{N=0}^{\infty} \Lambda^N$$

where $\Lambda^0 = \{\emptyset\}$ consists of a single point. Writing $\omega = \oplus_{N=0}^{\infty} \omega_N$, with $\omega_0(\{\emptyset\}) = 1$, we see that $Z_\Lambda^{g.c.}(T,\mu)$ is the normalization constant for the measure on Q_Λ whose density with respect to ω is $z^N e^{-\beta U_\Lambda(q)}$ on Λ^N. Write

$$\langle f \rangle_{\Lambda,T,\mu} = [Z_\Lambda^{g.c.}(T,\mu)]^{-1} \Sigma_{N=0}^{\infty} z^N \int_{\Lambda^N} f_N(q) e^{-\beta U_\Lambda(q)} d\omega_N(q)$$

for the normalized expectation of a function f on Q_Λ. Here $f_N = f|\Lambda^N$. The measure space determined by Q_Λ and this expectation is called the <u>configurational</u> ensemble. For any Borel set $A \subset \Lambda$ define a random variable n_A on Q_Λ by

$$n_A(\gamma) = \Sigma_{N=1}^{\infty} \Sigma_{k=1}^{N} \chi_A(\pi_k^N \gamma)$$

where $\pi_k^N \colon \Lambda^N \to \Lambda$ is the kth coordinate projection. For any point γ the sum on N has only one nonzero term. As in the discussion of the frequency measure, $\langle n_A \rangle_{\Lambda,T,\mu}$ has the interpretation of expected number of particles in A when the system has temperature T and chemical potential μ.

Here is the type of theorem that captures the idea of a thermodynamic limit for the measure spaces $(Q_\Lambda, \langle\ \rangle_{\Lambda,T,\mu})$.

Theorem. Under reasonable conditions on the potential $V(\vec{x})$ there exists an increasing sequence, Λ_n, of cubes in R^3 with union R^3 such that for any finite set, $A_1,\ldots,A_j$, of bounded Borel sets in R^3 the joint distributions of $n_{A_1},\ldots,n_{A_j}$ with respect to $(Q_{\Lambda_n}, \langle\ \rangle_{\Lambda_n,T,\mu})$ converge as $n \to \infty$. The limit determines a process $A \to n_A$, defined for bounded Borel sets A.

Remarks 17.1. The sense of convergence usually considered in the mathematical physics literature is convergence of the expectations of products; $\lim_{\Lambda_n \to R^3} \langle n_{A_1} \cdots n_{A_k} \rangle_{\Lambda_n,T,\mu}$.

17.2. The limit process may not be unique. That is, it may depend on the choice of cubes Λ_n. One expects non uniqueness to be associated with existence of coexisting phases at the temperature T and chemical potential μ. We shall discuss this in more detail in lattice models. For further discussion in the continuous case (i.e., space = R^3 rather than space = Z^3) see [Mi], [Pr], [Rl, Ch. 4].

17.3. An ideal gas is a gas whose particles do not interact with each other. If we put $V = 0$ in the model we have been discussing we obtain the mathematical model for an ideal gas of point particles (monatomic ideal gas). In this case the joint distributions of $n_{A_1},\ldots,n_{A_j}$ do not change once $\Lambda_n \supset \cup_{i=1}^{j} A_i$ and the existence of the thermodynamic limit of the configurational ensembles follows easily. The reader may recognize the process obtained in the thermodynamic limit as the Poisson process on R^3 with mean $\langle n_A \rangle_{T,\mu} = z\cdot\mathrm{vol}(A)$. Thus for an ideal gas the activity is precisely the density.

17.4. Even when the potential V is not zero the random variables n_A have their values in $\{0,1,2,\ldots\}$. The resulting process is therefore a point process. See e.g. Klaus Krickeberg in this volume.

17.5. A random field on R^m is, informally, a function φ on R^m whose values lie in a space $\mathfrak{M}(\Omega, X)$ of measurable functions on Ω to the measurable space X. Thus for t in R^m and ω in Ω, $\varphi(t)(\omega)$ is a point of X. This kind of definition captures in a technically correct way the general notion of random field provided one replaces in it R^m by Z^m. Indeed such a function φ on Z^m determines a measure σ on the product space X^{Z^m} and any measure σ on the product space determines a random field on Z^m by means of $\varphi(t) =$ "t^{th}" coordinate function on X^{Z^m}. It is exactly such random fields on Z^m that we shall consider in the remainder of this chapter. Thus we shall be concerned with understanding how statistical mechanics leads to the construction of interesting measures σ on a product space X^{Z^m}. For us, X will be compact.

But for a random field on R^m the preceding definition is too narrow to capture all of the random fields that arise in physics, in spite of our omission of any of the customary discussion of joint measurability of $(t,\omega) \to \varphi(t)(\omega)$. Proceeding informally, let us define, in case X is the real line,

$$\varphi(f)(\omega) = \int_{R^m} f(t)\varphi(t)(\omega)dt. \tag{17.3}$$

Then $\varphi(f)$ is a real valued random variable on Ω for each function f in some class, $\mathcal{T}$, of "test functions". For example in the case of the Poisson process just discussed one should identify n_A with $\varphi(\chi_A)$ and $\mathcal{T}$ can be taken to be the set of all characteristic functions, χ_A, of bounded Borel sets in R^m. In this case the integrand $\varphi(t)(\omega)$ doesn't really exist, but only its symbolic integral, $\int_A \varphi(t)(\omega)dt$. Similarly the random fields that arise in quantum field theory are "generalized random fields" in the sense that only the random variables $\varphi(f)$ exist but not the integrand in (17.3). In this case the test functions are usually taken to be in the linear space $C_c^\infty(R^m)$ or the Schwartz space $\mathcal{S}(R^m)$. For further discussion of random fields in the context of quantum field theory see [Si 3].

18. Lattice Gas Models.

Many of the interesting properties of the thermodynamics of gases (e.g. phase transitions, critical phenomena) are already visible in models in which the continuum R^3 is replaced by the lattice Z^3. The reason for this is that these phenomena are usually manifestations solely of the fact that there are many molecules present in matter. Mathematically these phenomena are concomitants of the operation of taking the thermodynamic limit. (Recall that in a finite volume the pressure $P_\Lambda(T,\mu)$ is analytic.) In the lattice models one can study the thermodynamic limit while avoiding certain technical continuum problems.

In order to motivate the customary lattice gas models let us return to the configurational ensemble described in the preceding section and note that in view of (17.2) the pressure, $P_\Lambda(T,\mu) \equiv (\beta V)^{-1}\log Z_\Lambda^{g.c.}(T,\mu)$, is determined by the normalization constant for the configurational ensemble. For this reason the configurational ensemble contains most of the interesting physics.

Lets assume that the integral $\int_{\Lambda^N} \exp[-\beta U_\Lambda(q)]dq_1\cdots dq_{3N}/N!$ can be approximated for all future purposes (including taking the thermodynamic limit) by a Riemann sum. (We won't justify this customary assumption, but it is probably valid under some reasonable conditions.) We choose a small lattice spacing $\varepsilon > 0$ and replace the integral by a sum over lattice points:

$$\int_{\Lambda^N} e^{-\beta U_\Lambda(q)}(N!)^{-1}dq_1\cdots dq_{3N} \sim \Sigma_{q\in A^N} e^{-\beta U_\Lambda(q)}(N!)^{-1}\,\varepsilon^{3N}$$

where $A = \Lambda \cap (\varepsilon Z^3)$.

Now $U_\Lambda(q) \equiv U_\Lambda(q_1,\ldots,q_{3N})$ is invariant under all of those permutations of $q_1,\ldots,q_{3N}$ which simply interchange the points $\vec{x}_1,\ldots,\vec{x}_N$. (Recall that q is short for $(\vec{x}_1,\ldots,\vec{x}_N)$.) Hence in the summand each term for which $\vec{x}_1,\ldots,\vec{x}_N$ are distinct occurs $N!$ times. If we write $U_\Lambda(X)$ for the common value of the potential $U_\Lambda(\vec{x}_1,\ldots,\vec{x}_N)$, where $X = \{\vec{x}_1,\ldots,\vec{x}_N\}$ is a subset of A of cardinality N, then the last sum may be written

$$\Sigma_{q\in A^N} e^{-\beta U_\Lambda(q)}(N!)^{-1}\epsilon^{3N} = \Sigma_{\substack{X\subset A\\ |X|=N}} e^{-\beta U_\Lambda(X)}\epsilon^{3N} + \epsilon^3 B.$$

The term $\epsilon^3 B$ arises from the terms on the left in which two or more of the vectors $\vec{x}_1,\ldots,\vec{x}_N$ coincide. It is not hard to see that if $e^{-\beta U_\Lambda(q)}$ is bounded then B remains bounded as $\epsilon \downarrow 0$, so that $\epsilon^3 B$ goes to zero in this limit. We shall, therefore, use only the first term in this lattice approximation to the continuum case. The partition function $Z_\Lambda^{g.c.}(T,\mu)$ is then approximately

$$\Sigma_{N=0}^{\infty} z^N \Sigma_{\substack{X\subset A\\ |X|=N}} e^{-\beta U_\Lambda(X)}(\epsilon^3)^N$$
$$= \Sigma_{X\subset A} (z\epsilon^3)^{|X|} e^{-\beta U_\Lambda(X)}.$$

Of course for large N the summands are zero because A is a finite set.

Lattice gas models are based on the preceding approximate expression for the partition function $Z_\Lambda^{g.c.}(T,\mu)$, which, as we have seen, determines the equilibrium thermodynamics of the system. Rather than let $\epsilon \downarrow 0$ one fixes a small positive ϵ and defines the lattice gas partition function by the last expression. It should be remarked that if the particles are viewed as hard spheres of diameter ϵ then the potential $V(\vec{x})$ must be taken as $+\infty$ if $|\vec{x}| < \epsilon$ so that $e^{-\beta U_\Lambda(\vec{x}_1,\ldots,\vec{x}_N)}$ is zero if $|\vec{x}_i-\vec{x}_j| < \epsilon$ for any i, j. This has the effect of making the terms $\epsilon^3 B$ much smaller and therefore more "negligible" for ϵ fixed. For this reason the lattice approximation is sometimes considered to correspond to an assumption of "hard core" particles.

Let us change notation in the last expression, using the indicator function, s, of the set X as the basic summation variable instead of X itself. Then since $|X| = \sum_{j\in A} s(j)$ we have

$$Z_\Lambda^{g.c.}(T,\mu) \sim \sum_{\{0,1\}^A} e^{-\beta U_\Lambda(s)+\alpha\sum_{j\in A} s_j}$$

where $s_j \equiv s(j)$ and $e^\alpha = z\varepsilon^3 = e^{\beta\mu}[\varepsilon(2\pi mkT)^{1/2}/h]^3$. The preceding arguments relied on the fact that $U_\Lambda(q)$ is a symmetric function of $\vec{x}_1,\ldots,\vec{x}_N$. The same argument applies therefore to the expectation $\langle f\rangle_{T,\mu,\Lambda}$ if f is symmetric. It therefore applies to the functions n_A discussed in the preceding section and to any functions of them. With this as motivation we now define a lattice gas model thus: We identify the lattice εZ^3 with Z^3, we use Λ for a finite subset of Z^3 instead of a bounded open set in R^3, and we assume that for each such Λ, $U_\Lambda(\cdot)$ is given as a function on 2^Λ. We then define the finite volume partition function by

$$(18.1) \qquad Z_\Lambda(T,\mu) = \sum_{s\in 2^\Lambda} e^{-\beta U_\Lambda(s)+\beta\mu\sum_{j\in\Lambda} s_j}$$

and the <u>finite volume Gibbs state</u> by

$$(18.2) \qquad \langle f\rangle_{\Lambda,T,\mu} = \frac{1}{Z_\Lambda(T,\mu)} \sum_{s\in 2^\Lambda} f(s)e^{-\beta U_\Lambda(s)+\beta\mu\sum_{j\in\Lambda} s_j}$$

where $f: 2^\Lambda \to R$.

Correspondingly, the <u>finite volume Gibbs measure</u> is the probability measure σ_Λ on $\{0,1\}^\Lambda$ whose expectation is given by (18.2). Note that we have written $\beta\mu$ instead of α in the exponential. They differ by an important function of temperature which is customarily ignored in studying lattice models. This has the effect of ignoring the contribution to the pressure from the momentum of the particles!! Nevertheless many qualitative features of thermodynamics are preserved. Of course the finite volume pressure is still defined as

$$P_\Lambda(T,\mu) = \beta^{-1}|\Lambda|^{-1}\log Z_\Lambda(T,\mu)$$

as in the grand canonical ensemble. Before discussing the thermodynamic limit of the two basic quantities P_Λ and σ_Λ we shall discuss another interpretation and generalization of this structure.

19. Molecules with Internal Structure; Crystals.

The orientation of a diatomic molecule composed of two different kinds of distinct atoms A and B at a fixed distance from each other may be specified by a unit vector pointing from atom A toward atom B. The configuration space for such a rigid molecule, with fixed center of gravity, is therefore the two sphere, S^2. If, on the other hand, the two atoms are identical and are regarded as indistinguishable (which is appropriate for quantum mechanics) then the direction of the unit vector is determined only up to sign, and the configuration space is therefore the projective plane P^2. For a triatomic molecule the configuration space is the rotation group $SO(3)$ if the three atoms are distinct, (compare the introduction to these notes) and, if some symmetries are present (e.g., three identical atoms at the vertices of an equilateral triangle) then the configuration space is the quotient of $SO(3)$ by some finite subgroup (a six element subgroup in the case of maximal symmetry). In any case the configuration space of the molecule is some compact (metrizable) Hausdorff space.

Consider a crystal composed of identical molecules fixed at points of a lattice. In the spirit of the configurational ensemble discussed in the preceding sections, we shall ignore the vibrational motions of the molecules and consider only those aspects of the thermodynamics of the crystal which result from the fact that the energy of the crystal depends on the relative <u>orientation</u> of its molecules. For a fixed number of molecules the volume is fixed since we shall assume that every lattice cite in the (say cubic) crystal is occupied. Since volume (equivalently density) is not a thermodynamic coordinate in this model, for a fixed number of particles, its dual coordinate, pressure, does not enter the theory.

We suppose that the molecules of the crystal are located on a subset, Λ, of Z^m, $m = 1,2,3,\ldots$. $m = 3$ is the case of primary physical interest. But thin film experiments relate to the case $m = 2$ and quantum field theory in the Euclidean region relates to the case $m = 4$. Let X denote the configuration space of an individual molecule. We assume that X is a compact metrizable Hausdorff space as in the preceding examples. The configuration space for the crystal as a whole is then the product X^Λ. The Liouville measure for the crystal has, in general, a factor corresponding to the momentum variables and a factor corresponding to the configuration variables. (Actually, in accordance with the general principles of mechanics, the Liouville measure is a measure on the cotangent bundle $T^*(X^\Lambda)$, assuming X is a manifold.) The configuration factor is a product $\nu \times \nu \times\cdots\times \nu$ where ν is a measure on the Borel sets of X. For example for a diatomic molecule ν would be Lebesgue measure on S^2 or P^2, while if $X = SO(3)$, ν would be Haar measure. We shall ignore the momentum factor and consider only the thermodynamics of the configurational factor. Thus if in a configuration $s \in X^\Lambda$ the energy of the crystal is $U_\Lambda(s)$ then the partition function in the configurational <u>canonical</u> ensemble (cf. Section 15) is

$$Z_\Lambda(T) = \int_{X^\Lambda} e^{-U_\Lambda(s)/kT}\, \nu^\Lambda(ds)$$

where $\nu^\Lambda = \nu \times \nu \times\cdots\times \nu$ is the product measure on X^Λ.

Recall that the canonical ensemble gives, not the pressure, but the free energy $F_\Lambda(T) = -kT \log Z_\Lambda(T)$ as the logarithm of the partition function (cf. Equation (15.4)). In this crystal model the number, N, of particles does not enter as an independent argument of F_Λ because every cite is occupied so that $N = |\Lambda|$ and the density is fixed.

Example (Ising model). Suppose that at each lattice cite the molecule forms a magnet which can point only up or down. In this case the configuration space, X, consists of just two points which we take to be ± 1. If the energy of interaction between the different magnets is $U_\Lambda(s)$, with $s \in \{-1,1\}^\Lambda$, and if the magnets also interact with an external magnetic field of strength h via the classical interaction energy $\Sigma_{j\in\Lambda} hs(j)$ then the partition function is

$$(19.1) \qquad Z_\Lambda = \Sigma_{s\in X^\Lambda} e^{-\beta[U_\Lambda(s)+h\Sigma_{j\in\Lambda}s(j)]} .$$

Note the resemblance of this partition function for the canonical ensemble of a crystal to the partition function (18.1) for the grand canonical ensemble of a lattice gas. In fact if one puts $s(j) = 2t(j)-1$ in (19.1) and lets t run over $\{0,1\}^\Lambda$ one sees that the partition function (19.1) is just the lattice gas partition function (18.1) times $e^{\beta|\Lambda|}$ if one puts $\mu = 2h$. Upon taking the logarithm and dividing by $|\Lambda|$ one sees that the free energy per unit volume in the present model differs from the negative of the pressure P_Λ in the lattice gas model by a constant. Consequently, many qualitative features of the two models (phase transitions, critical behavior) are similar. In many treatments of these lattice models (e.g. [R1], [R2], [Is2]) $|\Lambda|^{-1}\log Z_\Lambda$ is referred to simply as pressure, although the correct interpretation depends on how one interprets the points of X.

In keeping with the terminology of our two principle references [R2], [Is 2] we shall refer to $|\Lambda|^{-1}\log Z_\Lambda$ as pressure, ignoring the harmless (for qualitative discussions) missing factor β^{-1}, in front, with the understanding that the correct interpretation may be free energy per unit volume. The general formalism that we shall describe in the next section emphasizes the Legendre transform relation between the entropy and the pressure which we discussed for gases in

Section 12. Our objective in the next section is to describe the basic mathematical facts in a way that includes the preceding gas and crystal lattice models.

20. General Lattice Models.

We take physical space to be the m dimensional lattice Z^m. The configuration space of the particle associated to an individual lattice cite will be taken to be a compact metrizable Hausdorff space, X, with a given a priori finite Borel measure ν. The potential energy of interaction of the molecules may be described as follows.

Consider a finite subset $\Lambda \subset Z^m$. The configuration of the system in Λ is a point s in X^Λ. One says that the potential energy $U_\Lambda(s)$ comes from a pair interaction if for each pair $\{i,j\}$ of lattice cites there is a function $\varphi_{i,j}$ on $X \times X$ such that

$$(20.1) \qquad U_\Lambda(s) = \Sigma_{\{i,j\}\subset\Lambda}\, \varphi_{ij}(s_i, s_j)$$

where we have written s_i instead of $s(i)$. This is the most important case. It includes the special lattice gas model described in Section 18, which itself arose from a pair potential in classical mechanics. It also includes the Ising model in its classical form--the nearest neighbor interaction. But there are more general forms for U_Λ which are worthy of study for various reasons. Some of these more general "many body interactions" arise in lattice gauge theory, an important part of elementary particle theory. Secondly, some physical chemists regard classical mechanical many body interactions as useful approximations to quantum mechanical pair interactions. Thirdly, the renormalization group approach for studying critical phenomena requires many body interactions to be included in the formalism. And fourthly, from a purely mathematical point of view, the restriction to pair interactions in the study of general random fields would be an unnatural artifice.

A three body potential is an assignment to each three element set $A \equiv \{i,j,k\}$, in Z^m of a function φ_A on X^A. The corresponding contribution to the energy $U_\Lambda(s)$ would be a term $\Sigma_{\{i,j,k\}\subset\Lambda}\, \varphi_{\{i,j,k\}}(s_i,s_j,s_k)$ which is to be added to the pair interaction energy given in (20.1). Similarly an n body interaction is an assignment of a function φ_A on X^A to each set $A \subset Z^m$ of cardinality $|A| = n$. For the case $n = 1$, $\varphi_i(s_i)$ is interpreted as the energy of interaction of the molecule at cite i with an external field. Thus in general U_Λ will be given by

$$U_\Lambda(s) = \sum_{\emptyset \neq A \subset \Lambda} \varphi_A(s|A).$$

Since Λ is a finite set this is a finite sum. Now it happens that any family $\{U_\Lambda\}_{|\Lambda|<\infty}$ of functions $U_\Lambda\colon X^\Lambda \to R$ can be represented in this form. However, it will be necessary to control the behavior of U_Λ as $|\Lambda| \to \infty$, and we shall consequently impose size restrictions on the family $\{\varphi_A\}$. Moreover for the purposes of physics the translation invariant potentials are of primary importance. We consider then the following Banach spaces of potentials.

Let φ be a real valued function on the disjoint union $\bigcup_A X^A$ where A runs over all non-empty finite subsets of Z^m. We assume

i) $\varphi|X^A$ is continuous

ii) φ is translation invariant; that is

$$\varphi(s^j|A+j) = \varphi(s|A)$$

for all non-empty finite sets $A \subset Z^m$ and all j in Z^m. Here $s^j(i) = s(i-j)$. We have dropped the superfluous subscript A in φ_A.

Let k be a non-negative integer. Put

$$\|\varphi\|_k = \Sigma_{A \ni 0}\, |A|^{k-1} \sup\{|\varphi(s)|: s \in X^A\}, \quad k = 0,1,2,\dots .$$

We let

$$\mathscr{P}_k = \{\varphi\colon \varphi \text{ satisfies i) and ii) and } \|\varphi\|_k < \infty\}.$$

One sees easily that $\mathscr{P}_k$ is a Banach space. Moreover $\mathscr{P}_k \subset \mathscr{P}_j$ if $j \leq k$.

In Chapter 1 we saw that the pressure and (negative) entropy in thermodynamics are Legendre transforms of each other. In Chapter 2 we saw that thermodynamics is recovered by taking the thermodynamic limit in statistical mechanics. Our goal in this section is to show how the general features of the thermodynamic formalism can be recaptured in these very general lattice models. To this end we shall state four basic theorems. The first three define and assert the existence and basic properties of the pressure, states, and entropy respectively. The last one asserts that the pressure and entropy are Legendre transforms of each other.

<u>Pressure</u>. Let φ be a potential in $\mathscr{P}_0$. Define U_Λ by

$$U_\Lambda(s) = \sum_{\emptyset \neq A \subset \Lambda} \varphi(s|A) \qquad s \in X^\Lambda \tag{20.2}$$

and put

$$Z_\Lambda(\varphi) = \int_{X^\Lambda} e^{-U_\Lambda(s)} \nu^\Lambda(ds). \tag{20.3}$$

Define

$$P_\Lambda(\varphi) = |\Lambda|^{-1} \log Z_\Lambda(\varphi). \tag{20.4}$$

<u>Theorem 1</u>. $\lim P_\Lambda(\varphi)$ exists as Λ runs through any sequence of cubes in Z^m whose union is Z^m. The limit function $P(\varphi)$ is a convex Lip 1 function on $\mathscr{P}_0$, and in fact

$$|P(\varphi) - P(\psi)| \leq \|\varphi - \psi\|_0.$$

For the proof we refer the reader to [R1] or [R2] or [Is 2].

<u>Remark</u>. 20.1. We have not exhibited the temperature explicitly. Here and in most of the lattice formalism the temperature occurs in

the combination $(kT)^{-1}U_\Lambda$. We shall absorb the factor $(kT)^{-1}$ into the potential φ. Thus a derivative of the pressure with respect to the temperature is essentially synonymous with the "radial" derivative of $\varphi \to P(\varphi)$ on the Banach space $\mathcal{P}_0$.

States. Let $\Lambda \subset Z^m$ be a finite set, as usual. Let φ be in $\mathcal{P}_1$ and write

$$\sigma_\Lambda = Z_\Lambda(\varphi)^{-1} e^{-U_\Lambda(s)} \nu^\Lambda(ds).$$

Then σ_Λ is a probability measure on X^Λ. As we have seen in Sections 17 and 18, σ_Λ is a lattice analog of the configurational grand canonical ensemble, at least in the case $X = \{0,1\}$, and in this case the thermodynamic state coordinates U and N $(V = |\Lambda|)$ are expectations with respect to σ_Λ (cf. Equs. (15.12) and (15.13)). Instead of coordinatizing thermodynamic state space with these expectations we shall use the probability measure σ_Λ itself as the mathematical object corresponding to the physical state. Of course we wish first to take the thermodynamic limit, $\Lambda \to Z^m$. Now it is easy to show, using the compactness of each X^Λ together with a standard diagonalization argument, that for any sequence of cubes Λ_n whose union is all of Z^m, there is a subsequence n_j such that the measures $\sigma_{\Lambda_{n_j}}$ converge to a probability measure σ on X^{Z^m} in the sense that

$$\lim_{j\to\infty} \int_{X^{\Lambda_{n_j}}} f(s)\sigma_{\Lambda_{n_j}}(ds) = \int_{X^{Z^m}} f(s)\sigma(ds)$$

whenever f is a continuous function on X^{Z^m} depending on finitely many coordinates. (The latter condition ensures that the integrals on the left are well defined for large j.) The limit measure σ, which appears to depend on the choice of cubes, may not be unique, and in fact its non uniqueness is associated with the existence of phase transitions at the given (suppressed) temperature. (See Remark 20.1.) It happens that all of the possible limit measures satisfy a set of equations--the equilibrium equations of Dobrushin, Lanford and Ruelle--which we shall now describe. These equations (the DLR equations) provide a useful and elegant method of characterizing the set of limit states. They are the fundamental equations of the theory of Gibbs states.

We write henceforth

$$\Omega = X^{Z^m}.$$

Then Ω is a compact Hausdorff space in the product topology and we denote by $C(\Omega)$ the Banach space of real valued continuous functions on Ω.

Let $\Lambda \subset Z^m$ be a finite, non empty set and let φ be in $\mathcal{P}_1$. Put

$$W_\Lambda(s) = \Sigma_{A \cap \Lambda \neq \emptyset}\, \varphi(s|A), \quad s \in \Omega. \tag{20.5}$$

The sum converges uniformly in s and it is dominated by $\Sigma_{j \in \Lambda}\, \Sigma_{A \ni j}\, |\varphi(s|A)|$, which is no greater than $|\Lambda|\; \|\varphi\|_1$ in view of the translation invariance of φ and the definition of $\|\varphi\|_1$. It follows that W_Λ is a continuous function on Ω. Let $s \in \Omega$ and put $t = s|\Lambda^c$. (Λ^c = complement of Λ in Z^m.) Define a probability measure on X^Λ by

$$\mu_\Lambda(dy|s) = Z_\Lambda(t)^{-1}\, e^{-W_\Lambda(y \vee t)}\, \nu^\Lambda(dy). \tag{20.6}$$

$y \vee t$ denotes the function on Z^m which is y on Λ and t on Λ^c. $Z_\Lambda(t)$ is the normalization constant:

$$Z_\Lambda(t) = \int_{X^\Lambda} e^{-W_\Lambda(y \vee t)}\, \nu^\Lambda(dy).$$

Finally let τ_Λ be the operator on the space $C(\Omega)$ defined by

$$(\tau_\Lambda f)(s) = \int_{X^\Lambda} f(y \vee t)\mu_\Lambda(dy|s), \quad t = s|\Lambda^c. \tag{20.7}$$

The continuity of $s \to (\tau_\Lambda f)(s)$ follows from the continuity (and boundedness) of W_Λ.

<u>Definition</u>. A <u>Gibbs state</u> for the potential φ is a probability measure σ on the Borel sets of Ω such that

(20.8) (DLR Equations) $\sigma(\tau_\Lambda f) = \sigma(f)$

for all non empty finite sets $\Lambda \subset Z^m$ and all f in $C(\Omega)$.

The preceding definition can be stated in more probabilistic terms thus: A probability measure σ on the Borel sets of Ω is a <u>Gibbs state</u> for the potential φ if its conditional probabilities are given by the measures μ_Λ in accordance with

$$\sigma(B \times X^{\Lambda^c} \mid s = t \text{ on } \Lambda^c) = \mu_\Lambda(B|t) \quad \text{a.e. } [\sigma]$$

for all Borel sets $B \subset X^\Lambda$ and all non empty finite sets $\Lambda \subset Z^m$.

The equivalence of this more explicitly probabilistic definition with the preceding, more functional analytic, form is easy to establish from the definition of conditional probability. We leave the equivalence to the reader to verify. We shall have need later only for the first form.

In order to understand the connection between the DLR equations (20.8) and the limit process $\sigma_\Lambda \to \sigma$ discussed before let us consider a non empty finite set $\Lambda \subset Z^m$ and a very large finite set M containing Λ. We shall describe in detail the simple calculation which shows that

(20.9) $\sigma_M(\tau f) = \sigma_M(f) \quad f \in C(X^M)$

where τ is an operator depending on Λ and M and which converges to τ_Λ as $M \uparrow Z^m$. Once this is done the DLR equation (20.8) for the "infinite volume" limit measure σ follows from limit arguments (which we shall omit) upon letting M increase to Z^m.

The energy of the configuration s in X^M is

$$U_M(s) = \Sigma_{A\subset M}\ \varphi(s|A)$$
$$= \Sigma_{\substack{A\cap\Lambda\neq\emptyset \\ A\subset M}}\ \varphi(s|A) + \Sigma_{A\subset M-\Lambda}\ \varphi(s|A)$$
$$= W(s) + U_{M-\Lambda}(s|M-\Lambda)$$

where $W(s)$ is the first sum on the second line. Note for later use that $W(s|M)$ converges uniformly to $W_\Lambda(s)$ as $M\uparrow Z^m$ and $s\in\Omega$. But it should be realized that U_M and $U_{M-\Lambda}$ never converge as $M\uparrow Z^m$ unless $\varphi = 0$. Now define for $t\in M-\Lambda$ the following probability measure on X^Λ

$$\mu(dy|t) = Z(t)^{-1}\ e^{-W(y\vee t)}\ \nu^\Lambda(dy)$$

where $Z(t) = \int_{X^\Lambda} e^{-W(y\vee t)}\ \nu^\Lambda(dy)$.
The density of this measure with respect to ν^Λ clearly converges in total variation to that of $\mu_\Lambda(\ |t)$, uniformly for t in X^{Λ^c}. The reader may wish to use this to complete the omitted limit arguments.

Suppose that g is in $C(X^M)$, but depends only on the $X^{M-\Lambda}$ coordinates. Write $s = y\vee t$ with $y\in X^\Lambda$ and $t\in X^{M-\Lambda}$. Then

$$\sigma_M(g) = Z_M^{-1}\int_{X^M} g(s)e^{-U_M(s)}\nu^M(ds)$$
$$= Z_M^{-1}\int_{X^{M-\Lambda}} g(t)e^{-U_{M-\Lambda}(t)}\left(\int_{X^\Lambda} e^{-W(y\vee t)}\nu^\Lambda(dy)\right)\nu^{M-\Lambda}(dt)$$
$$= Z_M^{-1}\int_{X^{M-\Lambda}} g(t)Z(t)e^{-U_{M-\Lambda}(t)}\ dt.$$

On the other hand if f is any function in $C(X^M)$ then

$$\sigma_M(f) = Z_M^{-1}\int_{X^{M-\Lambda}}\int_{X^\Lambda} f(y\vee t)e^{-W(y\vee t)}e^{-U_{M-\Lambda}(t)}\nu^\Lambda(dy)\nu^{M-\Lambda}(dt)$$
$$= Z_M^{-1}\int_{X^{M-\Lambda}}\left(\int_{X^\Lambda} f(y\vee t)\mu(dy|t)\right)Z(t)e^{-U_{M-\Lambda}(t)}\nu^{M-\Lambda}(dt)$$
$$= \sigma_M(\tau f)$$

where τ is the integral operator on $C(X^M)$ determined by $\mu(\ |t)$ in the same way that μ_Λ determines τ_Λ. In the last line we have used the previous computation with $g = \tau f$. This proves (20.9).

We have now completed our sketch of the proof that any thermodynamic limit of the measures $\sigma_\Lambda \equiv Z_\Lambda^{-1} e^{-U_\Lambda(\cdot)} d\nu^\Lambda$ satisfies the DLR equations (20.8). But these may not be all the possible solutions to the DLR equations. Indeed, if one chooses any sequence of cubes Λ_n increasing to Z^m and any sequence of configurations $s_n \in \Omega$, then some subsequence of the sequence of measures $\mu_{\Lambda_n}(\ |s_n)$ will have a thermodynamic limit, ψ, by the same compactness argument used for the σ_Λ, and a computation similar to the preceding one also shows that ψ is a solution to the DLR equations. The measure $\mu_\Lambda(\ |s)$ is called a <u>finite volume Gibbs state</u> with boundary condition $t = s|\Lambda^c$. (Recall that $\mu_\Lambda(\ |s)$ depends only on t.) These are essentially all the solutions to the DLR equations--up to convex closures. Here is the precise statement.

<u>Theorem 2.</u> Let $\varphi \in \mathcal{P}_1$. The set $\mathcal{G}_\varphi$ of all Gibbs states for φ is a non empty weakly closed convex set in the dual space $C(\Omega)^*$. It is the closed convex hull of the set of all thermodynamic limits of finite volume Gibbs states for φ with boundary conditions. Moreover, $\mathcal{G}_\varphi$ is a Choquet simplex.

<u>Remarks</u>. 20.2. For the proof we refer the reader to [R2, Chapter 1] or [Is 2, Chapters 2 and 3].

20.3. Recall that a simplex in R^n is a non empty closed convex set any point of which is a <u>unique</u> convex combination of its extreme points. Thus a triangle and tetrahedron are simplices, but a square and cube aren't. A Choquet simplex is a generalization of simplex to arbitrary (possibly infinite) dimension. See [R2, Appendix A5] or [Is 2, Chapter 4] for a precise definition.

Now consider any probability measure σ on the Borel sets of Ω. For example σ may be one of the Gibbs states just described. If j is in Z^m and s is in Ω the translate of s by j is the configuration s^j given by $s^j(i) = s(i-j)$. If f is in $C(\Omega)$ its translate by j is $f^j(s) = f(s^j)$.

Definition. σ is translation invariant if $\sigma(f^j) = \sigma(f)$ for all j in Z^m and all f in $C(\Omega)$.

We shall see later (Theorem 4) that, for any φ in $\mathcal{P}_1$, $\mathcal{G}_\varphi$ contains at least one translation invariant Gibbs state. The non-translation invariant states enter in the description of phase separation, e.g. water and water vapor separated by a surface when these two phases coexist.

We must now understand the meaning of expected energy in the infinite volume limit. In accordance with the formalism of either the canonical or grand canonical ensembles, Equs. (15.7) or (15.12), the energy of the thermodynamic system modeled by our lattice system in a bounded region $\Lambda \subset Z^m$ is $\sigma_\Lambda(U_\Lambda)$. In order to discuss the energy in the thermodynamic limit it should be noted that for a homogeneous (i.e. translation invariant) medium the total energy is infinite if the energy per unit volume is not zero. It is therefore only the energy per unit volume, $|\Lambda|^{-1}\sigma_\Lambda(U_\Lambda)$, which can be expected to be meaningful in the thermodynamic limit when the limit state, $\sigma \equiv \lim \sigma_\Lambda$, is translation invariant. In order to motivate the desired expression for the energy per unit volume which we need, while avoiding some (actually non trivial) technicalities, let us consider instead $\lim|\Lambda|^{-1}\sigma(U_\Lambda)$ which in fact equals $\lim|\Lambda|^{-1}\sigma_\Lambda(U_\Lambda)$ when σ is translation invariant. We assume then that σ is a translation invariant probability measure on Ω.

Let

$$f_\varphi(s) = \sum_{A \ni 0} |A|^{-1}\varphi(s|A).$$

This series converges uniformly for φ in $\mathcal{P}_1$ (in fact even for φ in $\mathcal{P}_0$) and $|f_\varphi(s)| \leq \|\varphi\|_1$. We shall show that

$$\lim_{\Lambda \to Z^m} |\Lambda|^{-1}\sigma(U_\Lambda) = \sigma(f_\varphi). \tag{20.10}$$

The right side may then be used for the expected energy per unit volume in the thermodynamic limit. In order to prove (20.10) note first the combinatorial identity

$$\sum_{\emptyset \neq A \subset \Lambda} \varphi(s|A) = \sum_{j \in \Lambda} \sum_{\substack{A \ni j \\ A \subset \Lambda}} |A|^{-1} \varphi(s|A)$$

which holds because each term $\varphi(s|A)$ in the sum on the left occurs $|A|$ times on the right--once for each point j in A. Hence, using the translation invariance of σ in the last line below we have

$$\begin{aligned} |\Lambda|^{-1} \sigma(U_\Lambda) &= |\Lambda|^{-1} \sigma\Big(\sum_{\emptyset \neq A \subset \Lambda} \varphi(s|A)\Big) \\ &= |\Lambda|^{-1} \sum_{j \in \Lambda} \sigma\Big(\sum_{\substack{A \ni j \\ A \subset \Lambda}} |A|^{-1} \varphi(s|A)\Big) \\ &= |\Lambda|^{-1} \sum_{j \in \Lambda} \sigma\Big(\sum_{\substack{B \ni 0 \\ B \subset \Lambda - j}} |B|^{-1} \varphi(s|B)\Big). \end{aligned}$$

Now for each fixed j the integrand converges uniformly to f_φ, and if the convergence were also uniform in j then the normalized sum, $|\Lambda|^{-1}\Sigma_{j\in\Lambda}$ would clearly converge to $\sigma(f_\varphi)$. But the convergence is not uniform in j because for $j \in \Lambda$ and close to the boundary of Λ (we consider cubes only) the requirement $B \subset \Lambda - j$ may exclude some large terms from the series defining f_φ. One of the standard techniques of this subject enters here and we sketch it rather loosely. For j that are far from the boundary of the large cube Λ, say distance $\geq d$, the integrand is uniformly close to f_φ, while on the other hand the number of j which are close to $\partial\Lambda$ is on the order $d \cdot$ surface area of $\partial\Lambda$. Since the latter goes to zero after division by $|\Lambda|$ when $|\Lambda| \to \infty$, the normalized sum $|\Lambda|^{-1}\Sigma_{j\in\Lambda}\, \Sigma_{0\in B\subset\Lambda-j}|B|^{-1}\varphi(s|B)$ does actually converge uniformly to f_φ. This proves (20.10).

The natural bilinear pairing

$$\langle \sigma, \varphi \rangle = \sigma(f_\varphi)$$

that we have arrived at between $\mathcal{P}_1$ and the space $\mathcal{T}$ of translation invariant, real signed measures σ on Ω maps $\mathcal{T}$ into $\mathcal{P}_1^*$. The map is in fact one to one (see [Is 2, Lemma II.1.1]) and therefore the pressure (or free energy) is defined on a space, $\mathcal{P}_1$, which is more or less dual to the space, $\mathcal{T}$, in which the translation invariant states lie--as is in accordance with Sections 11 and 12.

Before discussing entropy, which in accordance with Chapter 1 should be a function on states, it may be illuminating to recall again, in the simplest case, that the domain of the pressure function is indeed coordinatized by temperature and chemical potential. To this end, let us put back the temperature and chemical potential into the lattice gas model of Section 18. For a pair potential of interaction $V(i-j)$ put

$$(20.11) \qquad \varphi(s|A) = \begin{cases} \beta V(i-j)s_i s_j & \text{if } A = \{i,j\} \quad i \neq j \\ \beta\mu s_i & \text{if } A = \{i\} \\ 0 & \text{otherwise.} \end{cases}$$

Here $\beta = (kT)^{-1}$ as usual. Thus β and $\beta\mu$ are linear coordinates in a two-dimensional subspace of $\mathcal{P}_1$, and the pressure $P(\varphi)$, restricted to this two dimensinnal subspace, is a function of temperature, T, and chemical potential, μ.

Entropy. For any translation invariant measure σ on Ω let σ^Λ be the probability measure on X^Λ defined by $\sigma^\Lambda(B) = \sigma(B\times X^{\Lambda^c})$ for any Borel set $B \subset X^\Lambda$. (σ^Λ is the marginal distribution of σ on the factor X^Λ.) If σ^Λ is absolutely continuous with respect to ν^Λ (recall $\nu^\Lambda = \prod_{j\in\Lambda} \nu$) define

$$(20.12) \qquad S_\Lambda(\sigma) = -\int_{X^\Lambda} (d\sigma^\Lambda/d\nu^\Lambda)[\log(d\sigma^\Lambda/d\nu^\Lambda)]d\nu^\Lambda.$$

Since $t \log t \geq t-1$ the integrand is bounded below by an integrable function, so the integral is well defined, although S_Λ may take the value $-\infty$. By way of motivation, the reader should recognize the resemblance of this definition to the equations (15.8) and (15.14).

In case σ^{Λ} is not absolutely continuous with respect to ν^{Λ} we put $S_{\Lambda}(\sigma) = -\infty$. The entropy per unit volume, $s_{\Lambda}(\sigma)$, is defined now as $|\Lambda|^{-1}S_{\Lambda}(\sigma)$.

<u>Theorem 3</u>. For any translation invariant probability measure σ on Ω the thermodynamic limit

$$\lim_{\Lambda \uparrow Z^m} s_{\Lambda}(\sigma) \equiv s(\sigma)$$

exists as a number in $[-\infty, \nu(X)]$. Moreover $s(a\sigma_1 + (1-a)\sigma_2) = as(\sigma_1) + (1-a)s(\sigma_2)$ for $0 \leq a \leq 1$ and any two translation invariant probability measures σ_1 and σ_2.

Here, as before, the thermodynamic limit means the limit as Λ increases to Z^m through an arbitrary sequence of cubes. We mention, however, that here, as well as in all preceding instances, the limit can be taken through more general sequences (convergence in the sense of Van Hove, see [R1], [R2] or [Is 2]).

<u>Pressure and Entropy are Legendre transforms</u>. At the end of Section 12, we saw that the Legendre transform of the negative entropy (per unit volume) with respect to U and N gives the pressure $P(-1/T, \mu/T)$. As we noted at the end of the paragraph on states, $-1/T$ and μ/T are to be regarded as <u>linear</u> coordinates in a two-dimensional subspace of $\mathcal{P}_1$. (Cf. Equ. (20.11).) Thus the following theorem generatizes this classical thermodynamic relation between pressure and entropy.

$\mathcal{T}_1$ denotes the convex set consisting of all translation invariant probability measures on Ω.

<u>Theorem 4</u>. Let φ be in $\mathcal{P}_1$. Then

a. $P(\varphi) = \sup_{\sigma \in \mathcal{T}_1} \{s(\sigma) - \sigma(f_{\varphi})\}$.

b. The set I_{φ} consisting of those measures σ in $\mathcal{T}_1$ such that

$$P(\varphi) = s(\sigma) - \sigma(f_\varphi)$$

is non-empty and is a Choquet simplex.

c. I_φ consists exactly of the translation invariant Gibbs states for φ.

For proofs of this deep theorem we refer the reader to [R2, Chapters 3 and 4] or [Is 2, Chapters 2 and 3].

An element of I_φ is called an equilibrium state by Ruelle and an invariant equilibrium state by Israel.

Remarks. 20.4. The equation a. agrees with Equation (12.3) except for the sign of the linear term. This discrepancy would disappear if we defined the duality between $\mathcal{P}_1$ and $\mathcal{T}$ by $\langle\sigma,\varphi\rangle = -\sigma(f_\varphi)$ as is done in [R2].

20.5. The meaning of I_φ can be understood in part from Figure 11.1. For a given slope α (which corresponds to the linear functional $\sigma \to -\sigma(f_\varphi)$) there may be more than one point of contact of the supporting plane $y = \alpha u - b$ with the graph of f (but only if f has flat spots.) f corresponds to our $-s$ and the u axis is $\mathcal{T}$. The abscissas, u_0, of such contact points form a closed interval which in our infinite dimensional analog is the Choquet simplex I_φ. For the physical interpretation of these simplices see the discussion of phase coexistence at the end of Section 9.

20.6. The preceding thermodynamic formalism is incomplete and perhaps somewhat misleading in some fundamental aspects. Not every translation invariant probability measure on Ω is a Gibbs state for some interaction φ. Hence $\mathcal{T}_1$ is not "really" the manifold of thermodynamic states of the system whose configuration space is $\Omega \equiv X^{Z^m}$. (See, however, the density theorem [R2, Sections 3.17 and 3.18].) Moreover a convex sum, $a\sigma_1 + (1-a)\sigma_2$, of two Gibbs states σ_1 and σ_2 for two different interactions φ_1 and φ_2 is not necessarily the Gibbs state for any interaction φ in $\mathcal{P}_1$. Thus the relation between the affine structure in $\mathcal{T}_1$ and the affine structure of the classical thermodynamic state space (exemplified by U, V, N) is not so clear.

At any rate it is possible to make a complete metric space $\mathcal{M}$ out of the set of all equilibrium states in a natural way, and it seems to me that several basic problems (Gibbs phase rule, critical points) might be more perspicuously formulated by using this metric space as a model for thermodynamic state space rather than the inordinately large space $\mathcal{T}_1$. In order to sketch the definition of $\mathcal{M}$ we shall have to enlarge the space $\mathcal{P}_1$ of potentials by the following reformulation.

Let s and t be two configurations in Ω which differ at only finitely many cites in Z^m. Although $\lim_{\Lambda\to Z^m} U_\Lambda(s)$ will not generally exist for a non zero potential φ in $\mathcal{P}_1$, nevertheless $H(s,t) \equiv \lim_{\Lambda\to Z^m}(U_\Lambda(s)-U_\Lambda(t))$ does exist and is given by the convergent sum

$$(20.13) \qquad H(s,t) = \sum_{\emptyset\neq A} (\varphi(s|A)-\varphi(t|A)).$$

The convergence is easy to show. H is called the <u>relative Hamiltonian</u> associated to φ. This function of two variables seems to have been used first by Pirogov and Sinai [Pi]. It is an easy exercise to show that the conditional probabilities $\mu_\Lambda(\ |s)$ can be defined explicitly in terms of H (and the a priori measure ν). Hence H determines all the Gibbs states associated to φ. Now define

$$\|H\|_1 = \sup\{|H(s,t)|: s = t \text{ except at } 0\}.$$

Let $\Delta = \{(s,t) \in \Omega \times \Omega: s = t$ except on a finite set$\}$ and let $\mathcal{H}_1'$ be the set of skew symmetric real valued continuous functions K on Δ satisfying $K(s,t) = K(s,r) + K(r,t)$ for all finitely distinct triples r, s, t. The function H described above is in $\mathcal{H}_1'$ and we let $\mathcal{H}_1$ be the closure in $\mathcal{H}_1'$ of the set of such functions H.

(I conjecture $\mathcal{H}_1 \neq \mathcal{H}_1^{!}$.) Then $\mathcal{H}_1$ is a Banach space and if H is given by (20.13), then $\|H\|_1 \leq 2\|\varphi\|_1$. Thus $\mathcal{P}_1$ is mapped continuously into $\mathcal{H}_1$ (but not onto, I think). Now let $\mathcal{M}$ be the set of all translation invariant Gibbs states corresponding to relative Hamiltonians in $\mathcal{H}_1$. That is, if $\mathcal{G}_H^0$ is the set of translation invariant Gibbs states associated to the relative Hamiltonian H in $\mathcal{H}_1$, then $\mathcal{M} = \cup_{H\in\mathcal{H}_1}\mathcal{G}_H^0$. $\mathcal{M}$ becomes a complete metric space with respect to the following metric. Let σ and τ be in $\mathcal{M}$ and let σ^Λ and τ^Λ be their marginal measures on X^Λ. Then if Λ is finite σ^Λ and τ^Λ are mutually absolutely continuous and $d\sigma^\Lambda/d\tau^\Lambda$ is a continuous function on X^Λ. Put

$$(20.14)\qquad d(\sigma,\tau) = \sup_{\Lambda\ni 0} \sup\{\log[\frac{d\sigma^\Lambda}{d\tau^\Lambda}(s)\frac{d\tau^\Lambda}{d\sigma^\Lambda}(t)]|: s = t \text{ except at } 0\}.$$

The outer supremum is taken over finite sets $\Lambda \subset Z^m$ which contain the origin of Z^m. Besides being a complete metric on $\mathcal{M}$, d also satisfies the inequalities

$$\|H-K\|_1 \leq d(\sigma,\tau) \leq \|H\|_1 + \|K\|_1$$

if σ is in $\mathcal{G}_H^0$ and τ is in $\mathcal{G}_K^0$. In particular the map $\sigma \to H$ (for $\sigma \in \mathcal{G}_H^0$) is continuous from $\mathcal{M}$ onto $\mathcal{H}_1$. I conjecture that $\mathcal{M}$ is a Banach Lip 1 manifold modeled on $\mathcal{H}_1$. This is closely related to the Gibbs phase rule. (See [R2] or [Is 2].)

20.7. Finally, corresponding to the smaller spaces $\mathcal{P}_k$, $k \geq 2$, there are smaller spaces $\mathcal{H}_k$ of relative Hamiltonians (see [Gr 1] for a definition of the $\mathcal{H}_2$ norm) and corresponding complete metric spaces $\mathcal{M}_k$ of translation invariant Gibbs states. The technical advantage of $\mathcal{P}_2$ ($\mathcal{H}_2$) over $\mathcal{P}_1$ (respectively $\mathcal{H}_1$) will be discussed in the next section.

21. The Dobrushin Uniqueness Theorem.

Experience shows that at high temperature thermodynamic systems exist in only one phase. For example in the ice-water-vapor system only the gaseous form exists above some critical temperature T_c, no matter how high the pressure is (see e.g. Figure 9.1). This fact of experience is reflected, in the context of the thermodynamic formalism of the preceding section, by various uniqueness theorems for Gibbs states. These theorems assert that the set of Gibbs states $\mathcal{G}_\varphi$, corresponding to an interaction φ, consists of at most one state (and therefore exactly one state) if φ is a "high temperature" interaction. In view of the way in which the temperature enters into φ (see Equ. (20.11)), "high temperature" means that φ is small in some sense. Here is an easy way to state a version of Dobrushin's very general uniqueness theorem. The notation and definition of the spaces $\mathcal{P}_k$ is given in the preceding section.

Theorem. If φ is in the open unit ball of $\mathcal{P}_2$, then φ has exactly one gibbs state.

Sketch of Proof. A simple proof, due to Vasershtein [V], uses the following machinery, which is useful for other purposes also. Order the points of Z^m in any way and label them 1, 2, ... Write τ_j for the operators $\tau_{\{j\}}$ associated to φ as in the previous section. It is only these operators τ_Λ for one point sets Λ which enter the proof. Now if f is in $C(\Omega)$ and depends only on finitely many coordinates then $\tau_j f = f$ for sufficiently large j. Hence $\lim_{n\to\infty} \tau_1\tau_2\cdots\tau_n f$ exists in sup norm if f is a continuous cylinder function on Ω. But each factor τ_j is a contraction on $C(\Omega)$ in sup norm because each measure, $\mu_j(\ |s)$, defining τ_j, is a probability measure. Hence the operator T, defined by

$$Tf = \lim_{n\to\infty} \tau_1\tau_2\cdots\tau_n f$$

for continuous cylinder functions f, is a densely defined contraction operator on $C(\Omega)$ and therefore has a unique extension to all of $C(\Omega)$. The extension, T, is also a contraction on $C(\Omega)$. Now if σ is a Gibbs state for φ then for all n and any continuous cylinder function f we have $\sigma(\tau_1 \cdots \tau_n f) = \sigma(f)$ by the DLR equations. Upon taking the limit as $n \to \infty$ and using the continuity of T we get

$$\sigma(Tf) = \sigma(f) \qquad f \in C(\Omega).$$

Thus any Gibbs state for φ is a fixed point of the adjoint, T^*.

In order to show that T^* has a unique fixed point we choose the right Banach space, in accordance with Vasershtein, [V], as follows. For any function f in $C(\Omega)$ put

$$\delta_j(f) = \sup\{|f(s)-f(t)|: s = t \text{ except at } j\}$$

and

$$|f|_1 = \Sigma_{j=1}^{\infty} \delta_j(f).$$

Let C^1 denote the set of f in $C(\Omega)$ such that $|f|_1 < \infty$. The semi norm $| \ |_1$ on C^1 has a kernel consisting only of constants. The quotient space, C^1/constants, is a Banach space which we call B. If μ is a real signed measure on Ω such that $\mu(\Omega) = 0$ then μ annihilates constants and so defines an element of B^*. For example if σ and τ are two probability measures on Ω then $\sigma - \tau$ is in B^*.

<u>Lemma</u>. If $\|\varphi\|_2 < 1$ then there exists a number $\alpha < 1$ such that

$$|Tf|_1 \leq \alpha |f|_1 \tag{21.1}$$

for all f in C^1.

We shall delay discussion of the proof of this lemma till later. Dobrushin's uniqueness theorem now follows because T, which leaves constants invariant, acts in B with norm less than one, by the lemma.

Hence if σ and τ are both Gibbs states for φ then $|\sigma-\tau|_1^* = |T^*(\sigma-\tau)|_1^* \leq \alpha|\sigma-\tau|_1^*$ where $|\ |_1^*$ is the dual norm. Since $\alpha < 1$ this forces $|\sigma-\tau|_1^* = 0$ and so $\sigma = \tau$.

The proof of the lemma consists of two parts. The first part, due to Vasershtein, asserts that if the number α defined by

$$\text{(21.2)} \qquad \alpha = (1/2)\Sigma_{j=2}^{\infty} \sup\{\| \mu_1(\ |s)-\mu_1(\ |t)\|_{var.}: s = t \text{ except at } j\}$$

is less than one then (21.1) holds with this α. The original form of Dubrushin's uniqueness theorem imposes a condition on the conditional probabilities $\mu_j(\ |s)$ rather than on the potentials and consists exactly of the condition $\alpha < 1$ where α is given by (21.2). (I have restricted the discussion to the case of translation invariant conditional probabilities although minor alterations in the hypothesis allow the general case to be handled with no substantial change.) For a proof that $\alpha < 1$ implies (21.1) see [V, Section 6] or [Gr 1, Corollary 3.2 with $d \equiv 0$]. An independent proof of a similar result was given by Lanford [L].

The second part of the proof of the lemma consists in showing that for φ small enough the constant α in (21.2) is indeed less than one. Dobrushin showed that if $\|\varphi\|_2$ is small enough then $\alpha < 1$. But the proof yielding the best general constant is that of Simon [Si 1] which shows, in a very simple manner, that $\alpha < 1$ if

$$\text{(21.3)} \qquad \sum_{A \ni 0} (|A|-1)\sup\{|\varphi(s)|: s \in X^A\} < 1.$$

Clearly (21.3) holds if $\|\varphi\|_2 < 1$. Using an example of R. Israel, Simon also pointed out [Si 1] that for any $\varepsilon > 0$ there is a potential φ with multiple Gibbs states such that the left side of (21.3) is less than $1 + \varepsilon$. Thus Dobrushin's uniqueness theorem is in a certain sense best possible.

There are several other high temperature uniqueness theorems. Whereas Dobrushin's theorem applies to a neighborhood of zero in $\mathcal{P}_2$

it is also true that uniqueness of Gibbs states holds for φ in a neighborhood of zero in the larger space $\mathcal{P}_1$, provided that the individual configuration space X is finite. We refer the reader to the definitive paper by Gallavotti and Miracle-Sole [Ga] for this result. The gap between this uniqueness theorem and Dobrushin's theorem is unexplored. For other general high temperature techniques we refer the reader to [Ia], [Ku], [Is 1] and, for non-compact configuration space X, to [D2] and their bibliographies.

Finally we mention the role of the more restricted spaces $\mathcal{P}_k$. As we noted in Remark 20.5, the translation invariant Gibbs states are support functionals for the pressure, since by Theorem 4, $P(\psi) \geq s(\sigma) - \sigma(f_\psi) = P(\varphi) - \sigma(f_\psi - f_\varphi)$ if σ is in I_φ. When φ has only one Gibbs state the pressure has a unique tangent plane at φ and is therefore Gateaux differentiable at φ. Thus in the Dobrushin uniqueness region $\mathcal{D}$, consisting of all those φ in $\mathcal{P}_2$ for which α, given by (21.2), is less than one, the pressure is automatically once differentiable. The techniques underlying the proof of Dobrushin's theorem have been used to prove that the pressure is in fact C^2 (Gateaux) in $\mathcal{D}$ [Gr 2] and analytic in a natural subset of $\mathcal{D} \cap (\cap_{k=1}^{\infty} \mathcal{P}_k)$, [Is 1].

<u>Concluding Remark</u>. In these notes I have attempted to give a birds eye view of the three topics in the title with communication of overall perspective on this very extensive subject as the number one priority. Not only have I omitted most of the technical proofs--which are collected anyway in just a few easily accessible references--but also I have not discussed at all any of the vast number of results which apply to special models. Many of the most physically interesting and mathematically deep results in the statistical mechanical equilibrium theory of lattice gases and crystals require rather specific assumptions about the form of the interaction energy U_Λ

and the structure of the single particle configuration space X. By contrast I have imposed only size restrictions on U_Λ. For mathematical expositions of these models see for example F. Spitzer [Sp] and the forthcoming book of Barry Simon [Si 2].

REFERENCES

[A] M. Aizenman, S. Goldstein, J.L. Lebowitz, Conditional equilibrium and the equivalence of microcanonical and grandcanonical ensembles in the thermodynamic limit, Comm. Math. Phys. 62 (1978) 279.

[B] Boltzmann, Analytical proof of the second law from the theorems on equilibrium of energy. Wien Ber. 63^2 (1871), p. 712.

[Br] S.G. Brush, The Kind of Motion we Call Heat, Books 1 and 2 in Studies in Statistical Mechanics, E.W. Montroll and J.L. Lebowitz, Eds., North Holland Pub. Co., Amsterdam, 1976.

[C] H.B. Callen, Thermodynamics, John Wiley and Sons, New York, 1960.

[D 1] R.L. Dobrushin, The description of a random field by means of conditional probabilities and conditions of its regularity, Theory of Probability and its Applications, 13 (1968) 197-224.

[D 2] ———, Prescribing a system of random variables by conditional distributions, Theory of Prob. and its Applic. 15 (1970), 458-486.

[DT] R.L. Dobrushin and B. Tirozzi, The central limit theorem and the problem of equivalence of ensembles, Commun. Math. Phys. 54 (1977) 173-192.

[Do] Domb and Green, Phase Transitions and Critical Phenomena, Vols. 1 to 6 Academic Press, New York, 1972-1976

[E] P. and T. Ehrenfest, The Conceptual Foundations of the Statistical Approach in Mechanics, Cornell Univ. Press, Ithaca, N.Y., 1959.

[Fa] G. Falk and H. Jung, Axiomatik der Thermodynamik, Handbuch der Physik III/2, 1959.

[Fi] M.E. Fisher, The theory of equilibrium critical phenomena, Reports on Progress in Physics, Vol. 30 (1967) 615-730.

[Fr] J. Fröhlich, The pure phases (harmonic functions) of generalized processes or: mathematical physics of phase transitions and symmetry breaking, Bull. A.M.S. 84 (1978) 165-193.

[Ga] G. Gallavotti and S. Miracle-Sole, Correlation functions of a lattice system, Commun. Math. Phys. 7 (1968) 274-288.

[Gi] R. Giles, Mathematical Foundations of Thermodynamics, The MacMillan Co., New York, N.Y., 1964.

[Gr 1] L. Gross, Decay of correlations in classical lattice models at high temperature, Commun. Math. Phys. 68 (1979) 9-27.

[Gr 2] ——————, Absence of second order phase transitions in the Dobrushin uniqueness region, J. Statistical Physics, 25 (1981) 57-72.

[Ia] D. Iagolnitzer and B. Souillard, On the analyticity in the potential in classical statistical mechanics, Commun. Math. Phys. 60 (1978) 131-152.

[Is 1] R.B. Israel, High temperature analyticity in classical lattice systems, Commun. Math. Phys. 50 (1976) 245-257.

[Is 2] ——————, Convexity in the Theory of Lattice Gases, Princeton Univ. Press, 1979.

[K] A.I. Khinchin, Mathematical Foundations of Statistical Mechanics, Dover, 1949.

[Ku] H. Kunz, Analyticity and clustering of unbounded spin systems, Comm. in Math. Phys. 59 (1978) 53-69.

[L] O.E. Lanford III, Entropy and equilibrium states in classical statistical mechanics, in Lecture Notes in Physics, Vol. 20, Statistical Mechanics and Mathematical Problems, Ed. A. Lenard, Springer-Verlag, N.Y., 1973.

[ML] A. Martin-Löf, Statistical Mechanics and the Foundations of Thermodynamics, Lecture Notes in Physics, Vol. 101, Springer-Verlag, N.Y., 1979.

[Mi] R.A. Minlos, Lectures on Statistical Physics, Russian Mathematical Surveys, 23 (1968), no. 1 (English Translation) 137-194.

[Mis] L. Mistura, Change of variables in thermodynamics, Il Nuovo Cimento, 51B, N.1 (1979) 125-146.

[Pi] S.A. Pirogov and Ya.G. Sinai, Phase diagrams of classical lattice systems, Theoretical and Mathematical Physics 25 (1975) 358-369 (in Russian).

[Pr] C. Preston, Random Fields, Lecture Notes in Mathematics No. 534, Springer-Verlag, N.Y., 1976.

[Rk] Rockafellar, R.T., Conjugate duality and optimization, S.I.A.M. Regional Conference Series in App. Math. No. 16 (1974) (especially pages 13-16).

[Ro] D. Roller, The early development of the concepts of temperature and heat, Harvard Univ. Press, 1950.

[R 1] D. Ruelle, Statistical Mechanics, W.A. Benjamin, Inc., N.Y., 1969.

[R 2] ————, Thermodynamic Formalism, Addison-Wesley Pub. Co, Reading, Mass., 1978.

[Se 1] J. Serrin, The concepts of thermodynamics, in Contemporary Developments in Continuum Mechanics and Partial Differential Equations de la Penha and Medeiros (Eds.), North-Holland Pub. Co., 1978.

[Se 2] ————, Conceptual Analysis of the Classical Second Laws of Thermodynamics, Arch. Rat. Mech., 1980.

[Se 3] ————, Foundations of Classical Thermodynamics, Univ. of Chicago, Dept. of Math., Lecture Notes in Mathematics, 1975.

[Si 1] B. Simon, A remark on Dobrushin's uniqueness theorem, Commun. Math. Phys. 68 (1979) 183-185.

[Si 2] ————, The Statistical Mechanics of Lattice Gases, to appear.

[Si 3] ————, The $P(\varphi)_2$ Euclidean (Quantum) Field Theory, Princeton Univ. Press, Princeton, N.J., 1974.

[Sp] F. Spitzer, Introduction aux processus de Markov a parametres dans Z^ν, in Lecture Notes in Mathematics, 390, Springer-Verlag, New York.

[Th] R.L. Thompson, Equilibrium States in Thin Energy Shells, Memoirs Amer. Math. Soc. 150 (1974).

[T] L. Tisza, Generalized Thermodynamics, M.I.T. Press, Cambridge, Mass., 1966.

[V] L.N. Vasershtein, Markov processes over denumerable products of spaces, describing large systems of automata, Problems in the Transmission of Information, Vol. 5, No. 3 (1969) 64-72 (English Translation).

[W] F.T. Wall, Chemical Thermodynamics, W.H. Freeman and Co., 1957.

[Z 1] M.W. Zemansky, Heat and Thermodynamics, 4th Edition, McGraw-Hill Book Co., New York, 1957.

[Z 2] ————, Heat and Thermodynamics, 5th Edition, McGraw-Hill Book Co., New York, 1968.

RANDOM FIELDS AND DIFFUSION PROCESSES

Hans Föllmer

Originally published in: *Ecole d'Eté de Probabilités de Saint-Flour XV–XVII – 1985–87*, Lecture Notes in Mathematics, Vol. **1362**, 101–203, DOI: 10.1007/BFb0086180, © Springer-Verlag Berlin Heidelberg 1988, Reprint by Springer-Verlag Berlin Heidelberg 2012

H. FOLLMER : "RANDOM FIELDS AND DIFFUSION PROCESSES"

INTRODUCTION

In these lectures, a random field P will be a probability measure on a product space S^I where S is some state space, for example $S = \{-1, +1\}$ or $S = C[0,1]$, and where I is some countable set of sites, for example the d-dimensional lattice Z^d.

We take the point of view of R.L. Dobrushin: A random field will be specified, but not necessarily uniquely determined, by a system of conditional probabilities π_V, where V is a finite subset of I, and where $\pi_V(\cdot|\eta)$ is the conditional distribution on S^V given the configural η outside of V. In that case, P will also be called a Gibbs measure with respect to the local specification (π_V). P is called a Markov field if $\pi_V(\cdot|\eta)$ only depends on the values $\eta(i)$ for the sites i in some finite "boundary" of V.

We want to give an introduction to some of the connections between Gibbs measures and infinite dimensional stochastic processes, and in particular between Markov fields and Markov processes. These connections appear on different levels:

(A) Markov fields as Markov processes

(B) Markov fields as invariant measures of Markov processes

(C) Markov processes as Markov fields.

To begin with, any Markov field on $S^I = (S^J)^Z$ with $I = Z^d$ and $J = Z^{d-1}$ can be viewed as the law of a stochastic process with state space S^J. This process may or may not be a Markov process: this is the problem of the "global" Markov property.

(B) has been a central topic in the study of time evolution, of interacting particle systems, and has been studied in various different contexts; cf., for example, [Li]. We are going to look at (B) for a class of infinite dimensional diffusion processes $X = (X_i)_{i \in I}$ of the form

$$dX_i = dW_i + b_i(X(t),t))dt \quad (i \in I)$$

where (W_i) is a collection of independent Brownian motions. Under some bounds on the interaction in the drift terms, the time reversed process is again of this form, and there is an infinite dimensional analogue to the classical duality equation

$$b + \hat{b} = \nabla \log \rho$$

which relates forward drift, backward drift and the density of the process at any given time. In the infinite dimensional case, the density is replaced by the system of conditional densities at each site, i.e., by a local specification. This leads to the description of invariant measures of the process as Gibbs measures and often as Markov fields. As to (C), note that the distribution of the infinite dimensional diffusion process above is a probability measure on $C[0,1]^I$ and can thus be regarded as a random field with state space $S = C[0,1]$ at each site $i \in I$. If we determine its local specification then we can apply random field techniques to the diffusion, and this may be useful, e.g., in view of large deviations or a central limit theorem not in the time but in the space direction.

At the Ecole d'Eté de Probabilités de Saint-Flour, it seemed natural to assume a very strong background in Stochastic Analysis, and to take a more introductory approach to random fields. Thus we begin with a self contained introduction to Gibbs measures in Ch. I, with special emphasis on Dobrushin's contraction technique and on the probabilistic limit theorems which are behind thermodynamical qunatities like energy and entropy. These topics are well known, and some excellent introductions are available, e.g. [Pr]. But we want to discuss various applications of the contraction technique, and the Shannon-McMillan theorem for the relative entropy with respect to a Gibbs measure will lead us to a more recent development, namely to the study of large deviations of the empirical field of a Gibbs measure.

In Ch. II we discuss some connections between Gibbs measures and infinite dimensional diffusion processes. In the spirit of (C), we first discuss some large deviations of the empirical field of an infinite collection of independent Brownian motions. Then we look at infinite dimensional diffusion processes from the point of view of time reversal. A first application is the description of invariant measures as Gibbs measures. But time reversal is also a useful tool for other purposes; for example, it can be used to study certain large deviations of the empirical field of the diffusion process. In the last section we apply Dobrushin's contraction technique to an infinite dimensional diffusion process, again in the spirit of (C).

Je voudrais remercier P.L. Hennequin de m'avoir invité à faire ce cours: c'était un grand plaisir pour moi de participer à l'Ecole d'Eté de Saint-Flour.

I. An introduction to random fields

A random field will be a probability measure μ on a countable product space S^I . In Dobrushin's approach, random fields are specified (but in general not uniquely determined) by a consistent system of conditional probabilities. This is analogous to the specification of a Markov process μ on S^Z by the semigroup of its transition kernels (which may or may not admit more than one entrance law, if any). But in a spatial setting, it is natural to replace conditional probabilities on the future given the past by conditional probabilities on the behavior in some finite subset $V \subset I$ given the situation outside of V . This leads to the specification of a random field by a consistent family of stochastic kernels (π_V) indexed by the finite subsets of I . In this way, the total order on the time axis is lost, and some of the usual techniques from the theory of Markov processes have to be modified or replaced accordingly.

Section 1 contains basic definitions and some general facts about the structure of the class of all random fields specified by a collection (π_V) of local conditional probabilities. In order to guarantee uniqueness, we have to introduce bounds on the interaction . In Section 2 we give a short introduction to Dobrushin's contraction technique which is based on estimates of the interaction between the different sites in I . This leads not only to a strong uniqueness theorem, but also to additional regularity properties of the unique Gibbs measure.

In Section 3 the local specification is given by a stationary interaction potential on a d-dimensional lattice, and this allows us to introduce "thermodynamical" quantities like specific energy. Our main purpose is to derive a Shannon-Mc Millan theorem for the relative entropy $h(\nu;\mu)$ of a stationary measure ν with respect to a stationary Gibbs measure μ . This is the key to Section 4 where we describe some joint work with S. Orey on large deviations of the empirical field of a Gibbs measure. In the second part of Section 4, the effect of a phase transition on large deviations is discussed in terms of "surface entropy".

1. Random fields and their local specification

In this section we introduce Dobrushin's description of random fields on S^I in terms of their local specification by a system (π_V) of conditional probabilities, indexed by the finite subsets V of I. The first part contains some basic definitions and remarks; in the second we collect some general results on the integral representation of the set $G(\pi)$ of all random fields specified by a given family (π_V).

1.1 Definitions

Let I be a countable set of sites, let S be a standard Borel space of states, and consider the product space $\Omega = S^I$ of all configurations $\omega : I \to S$. For $J \subseteq I$ we introduce the σ-field $\underline{\underline{F}}_J = \sigma(\omega(i)\ ;\ i \in J)$, and we write $F = \underline{\underline{F}}_I$. The restriction of ω to J will be denoted by ω_J.

Let $M(\Omega)$ denote the class of probability measures on $(\Omega,\underline{\underline{F}})$. A probability measure $\mu \in M(\Omega)$ will also be called a random field.

(1.1) Remark. $(\Omega,\underline{\underline{F}})$ is again a standard Borel space. Thus we can choose, for any $\mu \in M(\Omega)$ and any $V \subseteq I$, a regular conditional distribution of μ with respect to $\underline{\underline{F}}_{V^c}$, i.e., a stochastic kernel $\pi_V(\omega,dy)$ from $(\Omega,\underline{\underline{F}}_{V^c})$ to $(\Omega,\underline{\underline{F}})$ such that

$$(1.2) \qquad \pi_V(\omega,\cdot) = \delta_\omega \quad \text{on } \underline{\underline{F}}_{V^c}$$

and

$$(1.3) \qquad E_\mu[\varphi|\underline{\underline{F}}_{V^c}](\omega) = \int \pi_V(\omega,d\eta)\varphi(\eta) =: (\pi_V\varphi)(\omega)$$

for any $\underline{\underline{F}}$-measurable function $\varphi \geq 0$. These conditional distributions are consistent in the following sense: since $\underline{\underline{F}}_{W^c} \subseteq \underline{\underline{F}}_{V^c}$ for $W \supseteq V$, we have

$$(\pi_W\pi_V\varphi)(\omega) = E_\mu[E_\mu[\varphi|\underline{\underline{F}}_{V^c}]|\underline{\underline{F}}_{W^c}](\omega) = E_\mu[\varphi|\underline{\underline{F}}_{W^c}](\omega)$$

$$= (\pi_W\varphi)(\omega) \qquad \mu\text{-a.s.}$$

for any measurable $\varphi \geq 0$, hence

(1.4) $$(\pi_W \pi_V)(\omega,\cdot) = \pi_W(\omega,\cdot) \qquad \mu\text{-a.s.}$$

We are now going to prescribe the local conditional behavior of a random field by fixing a system of conditional distributions π_V for the finite subsets $V \subseteq I$. These conditional distributions will be consistent in a strict sense, i.e., without the intervention of null sets. We are then going to study the class of all random fields which are compatible with these kernels in the sense of (1.3)

(1.5) Definition. For each finite $V \subseteq I$, let π_V be a stochastic kernel from $(\Omega, \underline{\underline{F}}_{V^c})$ to $(\Omega, \underline{\underline{F}})$ such that

(1.6) $$\pi_V(\omega,\cdot) = \delta_\omega \quad \text{on } \underline{\underline{F}}_{V^c} .$$

The collection (π_V) is called a local specification if

(1.7) $$\pi_W \pi_V = \pi_W \qquad \text{for } V \subseteq W .$$

A random field μ is called a Gibbs measure with respect to the local specification (π_V) if, for any finite V , π_V is a conditional distribution of μ with respect to $\underline{\underline{F}}_{V^c}$ in the sense of (1.3).

From now on we fix a local specification (π_V) and denote by $G(\pi)$ the corresponding class of Gibbs measures.

(1.8) Lemma. A random field μ belongs to $G(\pi)$ if and only if

(1.9) $$\mu \pi_V = \mu .$$

for any finite $V \subseteq I$.

Proof. 1) For $\mu \in G(\pi)$ and finite V we have

$$\int \varphi d\mu = \int E_\mu[\varphi | \underline{\underline{F}}_{V^c}] d\mu = \int (\pi_V \varphi) d\mu = \int \varphi d(\mu \pi_V) ,$$

hence $\mu = \mu\pi_V$.

2) Condition (1.6) implies $\pi_V(\varphi\psi)(\omega) = \psi(\omega)(\pi_V\varphi)(\omega)$ for $\underline{\underline{F}}$-measurable $\varphi \geq 0$ and $\underline{\underline{F}}_{V^c}$-measurable $\psi \geq 0$. Thus, (1.9) implies

$$E_\mu[\varphi\psi] = E_\mu[\pi_V(\varphi\psi)] = E_\mu[\psi\pi_V\varphi] ,$$

hence (1.3).

(1.10) Remark on existence. Introducing a polish topology on S , we may view Ω as a polish space, and then $M(\Omega)$ becomes a polish space with respect to the weak topology. In particular,

$$(1.11) \qquad \mu_n \to \mu \quad \Leftrightarrow \quad \int f_k d\mu_n \to \int f_k d\mu \qquad (k = 1,2,\ldots)$$

for a suitably chosen countable family of bounded continuous functions on Ω . Now suppose that S is compact, and that (π_V) has the following Feller property:

$$(1.12) \qquad f \in C(\Omega) \quad \Rightarrow \quad \pi_V f \in C(\Omega)$$

for any finite V . Then $\mu_n \to \mu$ implies

$$\lim_n \int fd(\mu_n\pi_V) = \lim_n \int (\pi_V f)d\mu_n = \int (\pi_V f)d\mu = \int fd(\mu\pi_V)$$

for $f \in C(\Omega)$, i.e., $\mu \to \mu\pi_V$ is a continuous map on the compact convex space $M(\Omega)$. By Schauder-Tychonow, we have $\{\mu \in M(\Omega) | \mu\pi_V = \mu\} \neq \emptyset$, and since

$$G(\pi) = \bigcap_{V \text{ finite}} \{\mu \in M(\Omega) | \mu\pi_V = \mu\}$$

$$= \bigcap_n \{\mu \in M(\Omega) | \mu\pi_{V_n} = \mu\}$$

for any increasing sequence $V_n \uparrow I$, we see that $G(\pi)$ is a non-empty convex compact set. The assumption of a compact state space can be replaced by a tightness condition on the local specification; cf. [Pr].

(1.13) Remark on spatial homogeneity. Let I be the d-dimensional lattice Z^d, and let $\Theta_i : \Omega \to \Omega$ be the shift map defined by $(\Theta_i\omega)(k) = \omega(i+k)$. In this case, we denote by

$$M_s(\Omega) := \{\mu \in M(\Omega) | \mu \circ \Theta_i = \mu \ (i \in I)\}$$

the class of all spatially homogeneous random fields. A local specification is called spatially homogeneous if

$$\pi_V(\Theta_i\omega,\cdot) = \pi_{V+i}(\omega,\cdot) \circ \Theta_i$$

resp.

$$(\pi_V\varphi) \circ \Theta_i = \pi_{V+i}(\varphi \circ \Theta_i)$$

for $i \in I$ and finite $V \subseteq I$. For any $i \in I$, $\mu \to \mu \circ \Theta_i$ defines a continuous map on $G(\pi)$: in fact,

$$\begin{aligned}\int(\pi_V\varphi)d(\mu \circ \Theta_i) &= \int(\pi_V\varphi) \circ \Theta_i d\mu \\ &= \int\pi_{V+i}(\varphi \circ \Theta_i)d\mu \\ &= \int\varphi \circ \Theta_i d\mu = \int\varphi d(\mu \circ \Theta_i) ,\end{aligned}$$

and this implies $\mu \circ \Theta_i \in G(\pi)$ due to (1.8). If S is compact and (1.12) is satisfied, then it follows that the set

$$G_s(\pi) = G(\pi) \cap M_s(\Omega)$$

of spatially homogeneous Gibbs measures is non-empty, convex and compact; the argument is analogous to (1.10).

(1.14) Remark on the Markov property. Suppose that for each $i \in I$ there is a finite set of neighbours $N(i) \subseteq I - \{i\}$. We say that (π_V) has the local Markov property if

$$\pi_V(\omega,\cdot) = \pi_V(\eta,\cdot)$$

whenever

$$\omega = \eta \quad \text{on} \quad \partial V := \bigcup_{i \in V} N(i) - V .$$

For any $\mu \in G(\pi)$, this local Markov property takes the form

$$(1.15) \qquad E_\mu[\varphi|\underline{\underline{F}}_{V^c}] = E_\mu[\varphi|\underline{\underline{F}}_{\partial V}] \quad (\varphi \geq 0 \text{ , } \underline{\underline{F}}_V\text{-measurable})$$

whenever V is finite. In this case, we also say that μ is a Markov random field. In general, the local Markov property (1.15) does not imply the global Markov property, i.e., the validity of (1.15) for any (not necessarily finite) $V \subseteq I$; see, however, Section 2.2.

1.2. Integral representation

Suppose that $G(\pi) \neq \emptyset$. In general $G(\pi)$ contains more than one element, i.e., a Gibbs measure $\mu \in G(\pi)$ is not uniquely determined by its local specification. In that case one speaks of a "phase transition". If S is compact, then $G(\pi)$ is a compact convex set, and Choquet's theorem leads to an integral representation of $G(\pi)$ in terms of extremal Gibbs measures. But such an integral representation can also be obtained without any compactness assumptions, and in a more explicit form which exhibits the role of the tail field

$$\underline{\underline{A}} = \bigcap_{V \text{ finite}} \underline{\underline{F}}_{V^c}$$

of asymptotic events; cf. [Fö1], [Dy]. The point is that martingale convergence allows us to pass from (π_V) to a limiting kernel π_∞ with respect to $\underline{\underline{A}}$:

(1.16) Theorem. If $G(\pi) \neq \emptyset$ then there exists a stochastic kernel $\pi_\infty(\omega,dy)$ from $(\Omega,\underline{\underline{A}})$ to $(\Omega,\underline{\underline{F}})$ such that

$$(1.17) \qquad \pi_\infty(\omega,\cdot) \in G(\pi)$$

for any $\omega \in \Omega$, and such that, for any $\mu \in G(\pi)$,

(1.18) $$E_\mu[\varphi|\underline{\underline{A}}] = \pi_\infty\varphi \qquad \mu\text{-a.s.}$$

for $\underline{\underline{F}}$-measurable $\varphi \geq 0$.

<u>Proof.</u> 1) Fix a sequence $V_n \uparrow I$, a sequence of bounded functions f_k $(k = 1,2,\ldots)$ as in (1.11), and a Gibbs measure $\mu \in G(\pi)$. By martingale convergence,

$$\lim_n \pi_{V_n} f_k(\omega) = \lim_n E_\mu[f_k|\underline{\underline{F}}_{V_n^c}](\omega) = E_\mu[f_k|\underline{\underline{A}}](\omega)$$

$$= \int f_k(\eta)\pi_\mu(\omega,d\eta) \qquad \mu\text{-a.s.}$$

where π_μ denotes a conditional probability distribution for μ and $\underline{\underline{A}}$. Thus, the set

$$\Omega_o = \{\omega|\lim_n \pi_{V_n}(\omega,\cdot) \text{ exists}\} \in \underline{\underline{A}}$$

satisfies $\mu(\Omega_o) = 1$ for any $\mu \in G(\pi)$. Defining

$$\tilde{\pi}_\infty(\omega,\cdot) := \lim_n \pi_{V_n}(\omega,\cdot) \quad \text{if} \quad \omega \in \Omega_o$$

$$:= \mu_o \qquad \text{if} \quad \omega \notin \Omega_o$$

for some fixed $\mu_o \in G(\pi)$, we obtain a kernel from $(\Omega,\underline{\underline{A}})$ to $(\Omega,\underline{\underline{F}})$ which satisfies (1.18).

2) Let $\mu \in G(\pi)$. For any $A \in \underline{\underline{A}}$ and for any k ,

$$\int_A \tilde{\pi}_\infty f_k d\mu = \int_A E_\mu[f_k|\underline{\underline{A}}]d\mu = \int_A E_\mu[\pi_V f_k|\underline{\underline{A}}]d\mu = \int_A \tilde{\pi}_\infty \pi_V f_k d\mu .$$

This implies $\tilde{\pi}_\infty(\omega,\cdot) = (\tilde{\pi}_\infty\pi_V)(\omega,\cdot)$ μ-a.s. for any finite V , hence $\tilde{\pi}_\infty(\omega,\cdot) \in G(\pi)$ μ-a.s. due to (1.8). Since

$$\tilde{\Omega}_o = \{\omega|\tilde{\pi}_\infty(\omega,\cdot) \in G(\pi)\} \in \underline{\underline{A}} ,$$

the kernel defined by

$$\pi_\infty(\omega,\cdot) := \tilde{\pi}_\infty(\omega,\cdot) \quad \text{if} \quad \omega \in \tilde{\Omega}_o$$

$$:= \quad \mu_o \qquad \text{if} \quad \omega \notin \tilde{\Omega}_o$$

has properties (1.17) and (1.18).

In the language of [Dy], we have shown that $\underline{\underline{A}}$ is an "H-sufficient statistics" with respect to $G(\pi)$. Using the general construction in [Dy], we obtain the following integral representation:

(1.19) Corollary. 1) Each $\mu \in G(\pi)$ is of the form

$$\mu = \int_{G_e(\pi)} \nu \ \gamma_\mu(d\nu) \ , \tag{1.20}$$

with a unique probability measure γ_μ on

$$G_e(\pi) = \{\mu \in G(\pi) | \mu = 0 - 1 \ \text{ on } \ \underline{\underline{A}}\} \subseteq \{\lim_n \pi_{V_n}(\omega,\cdot) | \omega \in \tilde{\Omega}_o\} \ .$$

2) γ_μ is the image of μ under $\omega \to \pi_\infty(\omega,\cdot)$

3) For each probability measure γ_μ on $G_e(\pi)$, the measure μ defined by (1.20) belongs to $G(\pi)$.

(1.21) Remark. 1) $G_e(\pi)$ is the set of extremal points in $G(\pi)$.

2) By 2), two measures in $G(\pi)$ coincide as soon as they coincide on $\underline{\underline{A}}$.

In the spatially homogeneous case (1.13), an analogous construction, which combines martingale convergence and the ergodic theorem, leads to an integral representation of the class $G_s(\pi)$ of spatially homogeneous Gibbs measures. Here the role of $\underline{\underline{A}}$ is taken by the σ-field $\underline{\underline{J}}$ of shift-invariant sets:

(1.22) Theorem. If $G_s(\pi) \neq \emptyset$ then there exists a stochastic kernel $\pi_s(\omega,dy)$ from $(\Omega,\underline{\underline{J}})$ to $(\Omega,\underline{\underline{F}})$ such that

$$\pi_s(\omega,\cdot) \in G_s(\pi) \tag{1.23}$$

for any $\omega \in \Omega$, and

(1.24) $$E_\mu[\varphi|\underline{\underline{J}}] = \pi_s\varphi \qquad \mu\text{-a.s.}$$

for any $\mu \in G_s(\pi)$.

Proof. By the d-dimensional ergodic theorem, by the slight extension (2.29) of the martingale convergence theorem sometimes referred to as "Hunt's Lemma", and by (1.25) below,

$$\begin{aligned} E_\mu[f|\underline{\underline{J}}] &= E_\mu[E_\mu[f|\underline{\underline{J}}]|\underline{\underline{A}}] \\ &= \lim_n |V_n|^{-1} \sum_{i \in V_n} E_\mu[f \circ \theta_i | \underline{\underline{F}}_{V_n^c}] \\ &= \lim_n |V_n|^{-1} \sum_{i \in V_n} \int f d(\pi_{V_n}(\omega,\cdot) \circ \theta_i) \end{aligned}$$

if this limit exists and belongs to $G_s(\pi)$. We can now proceed as in the proof of (1.16).

(1.25) Lemma. For any $\mu \in M_s(\Omega)$, the σ-field $\underline{\underline{J}}$ of shift-invariant sets is contained in the μ-completion of $\underline{\underline{A}}$.

Proof. Let f be a bounded $\underline{\underline{J}}$-measurable function, and let $W \subseteq I$ be finite. Since $f = f \circ \theta_i$ and $\mu = \mu \circ \theta_i$,

$$\begin{aligned} \|f - E_\mu[f|\underline{\underline{F}}_W]\| &= \|f \circ \theta_i - E_\mu[f|\underline{\underline{F}}_W] \circ \theta_i\| \\ &\equiv \|f - E_\mu[f|\underline{\underline{F}}_{W+i}]\| \end{aligned}$$

due to (1.13), where $\|\cdot\|$ denotes the $L^1(\mu)$-norm. Thus,

$$\begin{aligned} \|f - E_\mu[f|\underline{\underline{A}}]\| \le\ & \|f - E_\mu[f|\underline{\underline{F}}_W]\| \\ & + \|E_\mu[f|\underline{\underline{F}}_{W+i}] - E_\mu[f|\underline{\underline{F}}_{V^c}]\| \\ & + \|E_\mu[f|\underline{\underline{F}}_{V^c}] - E_\mu[f|\underline{\underline{A}}]\| . \end{aligned}$$

Take $i \in I$ such that $W + i \subseteq V^c$. Then the second term on the right is dominated by

$$||E_\mu[f|\underline{\underline{F}}_{W+i}] - f|| = ||E_\mu[f|\underline{\underline{F}}_W] - f|| .$$

By martingale convergence, V and W can be chosen such that the first and the third (hence also the second) term is $\le \epsilon$. This shows $f = E_\mu[f|\underline{\underline{A}}]$.

(1.26) Corollary. $G_s(\pi)$ admits the integral representation

(1.27) $$\mu = \int_{G_{s,e}} \nu \; \gamma_{s,\mu}(d\nu)$$

where

$$G_{s,e} = \{\mu \in G_s(\pi) | \mu = 0 - 1 \text{ on } \underline{\underline{J}}\}$$

is the class of all ergodic measures in $G_s(\pi)$, and where $\gamma_{s,\mu}$ is the image of μ under $\omega \to \pi_s(\omega,\cdot)$.

(1.28) Remark. Without the additional Gibbs structure, the construction behind (1.22) and (1.26) leads to an explicit integral representation of a stationary measure $\mu \in M_s(\Omega)$ as a mixture of ergodic measures; cf. [Dy]. For this, we will use the notation

(1.29) $$\mu = \int \mu_\omega \mu(d\omega)$$

where μ_ω denotes the appropriate ergodic version of the conditional distribution with respect to $\underline{\underline{J}}$.

2. Dobrushin's contraction technique

In this section we give a short introduction to Dobrushin's contraction technique [Do1,2]; see also [Ro] and [DP]. This technique does not only provide a powerful uniqueness theorem, it also allows us to derive a number of additional regularity properties of the unique Gibbs measure. We illustrate this point with the global Markov property, with some covariance estimates, and with the almost sure convergence of multi-parameter martingales.

2.1. Dobrushin's comparison theorem

Let μ be a Gibbs measure on $\Omega = S^I$ with local specification (π_V). We denote by $\pi_k(\cdot|\eta)$ the conditional distribution of $\omega(k)$ given $\underline{\underline{F}}_{I-\{k\}}$, so that

$$\pi_{\{k\}}(\eta,\cdot) = \pi_k(\cdot|\eta) \times \prod_{i \neq k} \delta_{\eta(i)} \qquad (k \in I) .$$

Let us now measure the influence of site i on site k by

$$(2.1) \qquad C_{ik} := \sup \{\tfrac{1}{2} \,||\pi_k(\cdot|\omega) - \pi_k(\cdot|\eta)|| : \omega = \eta \text{ off } i\}$$

where $||\cdot||$ denotes the total variation of a signed measure on S. The matrix $C = (C_{ik})_{i,k\in I}$ will be called Dobrushin's interaction matrix; C^n denotes the n-th power of C. For any probability measure ν on Ω let us define the vector $b = (b_k)_{k\in I}$ with components

$$(2.2) \qquad b_k := \tfrac{1}{2} \int ||\pi_k(\cdot|\eta) - \nu_k(\cdot|\eta)|| \, \nu(d\eta)$$

where $\pi_k(\cdot|\eta)$ is a conditional distribution of $\omega(k)$ with respect to $\underline{\underline{F}}_{I-\{k\}}$. This is a slight modification of the definition in [Do2], which will be useful for the covariance estimates below.

In order to compare μ and ν, let us introduce the class $C(\Omega)$ of functions which can be approximated uniformly by bounded measurable functions depending only on finitely many coordinates. We say that a vector $a = (a_i)_{i\in I}$ is an estimate for μ and ν if

(2.3) $$\left|\int f d\mu - \int f d\nu\right| \le \sum_i a_i \delta_i(f) \qquad (f \in C(\Omega))$$

where

$$\delta_i(f) := \sup\{|f(\omega) - f(\eta)| : \omega = \eta \text{ off } i\}$$

denotes the oscillation of f at site $i \in I$. For example, $a_i \equiv 1$ is always an estimate since $|f(\omega) - f(\eta)| \le \sum_i \delta_i(f)$ for any $f \in C(\Omega)$. The comparison theorem will follow from a successive improvement of this initial estimate, and for this we need an additional continuity requirement:

(2.4) $$f \in C(\Omega) \Rightarrow \pi_{\{k\}} f \in C(\Omega) \quad \text{for any } k \in I .$$

(2.5) Lemma. If a is an estimate for μ and ν then the vector $aC + b$ is also an estimate.

Proof: It is enough to check (2.3), with $aC + b$ instead of a, for functions which depend only on finitely many coordinates. Therefore, it is enough to show that for any finite subset $J \subset I$ the vector a^J with components

$$a_i^J = \min(a_i, (aC + b)_i) \qquad (i \in J)$$

$$= a_i \qquad (i \notin J)$$

is an estimate for μ and ν. We prove this by induction on the cardinality of J. For $J = \emptyset$ the statement is true. Now assume that a^J is an estimate and take $K = J \cup \{k\}$; we have to show that a^K is also an estimate. For $f \in C(\Omega)$ we have

$$\begin{aligned} \left|\int f d\mu - \int f d\nu\right| &\le \int \left[\int f d\pi_k(\cdot|\eta)\right] (\mu-\nu)(d\eta) \\ &+ \int \left|\int f d\pi_k(\cdot|\eta) - \int f d\nu_k(\cdot|\eta)\right| \nu(d\eta) \\ &\le \sum_i a_i^J \delta_i\left(\int f d\pi_k(\cdot|\cdot)\right) + b_k \delta_k(f) . \end{aligned}$$

Since

$$\delta_i\left(\int f d\pi_k(\cdot|\cdot)\right) \le \delta_i(f) + C_{ik}\delta_k(f)$$

for $i \neq k$ and $=0$ for $i = k$, we obtain

$$\left|\int f d\mu - \int f d\nu\right| \le \sum_{i \neq k} a_i^J \, \delta_i(f) + (a^J C + b)_k \delta_k(f) \, .$$

But $a^J C \le aC$, and since a^J with $a_k^J \le a_k$ is also an estimate, we can replace the right side by $\sum_i a_i^K \delta_i(f)$.

Applying the lemma successively, we see that for each $n \ge 1$ the vector

$$aC^{n+1} + \sum_{m=0}^{n} bC^m$$

is an estimate. Letting $n \uparrow \infty$, taking $a_i \equiv 1$ and defining the matrix

$$D := \sum_{m=0}^{\infty} C^m \quad ,$$

we see that the vector bD is an estimate as soon as C satisfies the condition

(2.6) $$\lim_n \sum_i C_{ik}^n = 0 \qquad (k \in I)$$

Note that (2.6) is satisfied if, for example,

(2.7) $$c := \sup_k \sum_i C_{ik} < 1 \; ,$$

since $\sum_i C_{ik}^n \le c^n$ by induction.

This proves the following variant of Dobrushin's comparison theorem [Do2, Th. 3]; see also [Kü3].

(2.8) Comparison Theorem: Under condition (2.6) we have

$$\left|\int f d\mu - \int f d\nu\right| \le \sum_i (bD)_i \, \delta_i(f) \qquad \text{for any } f \in C(\Omega) \, .$$

2.2 Uniqueness and global Markov property

If $\nu \in G(\pi)$, i.e., if ν has the same local specification as μ , then we have $b_i = 0$ $(i \in I)$ in (2.8), and this implies $\mu = \nu$. This is Dobrushin's well-known

(2.9) Uniqueness theorem [Do1]: Under condition (2.6) there is at most one measure $\mu \in G(\pi)$.

In fact, condition (2.6) not only implies uniqueness but also conditional uniqueness in the following sense. For any $J \subseteq I$ and for any $\eta \in S^{I-J}$, define the conditional specification $\pi_V^{J,\eta}$ ($V \subseteq J$ finite) on S^J with

$$(2.10) \qquad \pi_V^{J,\eta}(\omega, \cdot\,) = \pi_V(\zeta, \cdot\,)$$

where $\zeta(i) = \eta(i)$ for $i \in I-J$ and $\zeta(i) = \omega(i)$ for $i \in J$. The corresponding Dobrushin matrix $C^{J,\eta}$ satisfies $C^{J,\eta}_{ik} \le C_{ik}$; in particular it inherits condition (2.6). For any Gibbs measure $\mu \in G(\pi)$, let $\mu_J(\cdot|\eta)$ be a conditional joint distribution of $\omega(i)$ $(i \in J)$ with respect to $\underline{\underline{F}}_{I-J}$ and μ . For μ-almost η , $\mu_J(\cdot|\eta)$ is compatible with the conditional specification $(\pi_V^{J,\eta})$, and thus coincides with the corresponding unique Gibbs measure, due to (2.9).

Suppose, in particular, that μ is a Markov random field, i.e., that μ resp. (π_V) has the local Markov property (1.15). In this case, (2.10) shows that $(\pi_V^{J,\eta})$, hence $\mu_J(\cdot|\eta)$, only depends on the values $\eta(j)$ for $j \in \partial J$. This implies the global Markov property, i.e.,

$$E_\mu[\varphi|\underline{\underline{F}}_{J^c}] = E_\mu[\varphi|\underline{\underline{F}}_{\partial J}] \qquad (\varphi \ge 0 \ \ \underline{\underline{F}}_J \text{ - measurable})$$

for arbitrary (not only finite) $J \subseteq I$. Consider, in particular, the case $I = Z^d$. Here the global Markov property implies that μ may be viewed as the distribution of a Markov chain on the infinite-dimensional state space $E = S^{Z^{d-1}}$. If we denote by $L(t)$ the line $\{(t,j): j \in Z^{d-1}\}$, then the transition probabilities of the chain are of the form

$$P_t(\xi,.) \;=\; \text{distribution of } \omega_{L(t)} \text{ under } \mu_{L(t-1)}(.|\omega_{L(t-1)} = \xi).$$

Let us also assume that S is finite so that E may be regarded as a compact space. Then condition (2.6) implies, in addition, that this Markov chain has the Feller property:

$$\xi_n \to \xi \quad \Rightarrow \quad P_t(\xi_n,\cdot) \to P_t(\xi,\cdot)$$

in the weak topology for probability measures on E. In fact, any limit point of the sequence $\mu_{L(t-1)}(\cdot|\xi_n)$ is compatible with the conditional specification induced by ξ, and so it must coincide with $\mu_{L(t-1)}(\cdot|\xi)$.

The restrictions μ_t of μ to $S^{L(t)}$ form an entrance law for the chain:

$$\mu_{t-1}P_t = \mu_t \qquad (t \in Z^1) .$$

Conversely, any entrance law induces a measure $\mu \in G(\pi)$, and so we have shown:

(2.11) <u>Corollary.</u> Suppose that $I = Z^d$, that S is finite, and that (π_V) has the local Markov property and satisfies (2.6). Then the unique Gibbs measure $\mu \in G(\pi)$ may be viewed as a Markov chain on $S^{Z^{d-1}}$, which has the Feller property and admits exactly one entrance law.

(2.12) <u>Remark</u>. We refer to [Is] for counter-examples to the conjecture that any extremal Gibbs measure with respect to a Markov specification might have the global Markov property. For an "attractive" interaction there are techniques based on monotonicity which allow to show, for example, that the + and - states of the Ising model have the global Markov property; cf. (3.47) below.

2.3 <u>Covariance estimates</u>

Let us now use the comparison theorem (2.8) in order to obtain estimates for certain covariances.

(2.13) <u>Theorem.</u> Under condition (2.6), the covariance of any two functions f and g in $C(\Omega)$ with respect to μ satisfies

$$|\mathrm{cov}_\mu(f,g)| \le \frac{1}{4} \sum_{i,k} \delta_i(f) D_{ki} \delta_k(g).$$

<u>Proof.</u> We may assume $g > 0$ and $\int g \, d\mu = 1$. But then we can write

$$\mathrm{cov}_\mu(f,g) = \int f \, d\nu - \int f \, d\mu \, , \, ,$$

where $d\nu = g \, d\mu$ is a probability measure whose conditional probabilities are given by

$$\nu_i(d\gamma|\eta) = g(\gamma) \, [\int g \, d\pi_i(\cdot|\eta)]^{-1} \, \pi_i(d\gamma|\eta) \, .$$

Applying variant (2.8) of Dobrushin's comparision theorem, we obtain

$$|\mathrm{cov}_\mu(f,g)| \le \sum_i (\sum_k b_k D_{ki}) \, \delta_i(f)$$

with

$$b_k \le \frac{1}{4} \delta_k(g) \int [\int g \, d\mu_k(\cdot|x)]^{-1} \, g(x) \, \mu(dx) = \frac{1}{4} \delta_k(g) \, ,$$

using the elementary estimate $\frac{1}{2} ||\mu-\nu|| \le \frac{1}{4} \delta(g)$.

Let us now illustrate how this estimate provides information on the rate of decay of correlation. Following L. Gross [Gr], we consider the case $I = Z^d$ and fix a translation invariant semimetric $d(\cdot,\cdot)$ on I. Let

$$|f|_o := \sum_i e^{d(i,o)} \, \delta_i(f)$$

and

$$\gamma := \sup_k \sum_i e^{d(i,k)} \, C_{ik} \, .$$

Note that condition $c < 1$ in (2.7) implies $\gamma < 1$ for a suitable multiple of the Euclidean metric $|i-k|$ if $C_{ik} = 0$ for large enough $|i-k|$. Let θ_i denote the shift map on $\Omega = S^I$ associated to $i \in I$.

(2.14) Corollary. If $\gamma < 1$ then

$$\sum_i |\text{cov}_\mu(f, g \circ \theta_i)| e^{d(i,0)} \leq \frac{1}{4} (1-\gamma)^{-1} |f|_0 |g|_0 \tag{2.15}$$

for f and g in $C(\Omega)$.

Proof. Applying (2.13) and the triangle inequality for $d(\cdot,\cdot)$ we obtain

$$|\text{cov}(f, g \circ \theta_i)| e^{d(i,0)}$$

$$\leq \frac{1}{4} \sum_{k,j} e^{d(j,k)} D_{jk} e^{d(k,0)} \delta_k(f) e^{d(i,j)} \delta_{j-i}(g) .$$

Summing over i,j,k in that order, we get (2.15).

(2.16) Remark. These covariance estimates are a slight improvement over some similar estimates in [Gr1] and [Kü3]; cf. [Fö4].

(2.17) Remark. We have introduced the contraction technique in its simplest form, based on total variation as a measure of the distance between two probability measures on S in (2.1). The technique becomes more flexible if we replace total variation by the (Kantorovic-Rubinstein-) Vasserstein metric induced by a metric $r(\cdot,\cdot)$ on S ; this extension will be used in Ch. II. For two probability measures μ and ν on S define

$$R(\mu,\nu) = \sup \frac{|\int f d\mu - \int f d\nu|}{\delta(f)} \tag{2.18}$$

where the supremum is taken over all Lipschitz functions f on S with

$$\delta(f) := \sup_{s \neq t} \frac{|f(s)-f(t)|}{r(s,t)} < \infty ,$$

cf. [Do2]. Note that (2.18) reduces to $\frac{1}{2}\ ||\mu-\nu||$ if the metric $r(\cdot,\cdot)$ is discrete. If S is polish then (2.18) admits the dual description

$$R(\mu,\nu) = \inf \int r d\gamma \tag{2.19}$$

where the infimum is taken over all measures γ on $S \times S$ with marginals μ and ν. The contraction coefficients in (2.1) and (2.2) are now replaced by

$$C_{ik} = \sup \{ \frac{R(\pi_k(\cdot|\omega),\pi_k(\cdot|\eta))}{r(\omega(i),\eta(i))} : w = \eta \text{ off } i \} \tag{2.20}$$

and

$$b_k = \int R(\pi_k(\cdot|\eta),\nu_k(\cdot|\eta))\nu(d\eta) ; \tag{2.21}$$

the matrices $C = (C_{ik})$ and $D = \sum\limits_{n\geq 0} C^n$ are defined as before. Assume (2.6), and in the Feller assumption (2.4) replace $C(\Omega)$ by the class $L(\Omega)$ of functions f on Ω which satisfy

$$|f(\omega) - f(\eta)| \leq \sum r(\omega(i),\eta(i))\delta_i(f) , \qquad \sum_i \delta_i(f) < \infty$$

where

$$\delta_i(f) = \sup_{s\neq t} \{ \frac{|f(s)-f(t)|}{r(\omega(i),\eta(i))} : w = \eta \text{ off } i \} .$$

A measure μ is called tempered if

$$\sup_i \int r(\omega(i),\eta(i))\mu(d\omega) < \infty \tag{2.22}$$

for some fixed $\eta \in \Omega$. For a tempered Gibbs measure μ and a tempered probability measure ν, we have a uniformly bounded estimate $a = (a_k)$ to start the argument in (2.5), namely $a_k = \int r(\omega(k),\eta(k))(\mu+\nu)(d\omega)$. This leads to Dobrushin's comparison theorem (2.8) for functions $f \in L(\Omega)$, and to a corresponding uniqueness theorem for tempered Gibbs measures. The covariance estimate (2.13) takes the form

(2.23) $$|\mathrm{cov}_\mu(f,g)| \le \sigma^2 \sum_{i,k} \delta_i(f) D_{ki} \delta_k(g)$$

for f and g in $L(\Omega)$, where μ is a Gibbs measure which satisfies

$$\sigma^2 := \sup_i \int r(\omega(i),\eta(i))^2 \mu(d\omega) < \infty .$$

In the translation invariant case $I = Z^d$, exponential decay of correlation follows as in (2.14): If $\gamma < 1$ then

(2.24) $$\sum_i |\mathrm{cov}(f, g\circ\theta_i)| e^{d(i,o)} \le \sigma^2 (1-\gamma)^{-1} |f|_o |g|_o$$

for functions f and g in $L(\Omega)$; cf. [Fö4] for details.

In the translation invariant case $I = Z^d$, covariance estimates of the form (2.15) resp. (2.24) are the key to a central limit theorem; cf. [DT], [Kü3]. Suppose that the local specification is spatially homogeneous so that $c_{k-i} := c_{i,k}$ only depends on the difference $i-k$. Suppose also, for simplicity, that $c_k = 0$ for large enough $|k|$, so that

$$\sum_{k\neq o} c_k |k|^d < \infty .$$

Then the covariance estimates allow us to apply a spatial central limit theorem, e.g. in the form of [Bo]:

(2.25) <u>Theorem.</u> If $c < 1$ then the distribution of

$$S_n^*(f) = |V_n|^{-1/2} \sum_{i\in Vn} [f\circ\theta_i - \int f d\mu]$$

under the unique (tempered) Gibbs measure μ converges to the centered normal law with variance

(2.26) $$\sigma^2(f) = \sum_k \mathrm{cov}_\mu(f, f\circ\theta_k) < \infty$$

for any $f \in L(\Omega)$.

2.4 Almost sure convergence of two-parameter martingales

We consider the case $I = Z^2$ and denote by $s \leq t$ the coordinate-wise ordering of Z^2. For a bounded random variable X on $(\Omega, \underline{\underline{F}})$, consider the two-parameter martingale

(2.27) $$X_t := E_\mu[X | \underline{\underline{F}}_t] \qquad (t \in Z^2_+),$$

where $\underline{\underline{F}}_t := \sigma(\omega(i); 0 \leq i \leq t)$. We are interested in almost sure convergence as $t \uparrow \infty$. By a theorem of Cairoli [Ca], we know that almost sure convergence does hold if the underlying random field μ on S^I satisfies the following independence condition:

(2.28) For each $t = (t_1, t_2) \in Z^2_+$, the two σ-fields

$$\underline{\underline{F}}^i_{t_i} := \sigma(\omega(s);\ s_i \leq t_i) \quad (i = 1,2)$$

are conditionally independent with respect to their intersection $\underline{\underline{F}}_t = \underline{\underline{F}}^1_{t_1} \cap \underline{\underline{F}}^2_{t_2}$.

In fact, condition (2.18) allows us to write

$$X_t = E_\mu[E_\mu[X | \underline{\underline{F}}^1_{t_1}] | \underline{\underline{F}}^2_{t_2}],$$

and then it is enough to apply the following two-parameter version of the martingale convergence theorem due to Blackwell and Dubins:

(2.29) Lemma [BD]. If (Y_n) converges μ-almost surely and satisfies $\sup |Y_n| \in L^1$, then

$$\lim_{n,m} E_\mu[Y_n | \underline{\underline{G}}_m]$$

exists μ-almost surely for any increasing (or decreasing) sequence of σ-fields $(\underline{\underline{G}}_m)$; the parameter n may run through any partially ordered index set.

From the point of view of random fields, condition (2.28) is very restrictive: in most cases there is some diagonal interaction between $\underline{\underline{F}}^1_{t_1}$ and $\underline{\underline{F}}^2_{t_2}$ which does not pass through $\underline{\underline{F}}_t$. If this interaction becomes too strong then almost sure convergence may break down; see, for example, the counterexample in [DP1] . Let us now sketch how Dobrushin's condition (2.7) allows us to control this effect. Define $X^{(0)}_t \equiv X$ and

$$X^{(N)}_t = E_\mu[E_\mu[X^{(N-1)}_t | \underline{\underline{F}}^1_{t_1}] | \underline{\underline{F}}^2_{t_2}] \qquad (t \in Z^2_+) .$$

Lemma (2.29), with n ranging in Z^2_+ , implies almost sure convergence of $X^{(N)}_t$ for any $N \geq 1$. But if μ is a Markov random field which satisfies condition (2.7) , then we can use the global Markov property and the comparison theorem (2.8) in order to show that

$$\lim_N \sup_t |X^{(N)}_t - X_t| = 0 \quad \text{a.s.}$$

This implies almost sure convergence of the martingale (X_t) :

(2.30) Theorem. If μ is a Markov random field which satisfies condition (2.7) then bounded two-parameter martingales converge μ-almost surely.

We refer to [Fö5] for a detailed proof, and for an example which shows that it is not enough to require that μ is uniquely determined by its conditional probabilities.

2.5 Time-inhomogeneous Markov chains and annealing

Dobrushin's contraction technique for random fields may be viewed as the spatial extension of a classical contraction method for Markov chains. In [Do3], this technique has been used systematically in order to study the asymptotic behavior of time-inhomogeneous Markov chains. The annealing algorithm is an important example in this context, and its convergence may be viewed as a special case of the following general convergence theorem.

Let $P_n(x,dy)$ $(n = 1,2,\ldots)$ be a sequence of transition kernels on some state space, and define the contraction coefficient of P_n as

$$c(P_n) := \sup_{x,y} \frac{1}{2} ||P_n(x,.) - P_n(y,.)|| .$$

For two probability measures μ and ν we have

$$||\mu P_n - \nu P_n|| \leqq c(P_n)\, ||\mu-\nu|| ,$$

and this shows that

(2.31) $$\prod_n c(P_n) = 0$$

is a sufficient condition for "asymptotic loss of memory", i.e.,

(2.32) $$\lim_n ||\mu P_1 \ldots P_n - \nu P_1 \ldots P_n|| = 0$$

for two initial distributions μ and ν ; cf. [Do3].

Now suppose that each kernel P_n has a unique invariant distribution μ_n , and that

(2.33) $$\sum_n ||\mu_{n+1} - \mu_n|| < \infty .$$

This implies the existence of a unique limiting measure $\mu_\infty = \lim \mu_n$ (in total variation). Let us also assume $c(P_n) > 0$ for all n .

(2.34) Corollary. Under (2.31) and (2.33),

$$\lim_n \| \nu P_1 \cdots P_n - \mu_\infty \| = 0$$

for any initial distribution ν .

Proof. For a fixed N ,

$$\begin{aligned} & \| \nu P_1 \cdots P_n - \mu_\infty \| \\ = \; & \| (\nu P_1 \cdots P_N - \mu_\infty) P_{N+1} \cdots P_n + \mu_\infty P_{N+1} \cdots P_n - \mu_\infty \| \\ \le \; & \prod_{K=N+1}^{n} c(P_k) + \| \mu_\infty P_{N+1} \cdots P_n - \mu_\infty \| . \end{aligned}$$

But

$$\mu_\infty P_{N+1} \cdots P_n - \mu_\infty = (\mu_\infty - \mu_{N+1}) P_{N+1} \cdots P_n + \mu_{N+1} P_{N+2} \cdots P_n - \mu_\infty$$

$$= (\mu_\infty - \mu_{N+1}) P_{N+1} \cdots P_n + \sum_{k=1}^{n-N-1} (\mu_{N+k} - \mu_{N+k+1}) P_{N+k+1} \cdots P_n + \mu_n - \mu_\infty ,$$

and so

$$\begin{aligned} & \sup_{n \ge N} \| \mu_\infty P_{N+1} \cdots P_n - \mu_\infty \| \\ \le \; & \sup_{n \ge N} 2 \| \mu_\infty - \mu_n \| + \sum_{j > N} \| \mu_j - \mu_{j+1} \| \end{aligned}$$

converges to 0 due to (2.33).

In this form, the contraction technique has been used to establish convergence of the following "annealing algorithm"; cf. [GG] and also [Gi]. Let $E(x)$ be a function on a product space of the form S^I , where I is a finite index set and S is some finite state space at each site $i \in I$. Our purpose is to find global minima of E . In order to avoid being trapped in one of the local minima, we use a randomized search procedure. For each $\beta > 0$ consider the Gibbs measure

(2.35) $$\mu_\beta(x) = Z(\beta)^{-1} \exp(-\beta E(x))$$

and note that μ_β converges for $\beta \uparrow \infty$ to the uniform distribution μ_∞ on the set of global minima of E. For a fixed β, the local specification of μ_β satisfies

$$\pi^\beta_{\{i\}}(x,.) = \pi^\beta_i(\cdot|x) \times \prod_{j \neq i} \delta_{x(j)}$$

with

(2.36) $$\pi^\beta_i(s|x) = Z_i(\beta)^{-1} \exp[-\beta E((s,x))]$$

where (s,x) is the configuration which coincides with x off i and has value s in i. Let $I = \{1,\ldots,N\}$ be some enumeration of I. Then μ_β is the unique invariant measure for the Markov chain with transition probability

(2.37) $$P_\beta = \pi^\beta_{\{1\}} \cdots \pi^\beta_{\{N\}}$$

on S^I. Thus, we can expect to be close to the global minima of the function E if we let the chain P_β run for a sufficient amount of time and for large enough β. Now the idea is to choose a sequence $\beta(n) \uparrow \infty$ such that the time-inhomogeneous Markov chain with $P_n = P_{\beta(n)}$ $(n = 1,2,\ldots)$ converges to μ_∞, i.e.,

(2.38) $$\lim_n ||\nu P_{\beta(1)} \cdots P_{\beta(n)} - \mu_\infty|| = 0$$

for any initial distribution ν. The appropriate rate at which $\beta(n)$ is allowed to go to infinity is computed in view of condition (2.31):

(2.39) Theorem [GG]: There is a constant γ such that (2.38) holds for

(2.40) $$\beta(n) \leq \gamma \log n .$$

Proof. For $s \in S$ and for any $x \in S^I$ we have

$$\pi^\beta_i(s|x) \geq |S|^{-1} \exp(-\beta\delta_i(E))$$

where $\delta_i(E)$ denotes the oscillation of E in the i-th coordinate. Thus,

$$\min_{x,y} P_\beta(x,y) \geq (|S|^{-1} \exp(-\beta\Delta))^N$$

with $\Delta = \max_i \delta_i(E)$. Since

$$\frac{1}{2} ||\mu-\nu|| = \sum_x (\mu(x)-\nu(x))^+$$

$$= 1 - \sum_x \min(\mu(x),\nu(x))$$

$$\geq 1 - \inf_x |S^I| \min(\mu(x),\nu(x))$$

for any two probability measures μ and ν on S^I , we obtain

$$c(P_\beta) \leq 1 - \exp(-\beta N\Delta) .$$

Thus, condition (2.31) is satisfied for

$$\beta(n) \leq (N\Delta)^{-1} \log n .$$

As to condition (2.33), note that for each $x \in S^I$ the sequence $\mu_{\beta(n)}(x)$ is either increasing or decreasing. Thus,

$$\sum_n ||\mu_{(n+1)}-\mu_{(n)}|| = \sum_{x\in S} \sum_n (\mu_{\beta(n+1)}(x)-\mu_{\beta(n)}(x))^+$$

$$= \sum_{x\in S} (\mu_\infty(x)-\mu_{\beta(1)}(x))^+ < \infty .$$

(2.41) <u>Remarks.</u> 1) See, e.g., [Ha] for further refinements concerning the constant γ .

2) Let P_ν be the distribution of the Markov chain with initial distribution ν and transition kernels $P_{\beta(n)}$ $(n = 1,2,\ldots)$. Under condition (2.40), P_ν coincides with P_μ on the tail field $\bigcap_n \sigma(X_n,X_{n+1},\ldots)$. But this does not yet imply ergodic behavior, i.e.,

(2.42) $$\lim_{n} \frac{1}{n} \sum_{i=1}^{n} f(X_i) = \int f d\mu_\infty$$

P_μ-almost surely for functions f on S^I (there are counterexamples to Theorem 1.3 in [Gi]). But, as observed by N. Gantert, (2.42) does hold if the constant used in the proof of (2.39) is divided by 2; this follows, e.g., from the laws of large numbers for time-inhomogeneous Markov chains in [IT].

3) In [GG] the annealing technique is applied to the restauration of distorted images. An image is described as a configuration $x \in S^I$, and it is viewed as the realization of some a priori distribution

$$\mu(x) = Z^{-1} \exp[E(x)]$$

on S^I; usually, μ is assumed to have a Markov property with respect to some graph structure on I, and $E(x)$ is specified in terms of some interaction potential as in the following section. Now suppose that we observe a distorted version y of x which is generated by some probability kernel $Q(x,y)$ on S^I, and denote by $\mu(\cdot|y)$ the corresponding a posteriori distribution on S^I; if both Q and μ have a local Markov property, then $\mu(\cdot|y)$ is again a Markov field. In any case, one can compute explicitely the function $E(\cdot|y)$ in the Gibbsian description

$$\mu(x|y) = Z(y)^{-1} \exp[E(x|y)]$$

of $\mu(\cdot|y)$. The Bayesian estimate with respect to the loss function

$$L(x,\hat{x}) = I_{\{x \neq \hat{x}\}}$$

is given by a picture $\hat{x}$ with maximal a posteriori probability $\mu(\cdot|y)$. This leads to the problem of finding the global minima of the function $E(\cdot|y)$, and here the annealing algorithm comes in.

3. Entropy, energy and the theorem of Shannon-Mc Millan

In Section 1 Gibbs measures were introduced in terms of their local specification. If the local specification is given by a stationary interaction potential on a d-dimensional lattice, then the variational principle of Lanford and Ruelle provides an alternative global characterization in terms of specific entropy and specific energy. The purpose of this section is to give a short introduction to the probabilistic limit theorems which are behind the existence of these "thermodynamical" quantities. In particular, we derive the d-dimensional Shannon-Mc Millan theorem for the specific entropy $h(\nu;\mu)$ of a stationary measure ν with respect to a Gibbs measure $\mu \in G_s(\pi)$; cf. [Fö2], [Pr]. In view of recent work on large deviations for lattice models, these results are of renewed interest. The Shannon-Mc Millan theorem is in fact the key tool in proving the lower bound (4.3) for large deviations of the empirical field of a Gibbs measure. We are going to discuss large deviations in Section 4, and so it seems reasonable to include a self-contained exposition. In view of Ch. II we admit a general state space S , otherwise we use the simplest setting of a bounded interaction on Z^d . For unbounded interactions we refer, e.g., to [Kü1], [Kü2], [Pi], [Gu], for a general amenable group I to [Te] and [Mo], for analogous results in the theory of interactive point processes to [NZ].

3.1 Specific entropy

Let μ and ν be two probability measures on some measurable space $(E,\underline{\underline{E}})$. We define the relative entropy of μ with respect to ν as

$$(3.1) \qquad H(\mu;\nu) := \int \log \varphi \, d\mu$$

if μ is absolutely continuous with respect to ν on $\underline{\underline{E}}$ with density φ , and $H(\mu;\nu) := +\infty$ else.

(3.2) Remarks. 1) By Jensen's inequality we have $H(\mu;\nu) \geq 0$, and $H(\mu;\nu) = 0 \iff \mu = \nu$ on $\underline{\underline{E}}$.

2) For $0 < \psi \in L^1(\nu)$, the inequality in 1) implies

$$H(\mu;\nu) = H(\mu;\tilde{\nu}) + \int \log \psi \, d\mu - \log \int \psi \, d\nu$$

$$\geq \int \log \psi \, d\mu - \log \int \psi \, d\nu$$

where $d\tilde{\nu} = \psi(\int \psi d\nu)^{-1} d\nu$. Thus,

$$(3.3) \qquad H(\mu;\nu) = \sup_{\psi} [\int \log \psi \, d\mu - \log \int \psi \, d\nu] .$$

If $\underline{\underline{E}}$ is the Borel σ-field with respect to some polish topology on E , then it is enough to take the supremum over bounded continuous functions which are bounded away from 0 ; cf. [Va]. This implies that $H(\cdot;\nu)$ is lower semicontinuous with respect to the weak topology.

3) The total variation $||\mu-\nu|| = \int |\varphi-1| d\nu$ can be estimated by

$$(3.4) \qquad ||\nu-\mu||^2 \leq 2 \, H(\mu;\nu) .$$

This follows from the elementary inequality $3(x-1)^2 \leq f(x)g(x)$ with $f(x) = 4 + 2x$ and $g(x) = x \log x - x + 1$ (cf. [Ke]):

$$3(\int |\varphi-1| d\nu)^2 \leq (\int \sqrt{f(\varphi)}\sqrt{g(\varphi)} d\nu)^2 \leq \int f(\varphi) d\nu \int g(\varphi) d\nu$$

$$= 6 \int \varphi \log \varphi \, d\nu .$$

We are going to use the following well-known

(3.5) <u>Lemma.</u> Let T be a partially ordered index set, let $\underline{\underline{A}}_t (t \in T)$ be an increasing directed family of σ-fields contained in $\underline{\underline{E}}$, and put $\underline{\underline{A}}_\infty := \sigma(\bigcup_t \underline{\underline{A}}_t)$. If

$$(3.6) \qquad \sup_t H_t(\mu;\nu) < \infty ,$$

where $H_t(\mu;\nu)$ denotes the relative entropy on $\underline{\underline{A}}_t$, then μ is absolutely continuous, with respect to ν on $\underline{\underline{A}}_\infty$, and

$$(3.7) \qquad H_\infty(\mu;\nu) = \sup_t H_t(\mu;\nu) .$$

Moreover, the densities $\varphi_t := \frac{d\mu}{d\nu}\big|_{\underline{\underline{A}}_t}$ and $\varphi_\infty := \frac{d\mu}{d\nu}\big|_{\underline{\underline{A}}_\infty}$ satisfy

$$\lim_t \log \varphi_t = \log \varphi_\infty \quad \text{in} \quad L^1(\mu) .$$

<u>Proof.</u> See, for example, [NZ] or [Or2].

Let us now go back to our product space S^I. We fix a reference probability measure λ_0 on S and denote by $\lambda = \prod_{i \in I} \lambda_0$ the corresponding product measure on S^I. For $\mu \in M(\Omega)$ and for finite $V \subseteq I$ let $H_V(\mu;\lambda)$ denote the relative entropy of μ with respect to ν on the σ-field $\underline{\underline{F}}_V$. From now on we assume that I is the d-dimensional lattice Z^d, and we write

$$V_n := \{t \in Z^d \mid 0 \le t_k \le n \quad (k=1,\dots,d)\} .$$

(3.8) <u>Definition:</u> For any $\mu \in M_s(\Omega)$, the specific entropy of μ with respect to λ is defined as

$$(3.9) \qquad h(\mu;\lambda) := \lim_n |V_n|^{-1} H_{V_n}(\mu;\lambda) = \sup_n |V_n|^{-1} H_{V_n}(\mu;\lambda) .$$

Note that (3.2) inplies that $h(\cdot;\lambda)$ is lower semicontinuous with respect to the weak topology on $M_s(\Omega)$. The existence of the limit and the equality in (3.8) follow from the fact that $H_V(\mu;\lambda)$ is superadditive in V; cf. [Pr] Th. 8.1. It is also contained in the following spatial version of the Shannon-McMillan theorem which states that there is L^1-convergence behind (3.8):

(3.10) <u>Theorem.</u> Suppose that $h(\mu|\lambda) < \infty$. Then the limit

$$(3.11) \qquad h_\mu(\omega) := \lim_n |V_n|^{-1} \log \frac{d\mu}{d\lambda}\Big|_{\underline{\underline{F}}_{V_n}}(\omega)$$

exists in $L^1(\mu)$ and satisfies

$$(3.12) \qquad h_\mu = E_\mu[H(\mu_0(\cdot|\underline{\underline{P}});\lambda_0)|\underline{\underline{J}}]$$

where $\underline{\underline{P}} := \sigma(\omega(i); i<0)$ denotes the σ-algebra of the "past" of $0 \in Z^d$ with respect to the lexicographical order on Z^d, and where

$\mu_0(\cdot|\underline{\underline{P}})$ is a conditional distribution of $\omega(0)$ with respect to $\underline{\underline{P}}$ and μ. In particular,

(3.13) $$h(\mu;\lambda) = \int H(\mu_0(\cdot|\underline{\underline{P}})(\omega);\lambda_0)\mu(d\omega).$$

2) We have

(3.14) $$h_\mu(\omega) = h(\mu_\omega;\lambda) \quad \mu\text{-a.s.}$$

if $\mu = \int \mu_\omega[\cdot]\mu(d\omega)$ is the ergodic decomposition of μ in (1.29).

<u>Proof.</u> 1) For $W \subseteq I - \{i\}$ we denote by $\mu_i(\cdot|\underline{\underline{F}}_W)(\omega)$ a conditional distribution of $\omega(i)$ with respect to $\underline{\underline{F}}_W$ and μ. Then we can write

(3.15) $$|V_n|^{-1} \log \frac{d\mu}{d\lambda}\Big|_{\underline{\underline{F}}_{V_n}}(\omega)$$

$$= |V_n|^{-1} \sum_{i\in V_n} [\log \frac{d\mu_0(\cdot|\underline{\underline{F}}_{W_{n,i}})(\omega)}{d\lambda}(\omega(0))] \circ \theta_i$$

where $W_{n,i} := \{j|j<0,\ j+i \in V_n\}$. But

$$\frac{d\mu_0(\cdot|\underline{\underline{F}}_W)(\omega)}{d\lambda_0}(\omega(o)) = \frac{d\mu}{d\nu}\Big|_{\underline{\underline{F}}_{W\cup\{o\}}}$$

if we define $\nu := \mu_{I-\{o\}} \times \lambda_0$. (3.17) below shows that assumption (3.6) is satisfied in our present case. By (3.5), the $L^1(\mu)$-convergence of (3.15) is thus reduced to the convergence of

$$|V_n|^{-1} \sum_{i\in V_n} \Big[\log \frac{d\mu_0(\cdot|\underline{\underline{P}})(\omega)}{d\lambda_0}(\omega(o))\Big] \circ \theta_i .$$

The spatial ergodic theorem yields the existence of the limit in (3.11) with

(3.16) $$h_\mu = E_\mu[\log \frac{d\mu_0(\cdot|\underline{\underline{P}})}{d\lambda_0}|\underline{\underline{J}}] .$$

and this implies (3.13).

2) Integrating (3.15) we obtain

$$|V_n|^{-1} H_{V_n}(\mu;\lambda) = |V_n|^{-1} \sum_{i\in V_n} H_{W_{n,i}}(\mu;\lambda) .$$

But $H_W(\mu;\lambda) \le H_{W'}(\mu;\lambda)$ for $W' \supseteq W$, and so we get, for a fixed $W = W_{n_o,i_o}$,

$$\text{(3.17)} \qquad H_W(\mu;\lambda) = \lim_n |V_n|^{-1} \sum_{\substack{i\in V_n \\ W_{n,i} \supseteq W}} H_W(\mu;\lambda)$$

$$\le \sup_n |V_n|^{-1} H_{V_n}(\mu;\lambda) = h(\mu;\lambda) < \infty .$$

3) The same argument as in Lemma (1.25) yields $\underline{\underline{P}} \supseteq \underline{\underline{J}}$ μ-a.s. This implies $\mu_o(\cdot|\underline{\underline{P}})(\omega) = (\mu_\omega)_o(\cdot|\underline{\underline{P}})(\omega)$ μ-a.s., hence

$$h_\mu(\omega) = E_\mu[\log \frac{d\mu_o(\cdot|\underline{\underline{P}})}{d\lambda_o}|\underline{\underline{J}}](\omega) = E_\mu[\log \frac{d(\mu_\omega)_o(\cdot|\underline{\underline{P}})}{d\lambda_o}|\underline{\underline{J}}](\omega)$$

$$= \int \log \frac{d(\mu_\omega)_o(\cdot|\underline{\underline{P}})}{d\lambda_o} d\mu_\omega = h(\mu_\omega) .$$

Our next purpose is to extend the Shannon-McMillan theorem to the case where the product measure λ is replaced by a Gibbs measure whose local specification is "nice".

3.2 Specific energy

For each finite $V \subseteq I$ let U_V be a $\underline{\underline{F}}_V$-measurable function on Ω. The collection (U_V) is called an interaction potential if

$$\text{(3.18)} \qquad U_{V+i} = U_V \circ \theta_i \qquad (i \in I)$$

$$\text{(3.19)} \qquad |U| := \sum_{O\in V} ||U_V|| < \infty$$

where $||U_V||$ is the supremum norm of U_V.

For $\omega,\eta \in \Omega$ let $(\omega,\eta)_V$ denote the configuration which coincides with ω on V and with η on V^c. The quantity

$$(3.20) \qquad E_V(\omega_V|\eta) := \sum_{A\cap V\neq\emptyset} U_A((\omega,\eta)_V)$$

will be called the conditional energy of ω on V given the environment η.

(3.21) <u>Theorem.</u> For any stationary $\mu \in M_s(\Omega)$, the pointwise specific energy

$$(3.22) \qquad e_U(\omega) := \lim_n |V_n|^{-1} \sum_{A\subseteq V_n} U_A(\omega)$$

exists, μ-almost surely and in $L^1(\mu)$, and satisfies

$$(3.23) \qquad e_U(\omega) = E_\mu[\sum_{0\in A} \frac{U_A}{|A|}|\underline{\underline{J}}] = \lim_n |V_n|^{-1} E_{V_n}(\omega_{V_n}|\eta)$$

for any η. In particular, the specific energy

$$(3.24) \qquad e_U(\mu) := \lim_n |V_n|^{-1} \int E_{V_n}(\cdot|\eta)\, d\mu = \int e_U(\cdot)\, d\mu ,$$

of μ exists, and

$$(3.25) \qquad e_U(\omega) = e_U(\nu_\omega) \qquad \nu\text{-a.s.}$$

if $\nu = \int \nu_\omega\, \nu(d\omega)$ is the ergodic decomposition of ν in (1.29).

<u>Proof.</u> For $g := \sum_{0\in A} \frac{U_A}{|A|}$ we have

$$\sum_{i\in V} g\circ\theta_i = \sum_{i\in V} \sum_{\substack{A:i\in A\\ A\subseteq V}} \frac{U_A}{|A|} + \sum_{i\in V} \sum_{\substack{A:i\in A\\ A\cap V^c\neq\emptyset}} \frac{U_A}{|A|}$$

$$= \sum_{A\subseteq V} \sum_{i\in V} I_A(i) \frac{U_A}{|A|} + \sum_{\substack{A\cap V^c\neq\emptyset\\ A\cap V\neq\emptyset}} \sum_{i\in V} I_A(i) \frac{U_A}{|A|} .$$

Since

$$E_V(\omega_V|\eta) = \sum_{A \subseteq V} U_A(\omega) + \sum_{\substack{A\cap V\neq\emptyset \\ A\cap V^c\neq\emptyset}} U_A((w,\eta)_V) ,$$

we obtain

$$|E_V(\omega_V|\eta) - \sum_{i\in V} (g\circ\theta_i)(\omega) \;| \le 2\Delta_o(V)$$

where

$$(3.26) \qquad \Delta_o(V) := \sum_{i\in V} \sum_{\substack{A:i\in A \\ A\cap V^c\neq\emptyset}} ||U_A|| \quad \le \quad |V - V_l|\,|U| + |V_l|\,\gamma_l ,$$

with

$$\gamma_l := \sum_{0\in A\not\subseteq N_l(0)} ||U_A||$$

and $V_l := \{i\in V|\ \text{dist}\,(i,V^c) > l\}$. Letting first $n \uparrow \infty$ and then $l \uparrow \infty$, we see that

$$(3.27) \qquad \lim_n \frac{\Delta_o(V_n)}{|V_n|} = 0 .$$

The theorem now follows from the d-dimensional ergodic theorem.

3.3 Specific entropy with respect to a Gibbs measure

For an interaction potential (U_V) as above we introduce the local specification

$$\pi_V(\eta,\cdot) = \pi_V(\cdot|\eta) \times \prod_{j\notin V} \delta_{\eta(j)}$$

where $\pi_V(\cdot|\eta)$ is the probability measure on S^V with density

$$(3.28) \qquad \frac{d\pi_V(\cdot|\eta)}{d\lambda_V}(\xi) = Z_V(\eta)^{-1} \exp(-E_V(\xi|\eta))$$

For $l \ge 1$ we define $\partial_l V$ and γ_l as in (3.26).

(3.29) Lemma. If $\eta = \eta'$ on $\partial_1 V$ then

$$\frac{d\pi_V(\cdot|\eta)}{d\pi_V(\cdot|\eta')} \le \exp(4|V|\gamma_1)$$

Proof. Since

$$E_V(\xi|\eta) = \sum_{\substack{A\cap V\neq\emptyset \\ A\subseteq V\cup\partial_1 V}} U_A((\xi,\eta)_V) + \sum_{\substack{A\cap V\neq\emptyset \\ A\not\subseteq V\cup\partial_1 V}} U_A((\xi,\eta)_V) ,$$

$\eta = \eta'$ on $\partial_1 V$ implies

$$E_V(\xi|\eta') \le E_V(\xi|\eta) + 2\,\Delta_1(V)$$

with

$$\Delta_1(V) := \sum_{\substack{A\cap V\neq\emptyset \\ A\not\subseteq V\cup\partial_1 V}} ||U_A|| \le |V|\cdot\gamma_1 ,$$

hence $Z_V(\eta') \ge Z_V(\eta)\exp(-2\,\Delta_1(V))$ and vice versa. Thus,

$$\frac{d\pi_V(\cdot|\eta)}{d\pi_V(\cdot|\eta')}(\xi) = \frac{Z_V(\eta')}{Z_V(\eta)}\exp[E_V(\xi|\eta') - E_V(\xi|\eta)] \le \exp(4|V|\,\gamma_1) .$$

Now suppose that $\mu\in G_s(\pi)$ is a stationary Gibbs measure with local specification (π_V) ; condition (3.19) implies in fact that $G_s(\pi)\neq\emptyset$, see [Pr] . By (3.29)

$$(3.30)\qquad \lim_n |V_n|^{-1}\sup_\xi\left|\log\frac{d\mu_{V_n}}{d\lambda_{V_n}}(\xi) - \log\frac{d\mu_{V_n}(\cdot|\eta)}{d\lambda_{V_n}}(\xi)\right| = 0$$

for any η , since $\lim_l \gamma_l = 0$ due to (3.19) . Thus ,

$$h(\mu;\lambda) = \lim_n \frac{1}{|V_n|}\left(-\log Z_{V_n}(\eta) - \int E_{V_n}(\cdot|\eta)\,d\mu\right)$$

due to (3.12) . The existence of $e_U(\mu)$ in (3.24) implies the existence of the "pressure"

$$(3.31)\qquad p_U = \lim \frac{1}{|V_n|} \log Z_{V_n}(\eta)$$

and the thermodynamical relation

$$(3.32)\qquad -p_U = h(\mu;\lambda) + e_U(\mu) .$$

We are now in a position to derive a Shannon-McMillan theorem for the specific entropy relative to a Gibbs measure $\mu \in G_s(\pi)$:

(3.33) Theorem. For any $\nu \in M_s(\Omega)$, the specific relative entropy with respect to μ

$$(3.34)\qquad h(\nu;\mu) := \lim |V_n|^{-1} H_{V_n}(\nu;\mu)$$

exists and satisfies

$$(3.35)\qquad h(\nu;\mu) = e_U(\nu) + h(\nu;\lambda) + p_U \geq 0 .$$

More precisely:

$$(3.36)\qquad h_{\nu;\mu}(\omega) := \lim_n |V_n|^{-1} \log \frac{d\nu_{V_n}}{d\mu_{V_n}}(\omega_{V_n})$$

exists in $L^1(\nu)$ and satisfies, ν-almost surely,

$$(3.37)\qquad h_{\nu;\mu}(\omega) = e_U(\omega) + h_\nu(\omega) + p_U = h(\nu_\omega;\mu) \geq 0 ,$$

where $\nu = \int \nu_\omega \, \nu(d\omega)$ is the ergodic decomposition of ν in (1.29). In particular,

$$(3.38)\qquad h(\nu;\mu) = \int h_{\nu;\mu}(\omega)\nu(d\omega).$$

Proof. (3.30) implies, for any $\eta \in \Omega$ and in $L^1(\nu)$,

$$\lim |V_n|^{-1} \log \frac{d\nu_{V_n}}{d\mu_{V_n}}(\omega_{V_n})$$

$$= \lim |V_n|^{-1} \log \frac{d\nu_{V_n}}{d\lambda_{V_n}}(\omega_{V_n}) - \lim_n |V_n|^{-1} \log \frac{d\mu_{V_n}(\cdot|\eta)}{d\lambda_{V_n}}(\omega_{Vn})$$

$$= h_{\nu}(\omega) + e_U(\omega) + p_U$$

due to (3.10), (3.23) and (3.31). Integrating with respect to ν we obtain (3.34) and (3.35). If $\nu = \int \nu_\omega \, \nu(d\omega)$ is the ergodic decomposition of ν, then (3.14), (3.25), and (3.35) imply

$$h_{\nu;\mu}(\omega) = h(\nu_\omega;\lambda) + e_U(\nu_\omega) + p_U = h(\nu_\omega;\mu) .$$

(3.35) shows that any $\nu \in M_s(\Omega)$ satisfies

$$e_U(\nu) + h(\nu;\lambda) \geq - p_U ,$$

and the minimal value $-p_U$ is in fact assumed for any $\nu \in G_s(\pi)$ due to (3.32). This is already one direction of the following variational principle due to Lanford and Ruelle [LR].

(3.39) <u>Theorem.</u> For $\nu \in M_s(\Omega)$ and $\mu \in G_s(\pi)$, the following properties are equivalent:

i) $\nu \in G_s(\pi)$

ii) $h(\nu;\mu) = 0$

iii) The function $e_U + h(\cdot;\lambda)$ assumes in ν its minimal value $-p_U$.

<u>Proof.</u> It only remains to show ii) $\Rightarrow$ i), and here we follow [Pr].

1) Let us fix $V = V_{n_o}$. We have to show that

(3.40) $$\nu_V(\cdot\,|\,\underline{\underline{F}}_{V^c})(\omega) = \pi_V(\cdot\,|\,\omega)$$

for ν-almost all ω. Lemma (3.29) implies

(3.41) $$\lim_{W \uparrow V^c} \frac{d\mu_V(\cdot\,|\,\underline{\underline{F}}_W)(\eta)}{d\lambda_V}(\xi) = \frac{d\pi_V(\cdot\,|\,\eta)}{d\lambda_V}(\xi)$$

uniformly in ξ, η . Lemma (3.5) implies, as in the proof of (3.10), that

$$(3.42) \qquad \lim_{W \uparrow V^c} \log \frac{d\nu_V(\cdot \mid \underline{\underline{F}}_W)(\omega)}{d\lambda_V}(\omega_V) = \log \frac{d\nu_V(\cdot \mid \underline{\underline{F}}_{V^c})(\omega)}{d\lambda_V}(\omega_V)$$

in $L^1(\nu)$. Thus,

$$(3.43) \qquad \lim_{W \uparrow V^c} \int H(\nu_V(\cdot \mid \underline{\underline{F}}_W)(\omega);\mu_V(\cdot \mid \underline{\underline{F}}_W)(\omega)\nu(d\omega)$$

$$= \int H(\nu_V(\cdot \mid \underline{\underline{F}}_{V^c})(\omega);\pi_V(\cdot \mid \omega)\nu(d\omega) .$$

In view of (3.40), we have only to show that the left side of (3.43) is 0 .

2) Let $B = V \cup \partial_1 V$. For $N = (n_0+2l)k$, V_N is the union of k^d disjoint translates $B(j)$ of B , each of the form $B(j) = V(j) \cup \partial_1 V(j)$ for some translate $V(j)$ of V . We can now write

$$H_{V_N}(\nu;\mu) = H_{V_N - \bigcup_j V(j)}(\nu;\mu)$$

$$+ \sum_j \int H(\nu_{V(j)}(\cdot \mid \underline{\underline{F}}_{C(j)})(\omega);\mu_{V(j)}(\cdot \mid \underline{\underline{F}}_{C(j)})(\omega))\nu(d\omega)$$

with $C(j) = V_N - \bigcup_{m \geq j} V(m)$, hence

$$(n_0+2l)^d \; |V_N|^{-1} H_{V_N}(\nu;\mu)$$

$$\geq \frac{1}{k^d} \sum_{j=1}^{k^d} \int H(\nu_V(\cdot \mid \underline{\underline{F}}_{W(j)})(\omega);\mu_V(\cdot \mid \underline{\underline{F}}_{W(j)})(\omega))\nu(d\omega)$$

with $\partial_1 V \subseteq W(j) \subseteq V^c$. But this shows that $h(\nu;\mu) = 0$ implies that the left side of (3.43) is 0 .

(3.44) <u>Remark.</u> In order to prove the Shannon-McMillan theorem for the relative entropy $h(\mu;\nu)$, we have used the local specification of ν and its thermodynamical description in terms of energy. For direct approach along the lines of the proof of (3.10), we would need a canonical choice of the conditional distribution $\nu_o(\cdot|\underline{\underline{P}})(\omega)$ which does not involve the null-sets of ν, and a continuity property which would guarantee the convergence

(3.45) $$\lim_{W \uparrow \{i;i<o\}} \frac{d\nu_o(\cdot|\underline{\underline{F}}_W)}{d\lambda_o} = \frac{d\nu_o(\cdot|\underline{\underline{P}})}{d\lambda_o}$$

with respect to μ. In that case, we would obtain the formula

(3.46) $$h(\mu;\nu) = \int \mu(d\eta)H(\mu_o(\cdot|\underline{\underline{P}})(\eta);\nu_o(\cdot|\underline{\underline{P}})(\eta))$$

in analogy to (3.13). If the local specification of ν satisfies the Dobrushin condition (2.6) and if S is finite, then (3.45) follows, as in the proof of the Feller property in (2.11), and so (3.46) does hold.

(3.47) Remark on the Ising model. Consider the ferromagnetic Ising model with $S = \{+1,-1\}$ and

$$U_V(\omega) = \beta\omega(i)\omega(j) \quad \text{if } V = \{i,j\} \text{ and } |i-j| = 1$$
$$= 0 \quad \text{else}$$

where $\beta > 0$. The limits $\lim_n \pi_{V_n}(\eta,\cdot)$ in (1.19) with $\eta \equiv +1$ resp. $\eta \equiv -1$ define two measures μ^+ and μ^- in $G_s(\pi)$; for large enough β, these two measures are different. For any measurable bounded function which is monotone with respect to the coordinatewise partial ordering on $\Omega = S^I$, and for any $\mu \in G(\pi)$,

(3.48) $$\int f d\mu^- \le \int f d\mu \le \int f d\mu^+$$

by the FKG inequality; cf. [Pr]. In particular,

(3.49) $$|G(\pi)| = 1 \iff \mu^+ = \mu^- .$$

Both μ^+ and μ^- have the global Markov property. In fact, for any $J \subseteq I$ the conditional distribution $\mu^+(\cdot|J)(\eta)$ can be defined as the "+-state" corresponding to the conditional specification on S^{I-J} determined by $\eta_{\partial J}$ as in (2.2); cf. [Fö3], [Go]. The monotonicity argument in [Fö3] shows that we also get (3.45), hence

(3.46), for $\nu = \mu^+$ and $\mu = \mu^-$. But the variational principle (3.39) implies $h(\mu|\mu^+) = 0$ for any $\mu \in G_s(\pi)$ and in particular for $\mu = \mu^-$. Thus, (3.46) implies

(3.50) $$\mu_0^+(\cdot|\underline{\underline{P}})(\eta) = \mu_0^-(\cdot|\underline{\underline{P}})(\eta)$$

for μ-almost all η, and for any $\mu \in G_S(\pi)$. Again by monotonicity, (3.50) implies $\mu^+(\cdot|\underline{\underline{P}})(\eta) = \mu^-(\cdot|\underline{\underline{P}})(\eta)$, and so (3.48), applied to the conditional system, leads to the following

(3.51) <u>Corollary.</u> For any $\mu \in G(\pi)$ and for μ-almost all η, the conditional specification on $J = \{j|j \not\leq 0\}$ determined by $\eta_{\partial J}$ admits no phase transition.

Thus, the Ising model can be viewed as a Markov chain as in 2.2, with a fixed transition probability, but with different invariant measures as soon as there is a phase transition.

4. Large deviations

Let μ be a stationary Gibbs measure with marginal distribution μ_0 , and suppose that μ is an extremal point in $G_s(\pi)$, hence ergodic by (1.26). The ergodic behavior of a typical configuration under μ can be described on various levels:

(1) convergence of average values for functions on S , i.e.,

$$\lim_n |V_n|^{-1} \sum_{i \in V_n} f(\omega(i)) = \int f d\mu_\omega ,$$

(2) convergence of the empirical distribution, i.e.,

$$\lim_n |V_n|^{-1} \sum_{i \in V_n} \delta_{\omega(i)} = \mu_0 ,$$

(3) convergence of the empirical field, i.e.,

$$\lim_n |V_n|^{-1} \sum_{i \in V_n} \delta_{\Theta_i \omega} = \mu .$$

On each level, we can look at large deviations from ergodic behavior. In the classical case where the random variables $\omega(i)$ $(i \in Z^d)$ are independent and identically distributed under μ , large deviations on level (1) are described by the theorem of Cramér and Chernoff, on level (2) by Sanov's theorem; cf., e.g., [Az]. Level (3) was introduced by Donsker and Varadhan in [DV2]. Here the situation is more subtle even in the classical case; in particular, the spatial structure of $I = Z^d$ may come in explicitely. For the one-parameter case with a "nice" stationary process μ , we refer to [DV2] and [Or2].

In this section we give an introduction to large deviations on level (3) for Gibbs measures in the multiparameter case $I = Z^d$, with special emphasis on the lower bound. In the first part, we state a general result obtained with S.Orey [FO], where the rate of convergence to 0 of a large deviation

$$\mu[\,|V_n|^{-1} \sum_{i \in V_n} \delta_{\omega(i)} \in A]$$

with $\mu \notin \bar{A}$ is described in terms of specific relative entropies $h(\nu;\mu)$. Here the Shannon-Mc Millan theorem (3.33) provides the technical key to the lower bound.

In the second part we discuss some joint work with M. Ort on the effect of a phase transition. Here it may happen that A contains another Gibbs measure $\nu \in G_S(\pi)$, and this will slow down the convergence to 0. More precisely, the order of convergence becomes exponential in the surface area $|\partial V_n|$ instead of the volume $|V_n|$, and the rate of convergence will now be described in terms of "specific surface entropies" $h_\partial(\nu;\mu)$.

4.1. Large deviations for the empirical field of a Gibbs measure

Consider a stationary Gibbs measure $\mu \in G_S(\pi)$ whose local specification is given by an interaction potential as in Section 3. For finite $V \subseteq I$ define the empirical field on V as the random element of $M_S(\Omega)$ given by

$$\rho_V(\omega) := |V|^{-1} \sum_{i \in V} \delta_{\theta_i \omega} .$$

In joint work with S. Orey [FO], it is shown that the sequence

$$\mu \circ \rho_{V_n}^{-1} \qquad (n=1,2,\dots)$$

satisfies a large deviation principle where the rate function is given by the specific relative entropy $h(\cdot;\mu)$:

(4.1) Theorem [FO]. If $A \subseteq M(\Omega)$ is open then

$$(4.2) \qquad \liminf_n |V_n|^{-1} \log \mu[\rho_{V_n} \in A] \geq - \inf_{\nu \in A \cap M_S(\Omega)} h(\nu;\mu) ,$$

and if $A \subseteq M(\Omega)$ is closed then

$$(4.3) \qquad \limsup_n |V_n|^{-1} \log \mu[\rho_{V_n} \in A] \leq - \inf_{\nu \in A \cap M_S(\Omega)} h(\nu;\mu)$$

F. Comets [Co] and S. Olla [Ol] have obtained the same result with a different method: they first study the case where μ is a product measure and then pass to the case of Gibbs measures by using Varadhan's abstraction of the Laplace method. In [FO] the case of Gibbs measures is analyzed directly. Here we show only how the lower bound follows from the thermodynamical limit laws in Section 3; for the proof of the upper bound see [FO].

Proof of the lower bound: 1) It is enough to show that

$$\liminf |V_n|^{-1} \log \mu[\rho_{V_n} \in G] \geq - h(\nu;\mu) \tag{4.4}$$

for any $\nu \in M_s(\Omega)$ and any open neighborhood G of ν. Let us first assume that ν is ergodic. It is no loss of generality to assume that G is of the form

$$G = \bigcap_{\kappa=1}^{n} \{ \tilde{\nu} \mid |\int f_k d\tilde{\nu} - \int f_k d\nu | < \epsilon \}$$

where $f_1,\dots,f_n$ are bounded uniformly continuous functions which are $\underline{\underline{F}}_{N_p(o)}$-measurable for some $p \geq 1$. In that case,

$$\{ \rho_{V_n} \in G \} \in \underline{\underline{F}}_{W_n} \tag{4.5}$$

where $W_n := V_n \cup \partial_p V_n$. Let φ_{W_n} denote the Radon-Nikodym density of ν with respect to μ on $\underline{\underline{F}}_{W_n}$. The Shannon-McMillan theorem (3.36) shows that

$$\lim_n |V_n|^{-1} \log \varphi_{W_n} = \lim_n |W_n|^{-1} \log \varphi_{W_n} = h(\nu;\mu) \tag{4.6}$$

in $L^1(\nu)$. Due to (4.5) we can write

$$\mu[\rho_{V_n} \in G] \geq \mu[\rho_{V_n} \in G , |V_n|^{-1} \log \varphi_{W_n} < h(\nu;\mu) + \epsilon] \tag{4.7}$$

$$\geq \exp(-|V_n|(h(\nu;\mu)+\epsilon)) \, \nu[\rho_{V_n} \in G , |V_n|^{-1} \log \varphi_{V_n} < h(\nu;\mu) + \epsilon].$$

Since ν was assumed to be ergodic, $\rho_{V_n} \to \nu$ ν-a.s.. This together with (4.6) implies that the last factor in (4.7) converges to 1 , and so we obtain (4.4)

2) For a general $\nu \in M_s(\Omega)$ take a sequence ν_n (n=1,2,...) of ergodic measures as in the following lemma. Since $\nu_n \in G$ for $n \geq n_0$, the left side of (4.4) is bounded from below by $-h(\nu;\mu)$.

The following lemma is a spatial version of a construction in [Or].

(4.8) <u>Lemma:</u> For any $\nu \in M_s(\Omega)$ there is a sequence of ergodic measures ν_n (n=1,2,...) such that $\nu_n \to \nu$ weakly and

$$\lim_n h(\nu_n;\mu) = h(\nu;\mu) . \tag{4.9}$$

<u>Proof:</u> 1) For $n \geq 1$ denote by ν_n' the measure which coincides with ν on each σ-field $\underline{\underline{F}}_{V_n+n\cdot k}$ $(k \in Z^d)$ and makes these σ-fields independent. The measure

$$\nu_n := |V_n|^{-1} \sum_{i \in V_n} \nu_n' \circ \theta_i^{-1}$$

is stationary. To show that ν_n is ergodic let $B \in \underline{\underline{J}}$ be an invariant event. Since $\underline{\underline{J}} \subseteq \underline{\underline{A}}$ mod ν_n by (1.25) , there exists $B^* \in A$ such that $B = B^*$ ν_n-a.s. , hence ν_n'-a.s. But ν_n' satisfies the Kolmogorov zero-one law, and so $\nu_n'(B) = \nu_n'(B^*) \in \{0,1\}$. Since B is invariant we obtain $\nu_n(B) = \nu_n'(B) \in \{0,1\}$.

2) For an $\underline{\underline{F}}_{V_p}$-measurable bounded function φ we have

$$\int \varphi \, d(\nu_n' \circ \theta_i^{-1}) = \int \varphi \, d\nu$$

if $i \in V_{n-p}$. Since $\lim_n |V_n|^{-1} |V_n - V_{n-p}| = 0$ we get

$$\lim_n \int \varphi d\nu_n = \int \varphi d\nu . \tag{4.10}$$

(4.10) implies weak convergence, and since $h(\cdot;\lambda)$ is lower semicontinuous we obtain

(4.11) $$\liminf_n h(\nu_n;\lambda) \geq h(\nu;\lambda) .$$

3) For $i \in V_n$ there are $(N-1)^d$ disjoint translates of V_n+i contained in $V_{N\cdot n}$, and this implies

$$H_{V_{N\cdot n}}(\nu_n' o\theta_i\ ;\lambda) \leq (N-1)^d\, H_{V_n}(\nu;\lambda) .$$

By convexity,

$$H_{V_{N\cdot n}}(\nu_n;\lambda) \leq |V_n|^{-1} \sum_{i\in V_n} H_{V_{N\cdot n}}(\nu_n' o\theta_i\ ;\lambda)$$

$$\leq (N-1)^d\, H_{V_n}(\nu;\lambda) .$$

Letting $N \uparrow \infty$, we obtain

$$h(\nu;\lambda) \geq |V_n|^{-1}\, H_{V_n}(\nu;\lambda) \geq h(\nu_n;\lambda) .$$

This together with (4.11) implies $\lim_n h(\nu_n;\lambda) = h(\nu;\lambda)$. But

$$h(\nu_n;\mu) = h(\nu_n;\lambda) + e_U(\nu_n) + p_U ,$$

and since

$$e_U(\nu_n) = \int \sum_{0\in A} \frac{U_A}{|A|}\, d\nu_n ,$$

we have

$$\lim_n e_U(\nu_n) = e_U(\nu)$$

due to (3.19) and (4.10). This implies (4.9) .

4.2. The effect of a phase transition

Let $\mu \in G_s(\pi)$ be a stationary Gibbs measure. If A is a open set on $M(\Omega)$ such that $\mu \notin \bar{A}$, then

$$\lim_n \mu[\rho_{V_n} \in A] = 0 . \tag{4.12}$$

If a phase transition $|G_s(\pi)| > 1$ occurs then we may have $A \cap G_s(\pi) \neq \emptyset$, hence

$$\inf_{\nu \in A \cap M_s(\Omega)} h(\nu;\mu) = 0$$

due to the variational principle (3.39). Thus, the lower bound (4.2) implies

$$\lim |V_n|^{-1} \log \mu[\rho_{V_n} \in A] = 0 , \tag{4.13}$$

i.e., the convergence in (4.12) is slower than exponential in the volume $|V_n|$.

Let us suppose that the local specification (π_V) has the Markov property. Then the conditional distributions of μ, $\nu \in G_s(\pi)$ on the boundary ∂V_n coincide, and so the relative entropy on $V_n \cup \partial V_n$ is given by

$$H_{V_n \cup \partial V_n}(\nu;\mu) = H_{\partial V_n}(\nu;\mu) .$$

But the relative entropy $H_{\partial V_n}(\nu;\mu)$ on the surface ∂V_n is of the order of $|\partial V_n|$. This suggests that in (4.13) the volume $|V_n|$ should be replaced by the surface area $|\partial V_n|$, and that the specific entropy $h(\nu:\mu)$ should be replaced by a "specific surface entropy" of the form

$$h_\partial(\nu;\mu) = \lim_n |\partial V_n|^{-1} H_{\partial V_n}(\nu;\mu) . \tag{4.14}$$

Here we scetch the argument for a lower bound in terms of surface entropy; cf. [Ort] and [FOr] for more details, and [Ort] for results on the upper bound. For the study of large deviations in terms of surface area $|\partial V_n|$ see also [Scho].

Let us consider the two-dimensional Ising model (3.47) in the presence of a phase transition $\mu^+ \neq \mu^-$. Restricted to the line $L = \{t \in Z^2; t_2 = 0\}$, μ^+ and μ^- may be viewed as stationary processes on $\{+1,-1\}^Z$, and we define $h_\partial(\mu^-;\mu^+)$ as the specific relative entropy of one process with respect to the other:

$$(4.15) \qquad h_\partial(\mu^-;\mu^+) := \int H(\mu_0^-(\cdot|\underline{\underline{P}}^1)(\omega);\mu_0^+(\cdot|\underline{\underline{P}}^1)(\omega))\mu^-(d\omega)$$

where $\underline{\underline{P}}^1 := \sigma(\omega(t) \; ; \; t \in L \; , \; t_1 < 0)$ denotes the one-dimensional "past". In (4.15), $\mu_0^+(\cdot|\underline{\underline{P}})(\omega)$ is defined, for any $\omega \in \Omega$, as the "+-state" which corresponds to the conditional specification determined by the restriction of ω to $\{t \in L; t_1 < 0\}$; cf. (3.47). In proving the lower bound below, the main problem is to show that (4.15) coincides with the surface entropy (4.14), and that there is a Shannon-Mc Millan theorem on surfaces behind the existence of the limit in (4.14).

(4.16) Theorem (FOr): Suppose that $A \subseteq M_s(\Omega)$ is an open set such that $\mu^+ \notin \bar{A}$ but $\mu^- \in A$. Then

$$(4.17) \qquad \liminf_n |\partial V_n|^{-1} \log \mu^+[\rho_{V_n} \in A] \geq h_\partial(\mu^-;\mu^+)$$

Proof. 1) We proceed as in the proof of the general lower bound (4.2), with $\mu = \mu^+$ and $\nu = \mu^-$, and so we have to control the densities

$$\mu^-[\omega_{V \cup \partial V_n}]/\mu^+[\omega_{V \cup \partial V_n}] = \mu^-[\omega_{\partial V_n}]/\mu^+[\omega_{\partial V_n}] \;.$$

The difference is that now we need a Shannon-Mc Millan theorem for

$$|\partial V_n|^{-1} \log(\mu^-[\omega_{\partial V_n}]/\mu^+[\omega_{\partial V_n}]) \;,$$

where we divide by the surface area $|\partial V_n|$ instead of the volume $|V_n|$. By martingale convergence, it is not hard to obtain

$$\lim_n |\partial V_n|^{-1} \log \mu^-[\omega_{\partial V_n}] = \int \log \mu_o^-(\cdot|\underline{\underline{P}}^1) d\mu^-$$

in $L^1(\mu^-)$. The proof of the $L^1(\mu^-)$-convergence of

$$\lim_n |\partial V_n|^{-1} \log \mu^+[\mu_{\partial V_n}] = \int \log \mu_o^+(\cdot|\underline{\underline{P}}^1) d\mu^- \tag{4.18}$$

is more subtle: this requires a combination of martingale techniques and monotonicity arguments. We argue separately on the four sides of ∂V_n. On the first side $S(1)$, we have

$$\frac{1}{n} \log \mu^+[\omega_{S(1)}] = \frac{1}{n} \sum_{t=0}^{n-1} \log \mu_o^+(\omega(0)|\omega_t) \circ \theta_t \tag{4.19}$$

where ω_t denotes the restriction of ω to $L_t := \{i \in L ; t \le i_1 < 0\}$. Let ω_t^+ be the configuration which coincides with ω_t on L_t and is $\equiv +1$ on $\{i \in Z^2; i_1 < -t\}$. By the FKG inequality (cf., e.g., [Pr]),

$$\mu_o^+(+1|\omega_t) \le \mu_o^+(+1|\omega_t^+) ,$$

hence

$$\limsup_t \mu_o^+(+1|\omega_t) \le \lim_n \mu_o^+(+1|\omega_t^+) = \mu_o^+(+1|\underline{\underline{P}}^1)(\omega) ; \tag{4.20}$$

the limit on the right exists for any ω by monotonicity. Since $\mu_o^+(+1|\underline{\underline{P}}^1)(\omega)$ is monoton increasing in ω, we obtain

$$\begin{aligned} \mu_o^+(+1|\omega_t) &= E_{\mu^+}[\mu_o^+(+1|\underline{\underline{P}}^1)|\omega_t] \\ &\ge E_{\mu^-}[\mu_o^+(+1|\underline{\underline{P}}^1)|\omega_t] , \end{aligned} \tag{4.21}$$

using the inequality (3.48) for the conditional system determined by ω_t. But by martingale convergence,

$$\lim_t E_{\mu^-}[\mu_o^+(+1|\underline{\underline{P}}^1)|\omega_t] = \mu_o^+(+1|\underline{\underline{P}}^1)(\omega) \tag{4.22}$$

μ^--almost surely. This together with (4.20) and (4.21) implies

$$\lim_{t} \mu_o^+(\omega(o)|\omega_t) = \mu_o^+(\omega(o)|\underline{\underline{P}}^1)(\omega) \qquad \mu^- \text{-a.s.} \tag{4.23}$$

Applying the ergodic theorem in (4.19) and using (4.23) we obtain

$$\lim_{n} |\partial V_n|^{-1} \log \mu^+[\omega_{S(1)}] = \frac{1}{4} \int \log \mu_o^+(\cdot|\underline{\underline{P}})d\mu^- .$$

On the remaining sides we have to argue "around corners", and monotonicity comes in in a more complicated manner. We refer to [FOr] for the complete argument.

II. Infinite dimensional diffusions

In this chapter we consider diffusion processes of the form

$$X(t) = (X_i(t))_{i \in I} \qquad (0 \le t \le 1)$$

with some countable index set I, which satisfy an infinite dimensional stochastic differential equation of the form

$$dX_i = dW_i + b_i(X(t),t)dt \qquad (i \in I) ;$$

$(W_i)_{i \in I}$ is a collection of independent Brownian motions. Such a process induces a probability measure P on $(C[0,1])^I$ and may thus be viewed as a random field with state space $S = C[0,1]$ at each site $i \in I$. This point of view is first illustrated in Section 1 where we look at some large deviations of an infinite dimensional Brownian motion.

In Section 2 we introduce a class of interactive diffusion processes of the above form where the interaction in the drift terms is controlled by an entropy condition. This condition allows us to describe the time reversed process by a stochastic differential equation of the same type, and to derive an infinite dimensional analogue to the classical duality equation which relates the forward description of a diffusion process to its backward description. In particular, this leads to the well known characterization of reversible invariant measures of interacting diffusion processes as Gibbs measures. But as in the classical duality theory of Markov processes, time reversal is also an important tool in the study of transient phenomena. This point is illustrated in Section 2.3 where we look at an explicit example of a infinite-dimensional Martin boundary and at a corresponding problem of large deviations.

In Section 3 we return to the description of an infinite dimensional diffusion process as a random field P on S^I with $S = C[0,1]$. In the interactive case, we have to determine the local specification of this random field if we want to apply the random field techniques of Ch. I. Thus, we have to look at the conditional diffusion at a site $i \in I$, given full information on the trajectories $X_j(t)$ $(0 \le t \le 1)$ for any $j \neq i$; this is related to a "smoothing" problem in non-linear filtering theory [Pa]. As an application, we show how Dobrushin's contraction technique in Section 2 of Ch. I can be applied to an infinite dimensional diffusion process; this leads to a spatial central limit theorem. In a similar way, and in view of large deviations in the interactive case, one could apply the "thermodynamical" techniques of Section 3 and 4 of Ch. I.

Section 2 is based on joint work with A. Wakolbinger [FW], Section 3 on recent work of J.D. Deuschel [De1,2,3]. It should be clear that this introduction to infinite dimensional diffusion processes is limited to some rather special topics; some of the central issues, for example the convergence of an interacting diffusion to an invariant Gibbs measure (e.g. [HS1,2]), are not even touched upon.

1. Some large deviations of infinite dimensional Brownian motion

Let $X_0, X_1, \ldots$ be a sequence of independent Brownian motions

$$X_i(t,\omega) \qquad (0 \le t \le 1 \ , \ i = 0,1,\ldots)$$

on R^d, each with initial distribution μ_0. We can take $X_0, X_1, \ldots$ to be the coordinate maps on $\Omega = S^I$, with $S = C([0,1],R^d)$ and $I = \{0,1,\ldots\}$, under the product measure

$$P = \prod_{i \in I} P_i$$

where P_i denotes Wiener measure on S with initial distribution μ_0. For large deviations of the empirical distribution

$$\frac{1}{n} \sum_{i=0}^{n-1} \delta_{X_i(\omega)} \in M(S) \quad ,$$

described by some subset A_0 of $M(S)$, the general results of Ch. I imply

$$(1.1) \qquad P\Big[\frac{1}{n}\sum_{i=0}^{n-1}\delta_{X_i}\in A_0\Big] \sim \exp\Big[-n \inf_{Q:Q_0\in A_0} h(Q;P)\Big]$$

where Q_0 denotes the marginal distribution of the stationary probability measure Q on S^I. For each n, the relative entropy $H_n(Q;P)$ with respect to P on $\underline{\underline{F}}_n := \sigma(X_0,\ldots,X_n)$ satisfies

$$H_n(Q;P) = H_n(Q;\prod_i Q_i) + \sum_{i=0}^{n}\int \log\frac{dQ_i}{dP_i}(X_i)\,dQ$$

$$\geq (n+1)\,H(Q_0;P_0)\,,$$

and this implies

$$h(Q;P) \geq h(\prod Q_i;P) = H(Q_0;P_0)$$

for any $Q \in M_S(\Omega)$ with marginal distribution Q_0 on S. Thus, (1.1) reduces to Sanov's theorem

$$(1.2) \qquad P\Big[\frac{1}{n}\sum_{i=0}^{n-1}\delta_{X_i}\in A_0\Big] \sim \exp\Big[-n \inf_{Q_0\in A_0} H(Q_0;P_0)\Big]\,.$$

Roughly speaking, (1.2) can be interpreted as follows. Under the condition that a large deviation described by A_0 does occur, the process $(X_0,X_1,\ldots)$ is most likely to behave like a collection of independent diffusions, each given by that measure $Q_0 \in A_0$ which minimizes the entropy $H(\cdot;P_0)$ with respect to Wiener measure P_0. Therefore, one would like to describe the minimizing measure Q_0 more explicitely. We are going to illustrate this point for some specific choices of A_0. The first three examples are classical, the last appears in recent work of Dawson and Gärtner [DG].

(1.3) Remark on the Girsanov transformation. 1) Let Q_0 be a probability measure on $S = C([0,1], R^d)$. We denote by $X(t)$ $(0 \leq t \leq 1)$ the coordinate process on S and by ν_t resp. μ_t the distribution of $X(t)$ under Q_0 resp. P_0. Suppose that Q_0 is absolutely continuous with respect to P_0. Then the theory of the (Cameron-Martin-Maruyama-) Girsanov transformation shows that there is an R^d - valued adapted drift process b_t $(0 \le t \le 1)$ on S such that

(1.4) $$\int_0^1 b_t^2 \, dt < \infty \quad Q_0\text{-a.s.}$$

(1.5) $$W^b(t) := X(t) - X(0) - \int_0^t b_s \, ds$$

is a Wiener process under Q_0,

(1.6) $$\frac{dQ_0}{dP_0} = \frac{d\nu_0}{d\mu_0} \exp[\int_0^1 b_t dW^b(t) + \frac{1}{2}\int_0^1 b_t^2 dt]$$

$$= \frac{d\nu_0}{d\mu_0} \exp[\int_0^1 b_t dX(t) - \frac{1}{2}\int_0^1 b_t^2 dt] ,$$

cf.,e.g., [IW]. Moreover, we have $H(Q_0;P_0) < \infty$ if and only if

$$H(\nu_0;\mu_0) + E_{Q_0}[\int_0^1 b_t^2 dt] < \infty ,$$

and in that case the entropy is given by

(1.7) $$H(Q_0;P_0) = H(\nu_0;\mu_0) + \frac{1}{2} E_{Q_0}[\int_0^1 b_t^2 dt] .$$

2) If, more generally, X is a Wiener process with respect to a filtration $(\underline{\underline{F}}_t)$ on some probability space $(\Omega,\underline{\underline{F}},P_0)$ and if Q_0 is a probability measure on $\underline{\underline{F}}$ with $H(Q_0;P_0) < \infty$ then there exists an adapted drift process with "finite energy"

$$E_{Q_0}[\int_0^1 b_t^2 dt] < \infty$$

such that (1.5) holds. In that case, the drift can be computed as a stochastic forward derivative in the sense of Nelson [Ne]: for almost all $t \in [0,1]$,

$$(1.8)\qquad b_t = \lim_{h\downarrow 0} \frac{1}{h} E_{Q_o}[X(t+h)-X(t)\,|\,\underline{\underline{F}}_t]$$

$$= \lim_{\alpha,\beta\downarrow 0} \frac{1}{\alpha+\beta} \int_{t-\alpha}^{t+\beta} b_s ds$$

in $L^2(Q_o)$; cf. [Fö6].

1.1. Large deviations of the average positions: Schilder's theorem

For $B \subseteq C([0,1],R^d)$ let us consider a large deviation of the form

$$(1.9)\qquad \frac{1}{n}\sum_{i=0}^{n-1} X_i \in B ;$$

since

$$P[\frac{1}{n}\sum_{i=0}^{n-1} X_i \in B] = P_o[n^{-1/2}X_o \in B] ,$$

this can also be regarded as a large deviation of a single Brownian motion with small variance. (1.9) means that in Sanov's theorem (1.2) we take $A_o = \{Q_o | \varphi_{Q_o} \in B\}$ where

$$\varphi_{Q_o}(t) := \int X(t)dQ_o \qquad (0 \leq t \leq 1) .$$

Let us assume that the initial distribution μ_o is concentrated on 0 , and let us fix $\varphi \in S$ with $\varphi(o) = 0$. Then the problem is reduced to finding a measure Q_o which minimizes $H(\cdot;P_o)$ under the constraint $\varphi_{Q_o} = \varphi$. The function φ is of interest only if there exists some Q_o with $\varphi_{Q_o} = \varphi$ and finite entropy $H(\cdot;P_o)$. But this implies, due to (1.5) and (1.7) ,

$$\varphi(t) = E_{Q_o}[\int_0^t b_s ds] = \int_0^t E_{Q_o}[b_s]ds ,$$

i.e., φ is absolutely continuous with derivative $\nabla\varphi(t) = E_{Q_o}[b_t]$ for almost all t . Moreover,

$$\frac{1}{2}\int_0^1 (\nabla\varphi)^2 dt \le \frac{1}{2}\int_0^1 E_{Q_o}[b_t^2]dt$$

$$= H(Q_o;P_o) < \infty ,$$

i.e. φ is an absolutely continuous function with "finite energy". For such a φ, the lower bound $\frac{1}{2}\int_0^1 (\nabla\varphi)^2 dt$ for the entropy $H(\cdot;P_o)$ is in fact attained if Q_o is the measure with $\nu_o = \delta_o$ and deterministic drift $b_t(\cdot) \equiv \nabla\varphi(t)$. In particular, Sanov's theorem (1.2) implies

(1.10) Schilder's theorem:

$$P[\frac{1}{n}\sum_{i=0}^{n-1} X_i \in B] \sim \exp[-n \inf_{\varphi\in B_o} \frac{1}{2}\int_0^1 (\nabla\varphi)^2 dt]$$

where B_o denotes the intersection of B with the class of absolutely continuous functions with finite energy; cf., e.g., [St].

1.2 Large deviations at the terminal time: h-path processes

If we are only interested in large deviations of the empirical distribution at the terminal time $t = 1$, the set A_o in (1.2) is of the form

$$A_o = \{Q_o | Q_o \circ X(1)^{-1} \in B\}$$

for some $B \subseteq M(R^d)$. This means that, for a given probability measure ν_1 on R^d, we have to find the measure Q_o with minimal entropy $H(\cdot;P_o)$ under the constraint that the marginal distribution at time 1 is given by ν_1. But we can write

$$(1.11) \qquad H(Q_o;P_o) = H(\nu_1;\mu_1) + \int \nu_1(dy) H(Q_o^y;P_o^y) ,$$

where Q_o^y and P_o^y denote the conditional distributions of Q_o and P_o on $S = C([0,1],R^d)$, given the terminal value $X(1,\cdot) = y$.

Thus, $H(\nu_1;\mu_1)$ is a lower bound, and this lower bound is attained by

$$(1.12)\qquad Q_o := \int \nu_1(dy)\, P_o^y \quad ,$$

i.e., by the distribution of Brownian motion starting with μ_o and conditioned to have terminal distribution ν_1 at time 1 . In particular,

$$(1.13)\qquad P[\frac{1}{n}\sum_{i=o}^{n-1} \delta_{X_i(1)} \in B] \sim \exp[-n \inf_{\nu_1 \in B} H(\nu_1;\mu_1)]$$

Of course, (1.13) is nothing else than Sanov's theorem applied to the classical case of a sequence of independent random variables with d-dimensional normal distribution. But on the higher level (1.1) we obtain additional information. For a given ν_1 with finite entropy $H(\nu_1;\mu_1)$, denote by $h(\cdot,1)$ the density of ν_1 with respect to μ_1 and by $h(x,t)$ the space-time harmonic function

$$(1.14)\qquad h(x,t) := \int h(y,1)\, p_{1-t}(x,y)\, dy$$

where $p_s(x,y)$ is the transition density of d-dimensional Brownian motion. Itô's formula, applied to $\log h$, shows that the measure Q_o in (1.12) is given by the density

$$\frac{dQ_o}{dP_o} = h(X(1),1)$$

$$= h(X(0),0) \exp[\int_0^1 b_t dX - \frac{1}{2}\int_0^1 b_t^2\, dt]$$

with drift

$$(1.15)\qquad b_t = \nabla \log h(X(t),t)$$

In other words, Q_o is an h-path process in the sense of J.L. Doob. The large deviation result (1.1) can now be interpreted as follows. Under a large deviation of the empirical distribution at the terminal time $t=1$, the process $(X_o, X_1,\ldots)$ behaves like a collection of independent h-path processes, where h is the space-time harmonic function induced by the measure ν_1 which has minimal entropy $H(\cdot;\mu_1)$ under the given constraint B.

1.3. Large deviations at the initial and the terminal time: Schrödinger bridges

Let us now look at large deviations of the empirical distribution both at time 0 and at time 1 ; here we assume that the initial distribution μ_o is equivalent to Lebesgue measure. This leads to the problem of finding the measure Q_o on S which minimizes $H(\cdot;P_o)$ under the constraint that the marginal distributions ν_o at time 0 and ν_1 at time 1 are fixed.

Let P_x^y denote the conditional distribution of Wiener measure on $C([0,1],R^d)$ given an initial value x at time 0 and a terminal value y at time 1 , i.e., P_x^y is the distribution of the Brownian bridge leading from x to y. Let μ and ν denote the joint distribution of $(X(0), X(1))$ under P_o and Q_o , and let us write

$$H(Q_o;P_o) = H(\nu;\mu) + \int \nu(dx,dy)\, H(Q_x^y;P_x^y) \quad .$$

This shows that $H(\cdot;P_o)$ is minimized by the measure

$$Q_o := \int \nu(dx,dy)\, P_x^y \;, \tag{1.16}$$

if ν is the measure on $R^d \times R^d$ which minimizes $H(\cdot;\mu)$ under the constraint that the marginals are given by ν_o and ν_1 . If the marginals admit any measure on the product space with finite entropy $H(\cdot;\mu)$, then a unique minimizing measure ν does exist, and its density is of the form

$$\frac{d\nu}{d\mu}(x,y) = f(x)\, g(y) \tag{1.17}$$

with $\log f \in L^1(\nu_o)$ and $\log g \in L^1(\nu_1)$; cf. [Cs] resp. the argument below.

The representation (1.17) implies the two equations

(1.18) $$\frac{d\nu_o}{d\mu_o}(x) = f(x) \int p_1(x,y)g(y)dy$$

$$\frac{d\nu_1}{d\mu_1}(y) = g(y) \int p_1(x,y)f(x)\frac{d\mu_o}{d\lambda}(x)dx \, \frac{d\mu_1}{d\lambda}(y)^{-1} .$$

It implies also that the associated measure Q_o in (1.16) has the Markov property. More precisely, Q_o is an h-path process with

$$h(x,t) := \int p_{1-t}(x,y)\, g(y)dy .$$

In fact we can write, using (1.18) and applying Itô's formula as in the previous case,

$$\frac{dQ_o}{dP_o} = f(X(0))\, g(X(1))$$

$$= \frac{d\nu_o}{d\mu_o}(X(0)) \exp[\int_0^1 b_t dX(t) - \frac{1}{2}\int_0^1 b_t^2 dt]$$

where the drift process (b_t) is given by

(1.19) $$b_t = \nabla \log h(X(t),t) .$$

(1.20) Remark. The dual equations (1.18), with $\mu_o = \mu_1 = \lambda$, appear in a paper by E. Schrödinger [Sch1], derived in a more heuristic manner, but clearly motivated by the question of large deviations: "Imaginez que vous observez un système de particules en diffusion, qui soient en équilibre thermodynamique. Admettons qu'à un instant donné t_o vous les ayez trouvées en répartition à peu près uniforme et qu'à $t_1 > t_o$ vous ayez trouvé un écart spontané et considérable par rapport à cette uniformité. On vous demande de quelle manière cet écart s'est produit. Quelle en est la manière la plus probable?" [Sch2]. The existence of solutions f and g is studied in [Fo], [Be]; it also follows by the probabilistic argument below. The associated "Schrödinger bridge" or "reciprocal process" Q_o in (1.16) is described in detail in [Ja]. Schrödinger comments in [Sch1] on "merkwürdige Analogien zur Quantenmechanik, die mir sehr des Hindenkens wert erscheinen"; a systematic account of the present role of Schrödinger bridges in Nelson's Stochastic Mechanics can be found in [Za].

The derivation of the crucial factorization (1.17) can be based on the following general argument, which will also be used in the next section; cf. [Cs].

Let μ be a probability measure on some polish space, and let f_i $(i=1,2,\ldots)$ be a sequence of bounded measurable functions. We want to minimize the entropy $H(\nu;\mu)$ under the constraints

$$\int f_i d\nu = c_i \quad (i=1,2,\ldots) . \tag{1.21}$$

Suppose that there exists some ν which satisfies (1.21) and has finite entropy $H(\nu;\mu) < \infty$. Then, and this is well known from the theory of exponential families (see, e.g., [Az]), there is a unique ν_n which minimizes $H(\cdot;\mu)$ under the first n constraints, and its density is of the form

$$\frac{d\nu_n}{d\mu} = Z_n^{-1} \exp\Big(\sum_{i=1}^{n} \lambda_i f_i\Big) . \tag{1.22}$$

(1.23) Lemma. There is a unique measure ν_∞ which minimizes $H(\cdot;\mu)$ under the constraints (1.21) . Moreover,

$$\lim_n H(\nu_n;\mu) = H(\nu_\infty;\mu) , \tag{1.24}$$

and

$$\lim_n H(\nu_\infty;\nu_n) = 0 . \tag{1.25}$$

In particular, ν_n converges to ν_∞ in total variation.

Proof. 1) Since $H(\nu_n;\mu)$ is monotone in n with

$$\sup_n H(\nu_n;\mu) \leq H(\nu;\mu) < \infty ,$$

and since

$$(1.26)\qquad H(\nu_m;\mu) = H(\nu_m;\nu_n) + \int \log \frac{d\nu_n}{d\mu}\, d\nu_m$$

$$= H(\nu_m;\nu_n) + H(\nu_n;\mu)$$

for any $m \geq n$, we obtain

$$\lim_n \sup_{m \geq n} H(\nu_m;\nu_n) = 0 .$$

But by Ch. I (3.3),

$$(1.27)\qquad \|\nu_m - \nu_n\| \leq (2\, H(\nu_m;\nu_n))^{1/2} ,$$

and so ν_n converges in total variation to some limit measure ν_∞ . Moreover,

$$H(\nu_\infty;\mu) \geq \lim_n H(\nu_n;\mu) \geq H(\nu_\infty;\mu)$$

where the last inequality follows from lower-semicontinuity of $H(\cdot;\mu)$ (or by Fatou's lemma, if one does not want to use topological assumptions). This shows (1.24), hence (1.25) if we use (1.26) with ν_∞ instead of ν_m .

Let us now apply the lemma to the situation above. The fact that ν on $R^d \times R^d$ has marginals ν_0 and ν_1 can be expressed by constraints of the form (1.21) where each f_i only depends on one of the two coordinates. Thus, the minimizing measure is given by $\nu_\infty = \lim \nu_n$ (in total variation) , and (1.22) shows that each ν_m is of the form $d\nu_n = h_n\, d\mu$ with $h_n(x,y) = f_n(x)\, g_n(y)$. Since h_n con-verges to $h_\infty = d\nu_\infty/d\mu$ in $L^1(\mu)$, we can now conclude that also h_∞ admits a factorization $h_\infty(x,y) = f(x)\, g(y)$; cf. Lemma 2.5 in [DV]. Moreover,

$$0 \leq H(\nu_\infty;\mu) = \int \log f d\nu_0 + \int \log g d\nu_1 < \infty$$

implies $\log f \in L^1(\nu_0)$ and $\log g \in L^1(\nu_1)$.

1.4 Large deviations of the flow of marginal distributions

In [DG], Dawson and Gärtner discuss large deviations of the whole flow of marginal distributions

$$\frac{1}{n}\sum_{i=1}^{n} \delta_{X_i(t,\omega)} \qquad (0 \le t \le 1) .$$

This leads to the problem of minimizing the entropy $H(Q_o;P_o)$ under the constraint that all the marginal distributions $\nu_t := Q_o \circ X(t)^{-1}$ $(0 \le t \le 1)$ are fixed in advance.

Suppose that the flow ν_t $(0 \le t \le 1)$ is admissible in the sense that there exists at least one measure Q_o on $C[0,1]$ with marginals ν_t and finite entropy $H(Q_o;P_o)$; some regularity properties of such admissible flows will be discussed in Section 2.1. Then we can write

$$H(Q_o;P_o) = H(\nu_o;\mu_o) + \frac{1}{2}\int [\int_0^1 b_t^2 \, dt] dQ_o$$

where (b_t) is the drift process associated to Q_o by (1.3). Introducing the Markovian drift

$$\tilde{b}_t(x) := E_{Q_o}[b_t | X_t = x]$$

and using the notation $\langle \nu_t, f\rangle = \int f d\nu_t$, we get

$$\text{(1.28)} \qquad H(Q_o;P_o) \ge H(\nu_o;\mu_o) + \frac{1}{2}\int_0^1 \langle \nu_t, \tilde{b}_t^2\rangle \, dt .$$

In order to obtain a lower bound which only involves (ν_t) , consider a smooth function f with compact support and note that Itô's formula implies

$$\langle \nu_t, f\rangle = \langle \nu_o, f\rangle + \int_0^t \langle \nu_s, \nabla f \, \tilde{b}_s\rangle ds + \frac{1}{2}\int_0^t \langle \nu_s, \Delta f\rangle ds .$$

Thus, for almost all t ,

$$\langle \dot{\nu}_t - \frac{1}{2}\Delta \nu_t, f\rangle = \langle \nu_t, \tilde{b}_t \nabla f\rangle$$

in the distributional sense, hence

$$(1.29)\qquad \frac{1}{2}\langle\nu_t,\tilde{b}_t^2\rangle = \frac{1}{2}\langle\nu_t,(\tilde{b}_t-\nabla f)^2\rangle + \langle\nu_t,\tilde{b}\nabla f\rangle - \frac{1}{2}\langle\nu_t,(\nabla f)^2\rangle \geq \langle\dot{\nu}_t-\frac{1}{2}\Delta\nu_t,f\rangle - \frac{1}{2}\langle\nu_t,(\nabla f)^2\rangle .$$

Defining the supremum of the right side over all smooth functions f with compact support as

$$||\dot{\nu}_t - \frac{1}{2}\Delta\nu_t||_{\nu_t} ,$$

we obtain the lower bound

$$(1.30)\qquad H(Q_o;P_o) \geq H(\nu_o;\mu_o) + \int_0^1 ||\dot{\nu}_t - \frac{1}{2}\Delta\nu_t||_{\nu_t} dt$$

due to (1.28) and (1.29). This lower bound is actually attained:

(1.31) Theorem ([DG]). There exists exactly one measure Q_o with marginals ν_t $(0 \leq t \leq 1)$ and

$$(1.32)\qquad H(Q_o;P_o) = H(\nu_o;\mu_o) + \frac{1}{2}\int_0^1 ||\dot{\nu}_t - \frac{1}{2}\Delta\nu_t||_{\nu_t} dt ,$$

and this measure has the Markov property.

Proof. The constraint can be written in the form (1.21) with a countable collection of bounded functions f_i $(i=1,2,\ldots)$ on $C[0,1]$, each of the form $f_i = \varphi_i(X(t_i))$. Let Q_n be the measure which minimizes $H(\cdot;P_o)$ under the constraints

$$\int f_i dQ = c_i \qquad (1 \leq i \leq n) .$$

It follows as in (1.19) that the drift process (b_t^n) associated to Q_n is of the form $b_t^n = \nabla f^n(X(t),t)$ for all but finitely many t . Now let $Q_o = \lim_n Q_n$ denote the limiting measure given by Lemma (1.23), and let (b_t) be the associated drift. (1.3) implies

$$H(Q_o;Q_n) = \frac{1}{2} E_{Q_o}[\int_0^1 (b_t-b_t^n)^2 dt] ,$$

and since $\lim H(Q_o;Q_n) = 0$ due to (1.25), we see that (b_t) is Markovian, i.e., $b_t = \tilde{b}_t(X(t))$ Q-a.s. for almost all t. It also follows that there is a sequence of smooth functions f_n with compact support such that

$$\lim_n \langle \nu_t,(\tilde{b}_t-\nabla f_n)^2\rangle = 0 .$$

(1.29) now implies

$$\frac{1}{2} \langle \nu_t,\tilde{b}_t^2\rangle = ||\dot{\nu}_t - \frac{1}{2} \Delta\nu_t||_{\nu_t} ,$$

hence (1.32), since

$$\begin{aligned} H(Q_o;P_o) &= H(\nu_o;\mu_o) + \frac{1}{2} \int[\int_0^1 b_t^2 dt]dQ_o \\ &= H(\nu_o;\mu_o) + \frac{1}{2} \int_0^1 \langle \nu_t,\tilde{b}_t^2\rangle dt . \end{aligned}$$

2. Infinite dimensional diffusions and Gibbs measures

Let P be a probability measure on $(C[0,1])^I$ and suppose that the coordinate process $X = (X_i)_{i\in I}$ satisfies a stochastic differential equation of the form

$$dX_1 = dW_i + b_i(X(t),t)dt \quad (i \in I) ,$$

where $(W_i)_{i\in I}$ is a collection of independent Wiener processes under P . We are going to derive a dual equation

$$dX_i = d\hat{W}_i + \hat{b}_i(X(t),t)dt \quad (i \in I) ,$$

for the behavior of X after time reversal, i.e., under the time reversed measure $\hat{P}$. In contrast to the finite dimensional case, such a dual equation may break down if the interaction between different sites via the drift terms (b_i) becomes too strong; cf. (2.35) below. In order to exclude such effects we use a condition of locally finite entropy.

In Section 2.1 this entropy technique is first illustrated in the finite dimensional case. Here it leads to regularity results for the distributions of the coordinate process under a measure which has finite entropy with respect to Wiener measure; in particular, the density functions are a.e. partially differentiable. These smoothness properties are derived by time reversal, not by Malliavin calculus: We use the computation of drifts as stochastic forward and backward derivates and a classical integration by parts in the resulting duality equation. As a digression, we show that an integration by parts on Wiener space, i.e., a Mallavian calculus argument, only comes in at a second stage: it is needed if we want to determine not only the the density but also the backward drift in the non-Markovian case.

In Section 2.2 we apply the entropy technique to the infinite dimensional case; this is based on joint work with A. Wakolbinger [FW]. We obtain an infinite dimensional version of the classical duality equation, which relates forward drift, backward drift and the local specification of the distribution μ_t of the process at time t . In particular, this leads to a description of invariant measures as Gibbs measures. But the duality equation is also useful for quite different purposes. This is illustrated in Section 2.3 where we look at a large deviation of the empirical field of an infinite dimensional Brownian motion.

2.1. Time reversal on Wiener space

Let P be a probability measure on $C([0,1],R^d)$ with initial distribution ν_o. We say that P has finite entropy with respect to Wiener measure if

$$H(P;P^*) < \infty \tag{2.1}$$

where P^* denotes Wiener measure with the same initial distribution $\mu_o^* = \mu_o$. It follows by (1.3) that the drift process (b_t) associated to P has finite energy:

$$\frac{1}{2} E[\int_0^1 b_t^2 dt] = H(P;P^*) < \infty . \tag{2.2}$$

As we have seen in Section 1, it is of interest to clarify the regularity properties of P which are implied by the finite entropy assumption (2.1). To begin with, the distribution μ_t of $X(t)$ under P is absolutely continuous with respect to Lebesgue measure, and we denote by $\rho_t(x)$ the density function. It turns out that this function is in fact a.e. smooth in the sense that it admits a.e. a gradient $\nabla\rho_t$:

(2.3) Proposition. For almost all t, the function ρ_t is absolutely continuous on almost all straight lines which are parallel to coordinate axes, and it satisfies

$$\int_\epsilon^1 [\int (\nabla \log \rho_t)^2(x) \rho_t(x) dx] dt < \infty$$

for any $\epsilon > 0$.

We are going to derive (2.3) as an exercise in time reversal. Let $\hat{P} = P \circ R$ denote the image of P under the pathwise time reversal R on $C([0,1],R^d)$ defined by

$$X(t,R(\omega)) = X(1-t,\omega) .$$

For the rest of this section, E and $\hat{E}$ will denote the expectation with respect to P and $\hat{P}$.

(2.4) Lemma. If P has finite entropy with respect to Wiener measure then $\hat{P}$ has finite entropy up to any time $t < 1$, i.e., there exists an adapted process $(\hat{b}_t)$ such that

$$(2.5) \qquad \hat{W}(t) := X(t) - X(o) - \int_0^t \hat{b}_s ds \qquad (0 \le t \le 1)$$

is a Wiener process under $\hat{P}$, and

$$(2.6) \qquad \hat{E}[\int_0^t \hat{b}_s^2 ds] < \infty$$

for any $t < 1$.

Proof. With respect to the time reversal $\hat{P}^* := P^* \circ R$ of P^* , the coordinate process $X(t)$ is a Brownian motion conditioned to have distribution μ_o at time $t = 1$. This implies that

$$W^*(t) := X(t) - X(o) - \int_0^t c_s ds$$

with $c_s := (1-s)^{-1}(X(1)-X(s))$ is a Wiener process with respect to $\hat{P}^*$ and the filtration given by $F_t^* := \underline{\underline{F}}_t V \sigma(X(1))$. Since

$$H(\hat{P};\hat{P}^*) = H(P;P^*) < \infty ,$$

(1.3), 2) implies that there is a process (a_t) adapted to $(\underline{\underline{F}}_t^*)$ with finite energy under $\hat{P}$ such that

$$W_t' := W_t^* - \int_0^t a_s ds \qquad (0 \le t \le 1)$$

is a Wiener process with respect to $\hat{P}$ and $(\underline{\underline{F}}_t^*)$. On the other hand,

$$\sup_s \hat{E}[X_1 - X_s)^2] = \sup_s E[(X_s - X_0)^2] < \infty$$

due to (2.2), and so we get

$$\hat{E}[\int_0^t (a_s + c_s)^2 ds] < \infty$$

for any $t < 1$. It follows that the process $(\hat{b}_t)$ defined as an optional version of the process

$$\hat{E}[a_t + c_t | \underline{\underline{F}}_t] \qquad (0 \le t \le 1) ,$$

satisfies (2.5) and (2.6).

Lemma (2.4) shows that under $\hat{P}$ the coordinate process satisfies a stochastic differential equation of the form $dX = d\hat{W} + \hat{b}dt$. Due to (1.8) and (2.6), we can compute the dual drift $(\hat{b}_t)$ as a stochastic forward derivative under $\hat{P}$ or, equivalently, as a stochastic backward derivative under P : for almost all $t \in [0,1]$,

(2.7)
$$\hat{b}_t = \lim_{h \downarrow 0} \frac{1}{h} \hat{E}[X(t+h) - X(t) | \underline{\underline{F}}_t]$$
$$= -\lim_{h \downarrow 0} \frac{1}{h} E[X(1-t) - X(1-t-h) | \hat{\underline{\underline{F}}}_{1-t}] \circ R$$

in $L^2(\hat{P})$, where $\hat{\underline{\underline{F}}}_{1-t} := R^{-1}(\underline{\underline{F}}_t)$ denotes the σ-field of events observable from time t on.

(2.8) <u>Theorem.</u> For almost all $t \in [0,1]$, ρ_t has the properties in (2.3), and its gradient $\nabla \rho_t$ satisfies the duality equation

(2.9)
$$\nabla \rho_t(x) = \rho_t(x) \; E[b_t + \hat{b}_{1-t} \circ R | X_t = x]$$

for almost all x .

<u>Proof.</u> 1) Let t_0 be a point in $(0,1)$ such that (1.8) holds for (b_s) and $t = 1-t_0$, and for $(\hat{b}_s)$ and $t = t_0$. Then, for any bounded $\underline{\underline{F}}_{t_0}$-measurable function F on $C([0,1],R^d)$

$$(2.10) \qquad E[(\hat{b}_{1-t_0} \circ R)\, F] = \hat{E}[\hat{b}_{1-t_0}(F \circ R)]$$

$$= \lim \frac{1}{h} \hat{E}[(X_{1-t_0+h} - X_{1-t_0})(F \circ R)]$$

$$= -\lim \frac{1}{h} E[(X_{t_0} - X_{t_0-h})F] .$$

2) For a smooth function f with compact support, Itô's formula implies

$$E[(X_{t_0} - X_{t_0-h})\, f(X_{t_0})] = E[(\int_{t_0-h}^{t_0} b_s ds)\, f(X_{t_0-h})]$$

$$+ E[(X_{t_0} - X_{t_0-h})(\int_{t_0-h}^{t_0} \nabla f(X_s) dX_s + \frac{1}{2} \int_{t_0-h}^{t_0} \Delta f(X_s) ds)] .$$

By (2.10) we get, using some straightforward estimates based on (1.8),

$$(2.11) \qquad E[(\hat{b}_{1-t_0} \circ R)\, f(X_{t_0})] = E[b_{t_0} f(X_{t_0})] - E[\nabla f(X_{t_0})] .$$

Since

$$E[\nabla f(X_{t_0})] = \int \nabla f(x) \rho_{t_0}(x) dx = -\langle \nabla \rho_{t_0}, f \rangle ,$$

(2.11) shows that the distributional gradient of ρ_{t_0} is in fact given by the right side of (2.9). This implies that ρ_{t_0} is absolutely continuous on almost all straight lines which are parallel to coordinate axes, and that the distributional gradient coincides with the usual gradient almost everywhere, cf., e.g., [Ma] 1.1.3. The integrability condition (2.3) now follows from (2.2) and (2.6).

(2.12) Remark. Suppose that (X_t) is a Markov process under P. Then the drift is of the form $b_t = b_t(X_t)$ with some measurable function $b_t(x)$. But the Markov property is preserved under time reversal. Thus, the dual drift is given by some measurable function $\hat{b}_t(x)$, and (2.9) reduces, but here without any regularity assumptions, to the classical duality equation

$$b_t(x) + \hat{b}_{1-t}(x) = \nabla \log \rho_t(x) \; ; \tag{2.13}$$

cf., e.g., [Na], [Ne], [HP]. In the case of a reversible Markov process, a smoothness property of the equilibrium density is derived in [Or1] and [Fu2].

In the general non-Markovian case, the duality equation (2.9) does not yet determine the dual drift, only its projection on the present state. For a complete description, we have to replace the functionals $F = f(X_{t_0})$ used in the proof of (2.8) by smooth functionals on $C([0,1],R^d)$ which depend on the full σ-algebra $\hat{\underline{F}}_{t_0}$. In particular, the integration by parts on R^d which was used in (2.11) has to be replaced by an integration by parts on Wiener space, i.e., by a Malliavin calculus argument.

(2.14) Remark on integration by parts on Wiener space. For a bounded predictable process (u_t) put $U_t = \int_0^t u_s ds$ and $X_t^{\epsilon,U} = X_t + \epsilon U_t$ $(0 \le t \le 1)$. Let us say that a function $F \in L^2(P^*)$ is P^*-smooth if there is a measurable process (φ_t) such that, for any bounded predictable process (u_t),

$$DF(.,U) = \lim_{\epsilon \downarrow 0} \frac{F(X^{\epsilon,U}) - F(X)}{\epsilon} = \int_0^1 u_s \varphi_s ds \tag{2.15}$$

in $L^2(P^*)$. If F is Fréchet-differentiable on $C([0,1],R^d)$ with bounded derivative $DF(\omega,dt)$, viewed as an R^d-valued measure on $[0,1]$, then (4.2) holds with $\varphi_s(\omega) = DF(\omega,[s,1])$. (2.15) is enough to use Bismut's derivation [Bi] of the basic integration by parts formula

$$(2.16)\qquad E^*[(\int_0^1 u_s dX_s)F] = E^*[DF(.,U)] = E^*[\int_0^1 u_s \varphi_s ds] .$$

If F is Fréchet-differentiable with bounded derivative $DF(.,dt)$ and G is P^*-smooth then (4.3), applied to FG, leads to

$$(2.17)\qquad E^*[FG \int_0^1 u_t dX_t] = E^*[DF(.,U)G] + E^*[FDG(.,U)] .$$

Let us now assume that the Girsanov density $G = dP/dP^*$ in (1.3) is P^*-smooth with

$$(2.18)\qquad DG(.,U) = \int_0^1 u_t \gamma_t dt .$$

If the drift b is a bounded smooth function on $C([0,1],R^d) \times [0,1]$ with bounded Fréchet derivatives $Db_t(.,ds)$, then G is indeed P^*-smooth and satisfies (2.18) with

$$(2.19)\qquad \gamma_t G^{-1} = b_t + \int_0^1 \beta_{t,s} dW_s^b$$

and $\beta_{t,s} := Db_s(.,[t,1])$; this follows, e.g., from Theorem (A.10) in [BJ] .

(2.20) Theorem. Suppose that $G = dP/dP^*$ is P^*-smooth and satisfies (2.18). Then the dual drift is given, for almost all $t \in (0,1)$, by

$$(2.21)\qquad \hat{b}_{1-t} = - E[b_t + a_t | \hat{\underline{F}}_t] \circ R$$

where

$$(2.22)\qquad a_t = \frac{1}{t} (W_t^b - \int_0^t \int_0^1 \beta_{s,r} dW_r^b ds) + \int_0^1 \beta_{t,s} dW_s^b .$$

Proof. We fix t_0 as in the proof of (2.10). Assume that F in (2.10) is Fréchet-differentiable with bounded derivative $DF(.,dt)$;

since F is $\hat{\underline{F}}_{t_0}$-differentiable, the measures $DF(.,dt)$ are concentrated on $[t_0,1]$. Note that (2.17) translates into

$$(2.23)\qquad E[F\int_0^1 u_t dX_t] = E[DF(.,U)] + E[FG^{-1}DG(.,U)]$$

with respect to P . For $u_t = I_{[t_0-h,t_0]}(t)$ we obtain

$$(2.24)\qquad E[(X_{t_0}-X_{t_0-h})F] = h\,E[DF(.,[t_0,1])] + E[FG^{-1}\int_{t_0-h}^{t_0} \gamma_t dt] .$$

This together with (2.10) implies

$$(2.25)\qquad -E[(\hat{b}_{1-t}\circ R)F] = E[DF(.,[t,1]] + E[FG^{-1}\gamma_t]$$

for almost all $t \in (0,1)$. Using again (2.24) with $h = t_0$ we obtain

$$E[DF(.,[t,1])] = \frac{1}{t}E[X_t F] - \frac{1}{t}E[F\int_0^t G^{-1}\gamma_s ds]$$

$$= E[F\,\frac{1}{t}(W_t^b - \int_0^t\int_0^1 \beta_{s,r}dW_r^b ds)]$$

Thus, we can write (2.25) as

$$E[(\hat{b}_{1-t}\circ R + b_t)F] = -E[a_t F] ,$$

and this implies (2.21).

(2.26) <u>Remark.</u> In the Markovian case with a smooth drift function $b_t(x)$ we have $\beta_{t,s} = \nabla b_s(X_s)I_{[t,1]}(s)$, and (2.21) becomes

$$(2.27)\qquad \hat{b}_{1-t}(x) = -b_t(x) + \frac{1}{t}E[\int_0^t (r\nabla b_r(X_r)-1)dW_r^b \,|\, X_t = x] .$$

Due to (2.13), this can also be read as a path space formula for $\nabla \log \rho_t(x)$.

2.2. Infinite dimensional diffusions and their time reversal

Let I be a countable index set, denote by

$$(2.28) \qquad X_i(t,\omega) \quad (0 \le t \le 1\ ,\ i \in I)$$

the coordinate processes X_i $(i \in I)$ on $\Omega = C[0,1]^I$, and let $(\underline{\underline{F}}_t)$ be the canonical right-continuous filtration on Ω . Let P be a probability measure on $(\Omega, \underline{\underline{F}}_1)$ such that the process

$$X(t) = (X_i(t))_{i \in I}$$

satisfies an infinite dimensional stochastic differential equation of the following form:

$$(2.29) \qquad dX_i = dW_i + b_i(X(t),t)dt \quad (i \in I)\ ,$$

where $W = (W_i)_{i \in I}$ is a collection of independent Wiener processes, and where $b_i(.,t)$ is a function defined on a suitable state space $E \subseteq R^I$ such that

$$P[X(t) \in E \quad (0 \le t \le 1)] = 1\ .$$

This means, in particular, that $X(t)$ $(0 \le t \le 1)$ can be viewed as an infinite dimensional Markov process with state space E .

(2.30) A class of examples. Let γ be some finite measure on I and define $E := L^2(\gamma)$. For each $i \in I$ let b_i be a function on $E \subseteq R^I$ which is locally Lipschitz in each coordinate and satisfies the following conditions:

$$(2.31) \qquad \sup|b_i(0)| < \infty\ ;$$

(2.32) There is a constant K such that

$$(x_i - y_i)(b_i(y)) \le K(x_i - y_i)^2$$

for any $i \in I$, and for $x,y \in E$ with $x = y$ off i;

(2.33) there is a continuous linear map $C = (C_{ij}) : L^1(\gamma) \to L^1(\gamma)$ such that

$$(b_i(x) - b_i(y))^2 \le \sum_j C_{ij}(x_j - y_j)^2$$

for any $i \in I$, and for $x,y \in E$ with $x_i = y_i$.

These conditions imply that, for a given collection $(W_i)_{i\in I}$ of independent Wiener processes, the stochastic differential equation (2.29) has a unique strong solution; cf. [ShSh], [DR]. If b is an E-valued Lipschitz continuous function on the Hilbert space $E = L^2(\gamma)$, then one can use general results on stochastic differential equations in Hilbert space; cf., c.g., [Mé]. But in view of applications in Statistical Mechanics as in (2.55) below, this assumption would be too restrictive. For a discussion of (2.29) in terms of stochastic differential equations in Hilbert space with "discontinuous drift" see [LR1,2].

In general, (2.29) means that $X(t)$ $(0 \le t \le 1)$ is a Markov process whose distribution P is "locally" absolutely continuous with respect to the model for an infinite dimen-sional Brownian motion considered in Section 1. This can be made precise in various ways, by a suitable extension of (1.3) to the infinite dimensional case. For our purpose, we want to make sure that the coordinate process $X(t)$ $(0 \le t \le 1)$ is a diffusion process of the form (2.29) not only under P , but also under the time reversed measure

(2.34) $$\hat{P} = P \circ R$$

where R denotes the pathwise time reversal on Ω defined by $X(t,R(\omega)) = X(1-t,\omega)$. In contrast to the finite dimensional case, this is not true in general. The point is that in infinite dimensions the present state $X(1,\omega)$ at time $t = 1$ may contain so much information that, looking backwards, the semimartingale property of the coordinate processes may be lost:

(2.35) Example. Take $I = \{0\} \cup \{(k,l) ; k,l = 1,2,\ldots\}$, $b_0 \equiv 0$ and $b_i(x,t) = f_k(t)x_0$ for $i = (k,l)$ where f_k $(k=1,2,\ldots)$ is a orthonormal basis in $L^2[0,1]$. For a given collection W_i $(i \in I)$ of independent Brownian motions, equation (2.29) has a unique strong solution given by $X_0 = W_0$ and

$$X_i(t) = W_i(t) + \int_0^t f_k(s)X_0(s)ds \tag{2.36}$$

for $i = (k,l)$. Let P be the corresponding measure on $C[0,1]^I$. By the law of large numbers,

$$\lim_{n\uparrow\infty} \frac{1}{n} \sum_{l=1}^{n} X_{(k,l)}(1) = \int_0^1 f_k(s)X_0(s)ds \tag{2.37}$$

P-almost surely for any $k \geq 1$. This shows that the whole path X_0 can be reconstructed if we know the state $X(1)$ at time $t=1$. In particular, X_0 cannot be a semimartingale with respect to $(\underline{\underline{F}}_t)$ and $\hat{P}$.

The loss of the semimartingale property in (2.35) is due to the long range interaction between site 0 and the sites $i \neq 0$. Let us now formulate an entropy condition which will exclude this effect. For $i \in I$ we introduce the σ-fields

$$\underline{\underline{F}}_t^i := \sigma(X(0) , X_j(s) ; j \neq i , 0 \leq s \leq t)$$

which contain the information on the initial state $X(0)$ and on the behavior of the coordinate processes X_j $(j \neq i)$ up to time t. Let P^i denote the probability measure on $(\Omega,\underline{\underline{F}}_1)$ which coincides with P on $\underline{\underline{F}}_1^i$ and makes $X^i(t) - X^i(0)$ $(0 \leq t \leq 1)$ a Wiener process which is independent of $\underline{\underline{F}}_1^i$; this corresponds to a decoupling of the i-th coordinate process.

(2.38) Definition. We say that P has locally finite entropy if

$$H(P;P^i) < \infty \quad (i \in I) , \tag{2.39}$$

where the entropy is computed on $\underline{\underline{F}}_1$. If (2.40) holds on $\underline{\underline{F}}_t$ then we say that P has locally finite entropy up to time t.

To (2.39) corresponds the following infinite dimensional analogue of remark (1.3) on the Girsanov transformation.

(2.40) Proposition. If P has locally finite entropy then there exist adapted processes b_i $(i \in I)$ such that

$$(2.41) \qquad W_i(t) := X_i(t) - X_i(0) - \int_0^t b_i(s)ds \qquad (i \in I)$$

defines a collection W_i $(i \in I)$ of independent Wiener processes under P. These drift processes satisfy the finite energy condition

$$(2.42) \qquad E[\int_0^1 b_i^2(s)ds] < \infty \qquad (i \in I)$$

and the locality condition

$$(2.43) \qquad E[\sum_{j \neq i} \int_0^1 (b_j(s) - E[b_j(s)|\underline{\underline{F}}^i])^2 ds] < \infty \qquad (i \in I)$$

Proof. For a fixed $i \in I$, the process $X_i(t) - X_i(0)$ $(0 \leq t \leq 1)$ is a Wiener process under P^i, and so the representation (2.41) and the finite energy condition (2.42) follows from (1.3), 2). The point is to show that these Wiener processes are independent, and that the locality condition (2.43) is satisfied; see [FW] for the details.

(2.44) Remarks. 1) If we would assume "globally" finite entropy, i.e.,

$$(2.45) \qquad H(P;P^*) < \infty$$

where P^* is the model for an infinite dimensional Brownian motion considered in Section 1, then the sum of the energy terms in (2.42) would converge; cf. [HW]. But (2.45) would be too strong for the class of examples introduces in (2.30).

2) The finite energy condition (2.42) implies that the drift processes can be computed as stochastic forward derivatives, i.e.,

$$(2.46) \qquad b_i(t) = \lim_{h \downarrow 0} \frac{1}{h} E[X_i(t+h) - X_i(t) | \underline{\underline{F}}_t]$$

in $L^2(P)$ for any $i \in I$, due to (1.3), Condition (2.43) says that the interaction is "local in the active sense": for a given $i \in I$, the drift of most coordinates $j \neq i$ is influenced only mildly by the behavior of the i-th coordinate.

3) Suppose that under P the coordinate process is the unique strong solution of the stochastic differential equation (2.29), where the drift functions satisfy the assumptions in (2.30). If, in addition,

$$|b^i(x)| \leq \text{const}(1+||x||^p)$$

for some $p \geq 1$, and if the initial distribution μ_o satisfies

$$\int ||x||^{2p} \mu_o(dx) < \infty ,$$

then P has locally finite entropy; cf. [FW].

4) The process in (2.35) is not of locally finite entropy. In fact, let $\bar{X}_o$ denote the process defined P-a.s. by the coordinates X_i $(i \neq 0)$ through equations (2.37). Since $\bar{X}_o = X_o$ P-a.s., $\bar{X}_o$ is a Brownian motion under P, hence under P^o since $P^o = P$ on $\underline{\underline{F}}^o$. But under P^o the two Brownian motions $\bar{X}_o$ and X_o are independent, and so P^o and P are singular to each other.

From now on we assume that $X(t)$ is a Markov process under P. We want to describe the structure of the time reversed process $X \circ R$ or, equivalently, the structure of the coordinate process X under $\hat{P} = P \circ R$.

By a straightforward extension of the argument in (2.4) to the infinite dimensional case, we see that locally finite entropy is essentially preserved under time reversal:

(2.47) <u>Lemma.</u> If P has locally finite entropy then $\hat{P}$ has locally finite entropy up to any time $t < 1$.

Since $\hat{P}$ is again a Markov process, (2.40) shows that X is again described by a stochastic differential equation of the form

$$(2.48) \qquad dX_i = d\hat{W}_i + \hat{b}_i(X(t),t)dt \quad (i \in I)$$

and that the drift functions $\hat{b}_i$ $(i \in I)$ can be computed as stochastic backward derivates in the manner of (2.46) resp. (2.7). This leads to the following infinite dimensional version of the classical duality equation (2.13).

Let μ_t denote the distribution of $X(t)$ under P, and let $\mu_{t,i}(\cdot|y)$ denote the conditional distribution of $X_i(t)$ under μ_t, given the outside configuration $y = (x_j)_{j \neq i}$. Condition (2.39) implies that $\mu_{t,i}$ is in fact absolutely continuous with some density function $\rho_{t,i}(x_i|y)$, and that these conditional densities have the following regularity properties:

(2.49) <u>Theorem (FW).</u> Suppose that P has locally finite entropy. Then, for any $i \in I$, for almost all $t \in [0,1]$ and for μ_t-almost all y, the conditional density $\rho_{t,i}(\cdot|y)$ is an absolutely continuous function, and its derivative is given by

$$(2.50) \qquad \nabla_i \rho_{t,i}(x_i|y) = \rho_{t,i}(x_i|y)\,(b_i(x,t) + \hat{b}_i(x,1-t))$$

for almost all $x_i \in R$.

<u>Proof.</u> We use the same argument as in (2.8), based on the computation of drifts as stochastic forward and backward derivatives. Taking $F = f(X_i(t))G$ with $G = g(X_j(t);\ j \in J)$ for finite $J \subseteq I - \{i\}$ in (2.10), we obtain

$$E[(b_i(X(t),t) + \hat{b}_i(X(t),1-t))\ f(X_i)\ G] = -E[\nabla_i f(X(t))\ G]\ .$$

This implies (2.5), as in the proof of (2.8).

(2.51) <u>Remark.</u> The crucial technical point is to guarantee a priori that the coordinate processes X_i $(i \in I)$ are semimartingales under the time reversed measure $\hat{P}$; then the derivation of the duality equation (2.50) is straightforward. It is clear that X_i is a semimartingale under $\hat{P}$ with respect to any "local" filtration $(\underline{\underline{F}}_t^J)$ generated by X_j $(j \in J)$ for some finite J. The problem is to control the limiting behavior as $J \uparrow I$, and this was our reason to introduce the condition of locally finite entropy.

(2.49) shows that, in the infinite dimensional case, the duality equation between forward drift, backward drift and the distribution μ_t at time t does not involve the distribution itself, only its local specification in the sense of Ch. I. Consider, in particular, the time homogeneous case $b_i(x,t) = b_i(x)$ and suppose that μ is an invariant measure of the process. Tn (2.50) shows that μ is a Gibbs measure in the sense that

$$\nabla_i \log \rho_i(x_i|y) = b_i(x_i,y) + \hat{b}_i(x_i,y) .$$

(2.52) Remark. Any locality properties of the conditional densities $\rho_i(x_i|\cdot)$ depend on the locality properties both of b and $\hat{b}$. Suppose, for example, that b has a spatial Markov property:

(2.53) $$x = y \text{ on } N(i) \Rightarrow b_i(x) = b_i(y)$$

for some finite neighbourhood $N(i)$ of i. Then the invariant measure μ is a Markov field if and only if also the dual drift $\hat{b}$ has a Markov property of the form (2.53). Note that we have always locality in the the extended sense of (2.43), both in the forward and in the backward direction.

Now suppose that the invariant measure μ resp. the stationary process μ resp. the stationary process P is reversible, i.e., $P = \hat{P}$. This implies in particular

$$b_i = \hat{b}_i \quad (i \in I) ,$$

and so μ is a Gibbs measure with respect to b, i.e.,

(2.54) $$\rho_i(\cdot|y) = z_i(y)^{-1} \exp[2 \int b_i(\cdot|y)] .$$

(2.55) Remark. Suppose that b is given by a pair potential (U_V), i.e., b_i is of the form

$$b_i(x) = -\frac{1}{2} \nabla_i (U_i(x_i) + \sum_{j \neq i} U_{ij}(x_i, x_j)) .$$

Then (2.54) means that μ is a Gibbs measure with respect to (U_V) in the sense of Ch. I.3. We refer to [ShSh] for conditions on (U_V) which guarantee that b fulfills the conditions in (2.30). If $|U_i'|$ is of polynomial growth then the condition in (2.44,3) is also satisfied, and so the preceding argument does apply.

Let us now suppose that, conversely, μ is an invariant measure which satisfies (2.54). Then (2.51) implies that both P and $\hat{P}$ are governed by the same stochastic differential equation (2.29). This is equivalent to reversibility of μ resp. of P if we assume uniqueness of solutions for (2.29):

(2.56) <u>Corollary.</u> Suppose that the infinite dimensional stochastic differential equation (2.29) has a unique weak solution on $E \subseteq R^I$, and that μ is an invariant measure concentrated on E. Then μ is reversible if and only if μ is a Gibbs measure in the sense of (2.54).

(2.57) <u>Remark.</u> 1) The characterization of reversible invariant measures of an interacting particle system as Gibbs measures has been studied in various different contexts; cf., e.g., [Li], [Lan], [HS1,2], [Fr], and in our present case [DR]. A Gibbsian description of non-reversible invariant measures appears, e.g., in [Kü4].

2) Under the assumptions in (2.56), Fukushima and Stroock [FS] used the following approach. For a Gibbs measure μ, (2.54) implies

$$\int fLg\,d\mu = \int gLf\,d\mu \tag{2.58}$$

for smooth functions f and g depending on finitely many coordinates, where

$$Lf = \sum \left(\frac{1}{2} \frac{\partial^2}{\partial x_i^2} + b_i \frac{\partial}{\partial x_i}\right) f$$

is the generator associated to (2.29). The problem is to show that (2.58) implies reversibility. Under additional regularity and compactness assumptions, Fukushima's reconstruction theorem implies that there is a reversible Markov process Q associated to L resp. to the corresponding Dirichlet form [Fu1]. But under the uniqueness assumption, Q must coincide with the measure P induced by μ, and this shows that P resp. μ is reversible. For the general non-reversible case (2.49), this Dirichlet space technique does not seem to be available. This was another reason to introduce the entropy condition.

2.3. Large deviations and Martin boundary: an infinite dimensional example

Let us return to the situation in Section 1 where

$$P = \prod_{i \in I} P_i$$

is the product of Wiener measures with initinal value 0. For a large deviation of the empirical field described by a subset A of $M(\Omega)$, the results of Ch. I imply

$$(2.58') \qquad P[|V_n|^{-1} \sum_{i \in V_n} \delta_{\theta_i \omega} \in A] \sim \exp[-\inf_{Q \in A \cap M_s(\Omega)} h(Q;P)]$$

Now suppose that, in analogy to Section 1.2, we are interested in a large deviation of the empirical field at the terminal time $t = 1$. Then (2.58') reduces to

$$(2.59) \qquad P[|V_n|^{-1} \sum_{i \in V_n} \delta_{\theta_i \omega(1)} \in B] \sim \exp[-\inf_{\nu \in B \cap M_s(E)} h(\nu;\mu)],$$

where B is a set of probability measures on $E = R^I$ and $\mu = \prod_{i \in I} N(0,1)$ is the marginal distribution of P at atime 1.

In fact, (2.59) follows also directly from I.(4.1), applied to a sequence of independent random variables with distribution $N(0,1)$. But in view of the behavior of the whole process (X_i) it is of interest, as in Section 1.2, to describe the measure $Q \in M_s(\Omega)$ which minimizes $h(Q;P)$ under the constraint that the marginal distribution at time 1 is given by $\nu \in M_s(E)$.

For $\eta = (\eta_i)_{i\in I} \in E$, let P^η denote the infinite dimensional Brownian bridge

$$P^\eta = \prod_{i\in I} P^{\eta_i}$$

where P^{η_i} is the measure on $C[0,1]$ corresponding to the one-dimensional Brownian bridge from 0 to η_i . Then

$$P^\nu = \int P^\eta \, \nu(d\eta) \tag{2.60}$$

may be viewed as infinite dimensional Brownian motion conditioned to have distribution ν at time 1 . We have

$$P_V^\nu = \int \prod_{i\in V} P^{\eta_i} \, \nu(d\eta) \quad ,$$

hence $H_V(P^\nu;P) = H_V(\nu;\mu)$ and

$$h(P^\nu;P) = h(\nu;\mu) \ .$$

Conversely, any $Q \in M_s(\Omega)$ with marginal ν at time 1 satisfies $h(Q;P) \geq h(\nu;\mu)$, and equality holds if and only if $Q = P^\nu$.

In contrast to Section 1.2, the minimizing measure $Q = P^\nu$ is now an interacting diffusion process, no longer a collection of independent diffusions, each described as an h-path process in the sense of (1.15). Let us use time reversal in order to derive the corresponding infinite dimensional stochastic differential equation. In fact, the structure of the time reversed process $\hat{Q}$ is clear: it is an infinite dimensional Brownian motion conditioned to go to $0 \in E$ at time 1 and starting with initial distribution ν . Thus, the time reversed drift $\hat{b} = (\hat{b}_i)$ is given by

$$\hat{b}_i(x,t) = -\frac{x_i}{1-t} \qquad (i \in I) \ .$$

Applying Theorem (2.49) we obtain the following

(2.61) Corollary. Under the minimizing measure $Q = P^{\nu}$, the process $X(t)$ $(0 \le t \le 1)$ satisfies an infite dimensional stochastic differential equation

$$dX_i = dW_i + b_i(X(t),t)dt \quad (i \in I) \tag{2.62}$$

as in (2.29), and the drift is given by

$$b_i(x,t) = \frac{x_i}{t} + \nabla_i \log \rho_{i,t}(x_i | x_j(j \ne i)) \quad (i \in I) \tag{2.63}$$

where $(\rho_{i,t})$ is the collection of conditional densities of $\nu_t = \int \nu(d\eta)P^{\eta} \circ X(t)^{-1}$.

Note that the process is not yet uniquely determined by (2.62) and (2.63) if there is a phase transition, i.e., if ν does not determine uniquely the local specification of ν_t $(0 \le t \le 1)$; cf. [FW] for an explicit example.

(2.64) Remark. $E = R^I$ may be regarded as the Martin boundary associated to P. If fact, (2.60) is an integral representation for the class of all probability measures Q on $C[0,1]^I$ with given conditional probabilities

$$Q[\cdot | X(s); t \le s < 1] = P[\cdot | X(s); t \le s < 1] \quad (0 < t < 1).$$

These conditional probabilities, defined consistently as Brownian bridges leading from $0 \in E$ to $X(t) \in E$ at time t, form a local specification in the sense of Ch. I. The Brownian bridges P^{η} $(\eta \in E)$ are the extremal points in the class of all probability measures compatible with this local specification, and the parametrizing space E may be called the Martin boundary; cf. [Dy], [Fö1]. In the finite dimensional case, each measure P^{ν} corresponds to a space-time harmonic function $h(x,t)$ as in (1.14), and (2.60) translates into an integral representation of space-time harmonic functions in terms of the classical Martin boundary R^d of space-time Brownian motion $(X(t),t)$ on $R^d \times [0,1)$. But in the infinite dimensional case this correspondence between functions and conditional processes is lost. The reason is that, other than in classical duality theory which relies on the existence of a reference measure [Dy], the measures P^{ν}

are in general not absolutely continuous with respect to P on $\underline{\underline{F}}_t = \sigma(X(s);\ s \le t)$. In particular, P^{ν} cannot be described as an h-path process where the space-time harmonic function $h(x,t)$ is given by the Radon-Nikodym densities

$$h(X(t),t) = \frac{dP^{\nu}}{dP}\Big|_{\underline{\underline{F}}_t} \qquad (0 \le t < 1) .$$

Thus, for a description of P^{ν} in the infinite dimensional case, we need an alternative to (1.15), and (2.61) shows that time reversal can be used for this purpose.

3. Infinite dimensional diffusions as Gibbs measures

Let P be the distribution of an infinite dimensional diffusion process

$$dX_i = dW_i + b_i(X)dt \quad (i \in I) . \tag{3.1}$$

with time-homogeneous drift. Since P is a probability measure on $C[0,1]^I$, it can be viewed as a random field with state space $S = C[0,1]$ at each site $i \in I$. In the case of an infinite dimensional Brownian motion, we have already taken this point of view in Section 1. But if we want to apply the random field techniques of Ch. I to the general case (3.1) of an interacting diffusion then it is important to derive its local specification, i.e., the conditional distribution of X_i on $C[0,1]$ given all the other coordinates $X_j \in C[0,1]$ $(j \neq i)$. This problem has been studied by J.D. Deuschel, and in this section we give an introduction to the results in [De1,2,3].

3.1 The local specification of an infinite dimensional diffusion

In order to simplify the exposition let us assume that the initial distribution is concentrated on some point $x \in R^I$. In that case, the σ-field generated by $Y = (X_j)_{j \neq i}$ can be identified with the σ-field $\underline{\underline{F}}^i$ introduced in Section 2.2. The conditional distribution $P_i(dX_i|\underline{\underline{F}}^i)$ of X_i with respect to $\underline{\underline{F}}^i$ is of the form

$$\pi_i(dX_i|Y) = c_i(Y)^{-1}(dP/dP^i)(X)\ P^*(dX_i) \tag{3.2}$$

for $X = (X_i, Y)$, where P^* denotes Wiener measure on $C[0,1]$ and where P^i is the decoupling of the i-th coordinate used in (2.38). The density dP/dP^i could be computed in the manner of (2.4); cf. the proof of (2.23) in [FW].

In order to obtain a more explicit description of $\pi_i(dX_i|Y)$, we assume that the drift is given by a smooth pair potential of finite range as in (2.55):

$$b_i(x) = \nabla_i E_i(x) \tag{3.3}$$

with

$$E_i(x) = U_i(x_i) + \sum_{j\neq i} U_{ij}(x_i,x_j)$$

where U_i and U_{ij} are smooth symmetric functions with bounded derivatives up to order 3, and $U_{ij} = 0 \quad (j \notin N(i))$ for some finite neighbourhood $N(i)$ of i. For a fixed $i \in I$, we can now use the following decoupling Q^i of the i-th coordinate. Define

$$(3.4) \qquad b_{ik} = b_k - \nabla_k E_i \qquad (k \in I)$$

so that $b_{ii} = 0$ and $b_{ik} = b_k$ for $k \notin N(i) \cup \{i\}$. Let Q^i denote the distribution of the infinite dimensional diffusion

$$(3.5) \qquad dX_k = dW_{ik} + b_{ik}\, dt \qquad (k \in I)$$

where $(W_{ik})_{k\in I}$ is a collection of independent Brownian motions. Under Q^i, the i-th coordinate X_i is a Wiener process which is independent of $\underline{\underline{F}}^i$, and so we can write

$$(3.6) \qquad \pi_i(dX_i|Y) = Z_i(Y)^{-1}(dP/dQ^i)(X)\, P^*(dX_i)$$

with $X = (X_i,Y)$ and some normalizing factor $Z_i(Y)$. In order to compute the density dP/dQ^i more explicitey, define

$$(3.7) \qquad g_i(x) = \sum_{k\in N(i)\cup\{i\}} b_{ik}(x)\nabla_k E_i(x) + \frac{1}{2}\nabla_k^2 E_i(x) + \frac{1}{2}(\nabla_k E_i(x))^2 .$$

Note that g_i may be regarded as a smooth function on $R^{N^2(i)\cup\{i\}}$ where

$$N^2(i) = N(i) \cup \bigcup_{k\in N(i)} (N(k)-\{i\})$$

denotes the second order neighbourhood of i.

(3.8) <u>Proposition.</u> The conditional distribution $\pi_i(dX_i|Y)$ is given by the conditional density

$$(3.9) \qquad Z_i(Y)^{-1} \exp[\Phi_i(X_i,Y)]$$

with respect to Wiener measure, where

$$(3.10) \qquad \Phi_i(X) = E_i(X(1)) - E_i(X(0)) - \int_0^1 g_i(X(s))ds .$$

Viewed as a function on $(C[0,1])^{N^2(i)\cup\{i\}}$, Φ_i is Fréchet differentiable, and the partial derivatives $D_k\Phi_i$ $(k \in N^2(i) \cup \{i\})$, viewed as signed measures on $[0,1]$, are bounded in total variation by

$$(3.11) \qquad ||D_k\Phi_i(X)|| \leq 2||\nabla_k E_i||_\infty + ||\nabla_k g_i||_\infty .$$

<u>Proof.</u> We have

$$(dP/dQ^i)(X) = \exp[\sum_{k\in N(i)\cup\{i\}} \int (b_k - b_{ik})dX_k - \frac{1}{2}\int (b_k^2 - b_{ik}^2)dt]$$

with

$$\frac{1}{2}(b_k^2 - b_{ik}^2) = \frac{1}{2}(\nabla_k E_i)^2 + b_{ik}\nabla_k E_i .$$

By Itô's formula,

$$\sum_k \int_0^1 (b_k - b_{ik})dX_k = \sum_k \int_0^1 \nabla_k E_i dX_k$$

$$= E_i(X(1)) - E_i(X(0)) - \frac{1}{2}\sum_k \int_0^1 \nabla_k^2 E_i(X(s))ds ,$$

and this implies (3.9). Our assumptions on $(b_k)_{k\in I}$ imply that Φ_i, viewed as a function on S^J with $S = C[0,1]$ and $J = N^2(i) \cup \{i\}$, is Fréchet differentiable. Its partial derivatives, viewed as signed measures on $[0,1]$, are of the form

$$D_k\Phi_i(X)(dt) = \nabla_k E_i(X(1))\delta_1(dt) - \nabla_k E_i(X(0))\delta_0(dt)$$
$$- \nabla_k g_i(X(t))(dt) ,$$

and this implies (3.11).

(3.12) Remark. 1) Since $\pi_i(\cdot|Y)$ is absolutely continuous with respect to Wiener measure P^* on $C[0,1]$, the coordinate process X_i is of the form

$$dX_i = dW_i^Y + b_i^Y(t)dt$$

with some conditional drift process $b_i^Y = b_i^Y(t)$ $(0 \le t \le 1)$ under $\pi_i(\cdot|Y)$. As in [BM], the conditional drift b_i^Y can be computed by Clark's formula [Cl]: For $\underline{\underline{G}}_t^i = \sigma(X^i(s); s \le t)$ and $\underline{\underline{F}}_t^i = \underline{\underline{F}}^i \cup \underline{\underline{G}}_t^i$, the martingale

$$M_i(t) = E_{Q^i}[\exp \Phi_i(X)|\underline{\underline{F}}_t^i]\,(X_i,Y)$$

$$= E^*[\exp \Phi_i(\cdot,Y)|\underline{\underline{G}}_t^i](X_i)$$

is of the form

$$M_i(t) = M_i(0) + \int_0^t H_i^Y(s)dX_i(s)$$

with

$$H_i^Y(s) = E^*[D_i\Phi_i(\cdot,Y)(s,1] \exp \Phi_i(\cdot,Y)|\underline{\underline{G}}_s^i] .$$

The drift is now given by

$$b_i^Y(t) = H_i^Y(t)/M_i(t) ; \tag{3.13}$$

cf., e.g., [DM]. This implies the uniform bound

$$||b_i^Y(t)||_\infty \le ||D_i\Phi_i||_\infty := \sup_X ||D_i\Phi_i(X,\cdot)|| \tag{3.14}$$

in particular,

$$\sigma_i^2 := \sup_Y \inf_{s\in S} \int ||X_i - s||^2 \pi_i(dX_i|Y) < \infty \tag{3.15}$$

where $||\cdot||$ denotes the supremum norm on $S = C[0,1]$. Due to the special additive structure of $\Phi_i(X)$ and the Markov property of P^*,

(3.13) also implies that, for a fixed Y, the drift process b_i^Y is of the form

(3.16) $$b_i^Y(\omega,t) = b_i^Y(X_i(\omega,t),t) ,$$

i.e., the conditional diffusion is actually a Markov process.

2) As an alternative to (3.13), the conditional drift function in (3.16) can also be computed as

(3.17) $$b_i^Y(x_i,t) = \nabla_i G_i^Y(x_i,t)$$

where G_i^Y is a solution of the partial differential equation

(3.18) $$\nabla_t G + \frac{1}{2}\nabla_i^2 G + \frac{1}{2}(\nabla_i G)^2 = g_i$$

with boundary condition

(3.19) $$G(x_i,1) = E_i(x_i) .$$

In fact, Itô's formula and (3.19) imply

$$E_i(X(1)) = G_i^Y(X(0),0) + \int_0^1 \nabla_i G_i^Y(X_i(t),t)dX_i(t)$$

$$+ \int_0^1 [\frac{1}{2}\nabla_i^2 G_i^Y + \nabla_t G_i^Y](X_i(t),t)dt ,$$

and so (3.18) shows that the densitiy in (3.9) is given by

$$\exp[\int_0^1 \nabla_i G_i^Y dX_i - \frac{1}{2}\int_0^1 (\nabla_i G_i^Y)^2 dt] .$$

This implies (3.17). Bounds and smoothness properties of the conditional drift can now be obtained by means of the Feynman-Kac formula. Moreover, the representation (3.17) leads to a characterization of the conditional drift as the solution of a stochastic control problem. Cf. [De2] for details.

3.2 Applying the contraction technique

Let us now look at the system of conditional probabilities (π_k) from the point of view of Dobrushin's contraction technique in Section I.2. For two probability measures μ and ν on $S = C[0,1]$, let

$$R(\mu,\nu) = \sup \frac{|\int f d\mu - \int f d\nu|}{\delta(f)}$$

denote the Vasserstein distance as in Ch. I (2.18); the Lipschitz constant $\delta(f)$ is defined with respect to the supremum norm on $C[0,1]$. For $\eta \in \Omega = S^I$ we use the notation

$$\pi_k(\cdot|\eta) = P_k(\cdot|\underline{\underline{F}}^k)(\eta) .$$

Recall the definition of σ_k^2 in (3.15), and put

$$||D_i\Phi_k||_\infty = \sup_{\omega\in\Omega} ||D_i\Phi_k(\omega,\cdot)||$$

as in (3.14).

(3.20) Theorem. For ω and η in $\Omega = S^I$, the Vasserstein distance between the conditional diffusions $\pi_k(\cdot|\omega)$ and $\pi_k(\cdot|\eta)$ satisfies

$$(3.21) \qquad R(\pi_k(\cdot|\omega),\pi_k(\cdot|\eta)) \le \sum_{i\in N^2(k)} c_{ik} \, ||\omega(i)-\eta(i)||$$

with

$$(3.22) \qquad c_{ik} \le \sigma_k \, ||D_i\Phi_k||_\infty$$

Proof. Take $k \in I$, $i \in N(k)$ and $\omega, \eta \in S^I$ with $w = \eta$ off i. Let f be a Lipschitz function on S with Lipschitz constant $\delta(f)$.
1) For $\alpha \in [0,1]$ define $\eta_\alpha = \omega + \alpha(\eta-\omega)$. Using the notation $\mu_\alpha = \pi_k(\cdot|\eta_\alpha)$ and

$$Z_\alpha = Z_k(\eta_\alpha) = \int \exp[\Phi_k(\cdot,\eta_\alpha)dP^* ,$$

we have

$$\frac{d}{d\alpha}\log Z_\alpha = \int \frac{d}{d\alpha}\Phi_k(\cdot,\eta_\alpha)d\mu_\alpha ,$$

hence

$$\frac{d}{d\alpha}\int f d\mu_\alpha = \frac{d}{d\alpha}(Z_\alpha^{-1}\int f \exp[\Phi_k(\cdot,\eta_\alpha)]dP^*) =$$

$$= \int f \left[\frac{d}{d\alpha}\Phi_k(\cdot,\eta_\alpha) - \frac{d}{d\alpha}\log Z_\alpha\right]d\mu_\alpha$$

$$= \int [f-f(s)]\left[\frac{d}{d\alpha}\Phi_k(\cdot,\eta_\alpha) - \frac{d}{d\alpha}\log Z_\alpha\right]d\mu_\alpha$$

for any $s \in S = C[0,1]$. This implies

$$\left|\frac{d}{d\alpha}\int f d\mu_\alpha\right|^2 \leq \delta(f)^2\sigma_k^2 \operatorname{var}_{\mu_\alpha}\left(\frac{d}{d\alpha}\Phi_k(\cdot,\eta_\alpha)\right) .$$

But

$$\left|\frac{d}{d\alpha}\Phi_k(\cdot,\eta_\alpha)\right| \leq ||D_i\Phi_k(\cdot,\eta_\alpha)||\ ||\omega(i)-\eta(i)|| ,$$

and so we get the estimate

$$\left|\frac{d}{d\alpha}\int f d\mu_\alpha\right| \leq \delta(f)\sigma_k\ ||D_i\Phi_k||_\infty ||\omega(i)-\eta(i)|| . \tag{3.23}$$

2) By (3.23)

$$\left|\int f d\pi_k(\cdot|\omega) - \int f d\pi_k(\cdot|\eta)\right| = \left|\int_0^1\left(\frac{d}{d\alpha}\int f d\mu_\alpha\right)d\alpha\right|$$

$$\leq \delta(f)\sigma_k ||D_i\Phi_k||_\infty ||\omega(i)-\eta(i)|| .$$

This implies (3.21) and (3.22).

By (3.8), (3.15) and (3.20), the distribution P of the diffusion (3.1) can be viewed as a tempered Gibbs measure on S^I in the sense of (2.17) in Ch. I. (3.9) and (3.10) show that this random field has a spatial Markov property. Due to (3.22) and (3.11), one can derive bounds on the pair potential which guarantee the Dobrushin condition

$$\sup_k \sum_i C_{ik} < 1 ; \tag{3.24}$$

cf. [De3] for an explicit computation. This allows to proceed as in Section 2 of Ch. I. In particular, P is the unique tempered Gibbs measure compatible with the system of conditional distributions (π_i).

Let us now consider the translation invariant case, with $I = Z^d$ and a spatially homogeneous pair potential. Then we have exponential decay of correlations in the sense of (2.24) in Ch. I. For any function f in the class $L(S^I)$ the central limit theorem in I (2.25) holds, i.e., the distribution of

$$S_n^*(f) = |V_n|^{-1/2} \sum_{k \in V_n} [f \circ \theta_k - \int f dP]$$

converges in law to a centered Gaussian random variable with variance

$$\sigma^2(f) = \sum_k \text{cov}(f, f \circ \theta_k) .$$

Following K. Itô [It1], we are led to weak convergence of the process $Y^{(n)}$ defined by

$$Y_t^{(n)}(\varphi) = S_n^* (\varphi \circ X(t)) \qquad (0 \le t \le 1)$$

for a suitable class S_∞ of test functions φ on R^I. Viewed as a process with values in the corresponding space S'_∞ of tempered distributions, $Y^{(n)}$ converges in law to a continuous S'_∞-valued process $Y = (Y_t)_{0 \le t \le 1}$ with variances

$$\text{var}(Y_t(\varphi)) = \sum_k \text{cov}(\varphi(X(t)), \varphi(X(t) \circ \theta_k)) .$$

Moreover, the evolution of this process can be described by a linear stochastic differential equation of the form

$$(3.25) \qquad Y_t(\varphi) = Y_0(\varphi) + \int_0^t Y_s(L\varphi) ds + \int_0^t dB_s(D\varphi)$$

where

$$L\varphi = \sum_k (b_k \nabla_k + \frac{1}{2} \nabla_k^2)\varphi$$

$$D\varphi = \sum_k \nabla_k \varphi \circ \theta_{-k} ,$$

and where $B = (B_t)_{0 \le t \le 1}$ is an S'_∞-valued Brownian motion with quadratic variation

$$\mathrm{var}(B_t(\varphi)) = \int_0^t E[\varphi^2(X(s))]ds .$$

We refer to [De3] for details, and also to [De5] for an alternative approach to (3.25) which does not pass through the description of P as a random field.

References

[AH] Albeverio,S ., Høegh-Kron, R.: Uniqueness and global Markov property for Euclidean fields and lattice systems. In: Quantum fields-algebras, processes (ed. Streit, L.) 303-330, Springer (1980)

[Az] Azencott, R.: Grandes deviations et applications. In: Ecole d'Eté de Probabilités de Saint-Flour VIII, Springer Lecture Notes Math. 774, 1-176 (1980)

[Be] Beurling, A.: An automorphism of product measures. Ann. Math. 72, 189-200 (1960)

[BJ] Bichteler, K., Jacod, J.: Calcul de Malliavin pour les diffusions avec sauts. Sém. Probabilités XVII, Lecture Notes Math. 986, Springer (1983)

[Bi] Bismut, J.M.: Martingales, the Malliavin Calculus and Hypolellipticity under general Hörmander's conditions. Z. Wahrscheinlichkeitstheorie verw. Geb. 56, 469-505 (1981)

[BM] Bismut, J.M., Michel, D.: Diffusions conditionelles I, II. J. Funct. Anal. 44, 174-211 (1981), 45, 274-292 (1982).

[BD] Blackwell, D. and Dubins, L.: Merging of opinions with increasing information. Ann. Math. Statist. 33, 882-886 (1962)

[Bo] Bolthausen, E.: On the central limit theorem for stationary mixing random fields. Ann. Prob. 10, 1047-1050 (1982)

[Ca] Cairoli, R.: Une inégalité pour martingales à indices multiples et ses applications. Sém. Prob. IV, Springer Lecture Notes in Math. 124, 1-27 (1970)

[Car] Carlen, E.A.: Conservative Diffusions: A constructive approach to Nelson's stochastic mechanics. Thesis, Princeton (1984)

[Cl] Clark, J.M.C.: The representation of functionals of Brownian motion by stochastic integrals. Ann. Math. Stat. 41, 1281-1295 (1970), 42, 1778 (1971)

[Co] Comets, F.: Grandes déviations pour des champs de Gibbs sur Z^d . C.R. Acad. Sc. Paris 303, sér. 1, n° 11, 511 (1986)

[Cs] Csiszár, I.: I-Divergence Geometry of Probability Distributions and Minimization Problems. Annals Prob. 3, No. 1, 146-158 (1975)

[DG] Dawson, D.A., Gärtner, J.: Long time fluctuations of weakly interacting diffusions. Preprint (1986).

[DM] Dellacherie, C., Meyer, P.A.: Probabilités et potential, Ch. V-VIII. Hermann (1980)

[De1] Deuschel, J.D.: Lissage de diffusion à dimension infinie et leurs propriétés en tout que mesure de Gibbs. Thèse ETH Zürich 7823 (1985)

[De2] Deuschel, J.D.: Non-linear Smoothing of Infinite-dimensional Diffusion Processes. Stochastics 19, 237-261 (1986)

[De3] Deuschel, J.D.: Infinite-dimensional diffusion processes as Gibbs measures on $C[0,1]^{\mathbb{I}}$. Preprint ETH Zürich (1986)

[De4] Deuschel, J.D.: Représentation du champ de fluctuation de diffusions indépendentes par le drap brownien. Sém. Prob. XXI, Springer LN Math. 1247, 428-433 (1987)

[De5] Deuschel, J.D.: A Central limit theorem for an infinite lattice system of interacting diffusion processes. Ann. Prob. (to appear)

[Do1] Dobrushin, R.L.: Description of a random field by means of conditional probabilities and the conditions governing its regularity. Theory Probab. Appl. 13, 197-224 (1968)

[Do2] Dobrushin, R.L.: Prescribing a system of random variables by conditional distributions. Theor. Probab. Appl. 15, 458-486 (1970).

[Do3] Dobrushin, R.L.: Central limit theorems for nonstationary Markov chains. Theory Probab. Appl. 1, Nr. 4, 329-383 (1956).

[DP] Dobrushin, R.L., Pecherski, E.A.: Uniqueness conditions for finitely dependent random fields. In: Coll. Math. Soc. J. Bolyai 27, Random Fields, 223-261, North-Holland (1981)

[DT] Dobrushin, R.L., Tirozzi, B.: The central limit theorem and the problem of equivalence of ensembles. Comm. Math. Phys. 54, 173-192 (1977)

[DV1] Donsker, M.D. and S.R.S. Varadhan: Asymptotic evaluation of certain Markov expectations for large time, III. Comm. Pure Appl. Math, 29, 389-461 (1976)

[DV2] Donsker, M.D. and S.R.S. Varadhan: Asymptotic evaluation of certain Markov expectations for large time IV, Comm. Pure Appl. Math. 36, 183-212 (1983)

[DR] Doss, H., Royer, G.: Processus de diffusion associés aus mesures de Gibbs. Z. Wahrscheinlichkeitstheorie 46, 125-158 (1979)

[DP1] Dubins, L., Pitman, J.W.: A divergent, two-parameter, bounded martingale. Proc. Amer. Math. Soc. 78, 414-416 (1980)

[Dy] Dynkin, E.B.: Sufficient statistics and extreme points. Ann. Probability 6, 705-730 (1978)

[El] Ellis, R.S.: Entropy, Large Deviation and Statistical Mechanics. Grundlehren 271, Springer (1975)

[Föl] Föllmer, H.: Phase transitions and Martin boundary, in: Sém. Prob. IX, Lecture Notes in Mathematics 465, 305-318, Springer (1975)

[Fö2] Föllmer, H.: On entropy and information gain in random fields. Z. Wahrsch. verw. Geb. 26, 207-217 (1973)

[Fö3] Föllmer, H.: On the global Markov property. In: Quantum fields, algebras, processes (ed. Streit, L.) 293-302, (1980)

[Fö4] Föllmer, H.: A covariance estimate for Gibbs measures. J. Funct. Anal. 46, 387-395 (1982)

[Fö5] Föllmer, H. Almost sure convergence of multiparameter martingales for Markov random fields. Ann. Probability 12, 133-140 (1984)

[Fö6] Föllmer, H.: Time reversal on Wiener space, Proceedings of the BiBoS-Symposium "Stochastic processes - Mathematics and Physics", Lecture Notes in Mathematics 1158, 119-129 (1986)

[FO] Föllmer, H, Orey, S.: Large deviations for the empirical field of a Gibbs measure. Ann. Prob. (to appear).

[FOr] Föllmer, H., Ort, M.: Large deviations and surface entropy Astérisque, Actes du Colloque Paul Lévy (to appear).

[FW] Föllmer, H., Wakolbinger, A.: Time reversal of infinite-dimensional diffusions. Stochastic Processes Appl. 22, 59-77 (1986)

[Fo] Fortet, R.: Résolution d'un système d'équations de M. Schrödinger. J. Math. Pures Appl. IX, 83-105 (1940)

[Fr1] Fritz, J.: Generalization of McMillan's theorem to random set functions. Studia Sci. Math. Hungar. 5, 369-394 (1970)

[Fr2] Fritz, J.: Stationary measures of stochastic gradient systems, Infinite lattice models. Z. Wahrscheinlich-keitsstheorie 59, 479-490 (1982)

[Fu1] Fukushima, M.: Dirichlet forms and Markov Processes. North Holland (1980)

[Fu2] Fukushima, M.: On absolute continuity of multidimensional symmetrizable diffusion. In Functional Analysis in Markov Processes, Springer LN in Math. 923, 146-176 (1982)

[FS] Fukushima, M., Stroock, D.W.: Reversibility of solutions to martingale problems. Preprint (1980)

[GG] Geman, S. and Geman, D.: Stochastic Relaxation, Gibbs Distributions and the Bayesian Restoration of Images. IEEE Trans. on Pattern Analysis and Machine Intelligence, Vol. 6, 721-741 (1984)

[Ge] Georgii, H.O.: Canonical Gibbs measures. Springer Lecture Notes Math. 760 (1979)

[Gi] Gidas, B.: Nonstationary Markov chains and the Convergence of the Annealing Algorithm. Journal of Stat. Physics 39, 73-130 (1985)

[Go] Goldstein, S.: Remarks on the global Markov property. Comm. Math. Phys. 74, 223-234 (1980)

[Gr1] Gross, L.: Decay of correlations in classical lattice models at high temperatures. Comm. Math. Phys. 68, 9-27 (1979)

[Gr2] Gross, L.: Thermodynamics, Statistical Mechanics, and Random Fields. In: Ecole d'Eté de Probabilités de Saint-Flour X, Springer Lecture Notes in Math. 929, 101-204 (1982)

[Gu] Gurevich, B.M.: A variational Characterization of one-dimensional Countable state Gibbs random fields. Z. Wahrscheinlichkeitstheorie verw. Geb. 68, 205-242 (1984)

[Ha] Hayek, B.: A tutorial survey of theory and applications of simulated annealing. In: Proc. 24th Conference of Decision and Control, Dec. 85 (to appear)

[HP] Haussmann, U., Pardoux, E.: Time reversal of diffusion processes. In: Proc. 4th IFIP Workshop on Stochastic Differential Equations, M. Métivier and E. Pardoux Eds., Lecture Notes in Control and Information Sciences, Springer (1985)

[HS] Holley, R., Stroock, D.W.: Applications of the stochastic Ising model to the Gibbs states. Comm. math. Phys. 48, 249-265 (1976)

[HS1] Holley, R., Stroock, D.W.: Diffusions on an infinite dimensional Torus. J. Funct. Anal. 42, 29-63 (1981)

[HS2] Holley, R., Stroock, D.W.: Logarithmic Sobolev inequalities and stochastic Ising models. Preprint M.I.T. (1986)

[HW] Hitsuda, M. Watanabe, H.: On stochastic integrals with respect to an infinite number of Brownian motions and its applications. In: Stochastic Differential Equations (ed. K. Itô), Kyoto (1976)

[It1] Itô, K.: Distribution-valued processes arising from independent Brownian motions, Math. Z. 182, 17-33 (1983)

[It2] Itô, K.: Foundations of stochastic differential equations in infinite dimensional spaces. CBMS-NSF Ser. in Appl. Math. 4 SIAM, Philadelphia (1984)

[IT] Iosifescu, M., Theodorescu, R.: Random processes and learning, Grundlehren der math. Wissenschaften, Bd. 150, Springer (1969)

[IW] Ikeda, N., Watanabe, S.: Stochastic differential equations and diffusion processes. North Holland (1981)

[Ja] Jamison, B.: Reciprocal processes, Z. Wahrscheinlichkeitstheorie verw. Geb. 30, 65-86 (1974)

[Ke] Kemperman, J.H.B.: On the optimum rate of transmitting information. In: Probability and Information theory, Springer Lecture Notes in Mathematics 89, 126-169 (1967)

[Ko] Kolmogorov, A.N.: Zur Umkehrbarkeit der statistischen Naturgesetze. Math. Ann. 113, 766-772 (1937)

[Kü1] Künsch, H.: Almost sure entropy and the variational principle for random fields with unbounded state space. Z. Wahrscheinlichkeitstheorie verw. Geb. 58, 69-85 (1981)

[Kü2] Künsch, H.: Thermodynamics and Statistical Analysis of Gaussian Random fields, Z. Wahrscheinlichkeitstheorie verw. Geb. 58, 407-421 (1981)

[Kü3] Künsch, H.: Decay of correlation under Dobrushin's uniqueness condition and its application. Comm. Math. Physics 84, 207-222 (1982)

[Kü4] Künsch, H.: Non Reversible Stationary Measures for Infinite Interacting Particle Systems. Z. Wahrscheinlichkeitstheorie verw. Geb. 66, 407-424 (1984)

[La] Lanford, O.E.: Entropy and Equilibrium states in classical statistical mechanics. Springer Lecture Notes in Physics 20, 1-113 (1973)

[LR] Lanford, O.E., Ruelle, D.: Observables at infinity and states with short range correlations in statistical mechanics. Comm. Math. Phys. 13, 194-215 (1969)

[Lan] Lang, R.: Unendlich-dimensionale Wienerprozesse mit Wechselwirkung. Z. Wahrscheinlichkeitstheorie verw. Geb. 39, 277-299 (1977)

[Li] Liggett, T.M.: Interacting Particle Systems. Springer (1985)

[LR1] Leha, G., Ritter, G.: On diffusion processes and their semigroups in Hilbert space with an application to interacting stochastic systems. Ann. Prob. 12, 1077-1112 (1984)

[LR2] Leha, G. Ritter, G.: On solutions to stochastic differential equations with discontinuous drift in Hilbert space. Math. Ann. 270, 109-123 (1985.

[LS] Liptser, R.S., Shiryaev, A.N.: Statistics of Random processes I. Springer (1977)

[Ma] Maz'ja, V.G.: Sobolev spaces. Springer (1985)

[Mé] Métivier, M.: Semimartingales. de Gruyter (1982)

[Mo] Moulin Ollagnier, J.: Ergodic Theory and Statistical Mechanics, Springer Lecture Notes in Math. 1115 (1985)

[Na] Nagasawa, M.: The Adjoint Process of a Diffusion with Reflecting Barrier, Kodai Math. Sem. Report 13, 235-248 (1961)

[Ne] Nelson, E.: Dynamical theories of Brownian motion. Princeton University Press (1967)

[NZ] Nguyen, X.X., Zessin, H.: Ergodic theorems for spatial processes. Z. Wahrscheinlichkeitstheorie verw. Geb. 48, 133-158 (1979)

[Ol] Olla, S.: Large deviations for Gibbs random fields. Preprint Rutyers University (1986)

[Or1] Orey, S.: Conditions for the absolute continuity of two diffusions. Trans. Amer. Math. Soc. 193, 413-426

[Or2] Orey, S.: Large deviations in ergodic theory, Seminar on Probability, Birkhäuser (1984)

[Ort] Ort, M.: Grosse Abweichungen und Oberflächenentropie. Diss. ETHZ (to appear)

[OW] Ornstein, D, Weiss, B.: The Shannon-Mc Millan-Breimann Theorem for a class of amenable groups. Isr. JM 44, 53-60 (1983)

[Pa] Pardoux, E.: Equations du filtrage non-linéaire, de la prédiction et du lissage. Stochastics 6, 193-231 (1982)

[Pi] Pirlot, M.: A strong variational principle for continuous spin systems. J. Appl. Prob. 17, 47-58 (1980)

[Pr] Preston, C.: Random Fields. Lecture Notes in Mathematics. Springer (1976)

[Ro] Royer, G.: Etude des champs euclidéens sur un réseau. J. Math. Pures Appl. 56, 455-478 (1977)

[Ru] Ruelle, D.: Thermodynamical Formalism. Encyclopedia of Math. and Appl. 5, Addison Wesley (1978)

[Schi] Schilder, M.: Some asymptotic formulae for Wiener integrals. Trans Amer. Math. Soc., 125, 63-85 (1966)

[Sch1] Schrödinger, E.: Ueber die Umkehrung der Naturgesetze, Sitzungsberichte Preuss. Akad. Wiss. Berlin, Phys. Math. 144, 144-153 (1931)

[Sch2] Schrödinger, E.: Sur la théorie relativiste de l'électron et l'interprétation de la mécanique quantique", Ann. Inst. H. Poincaré 2, 269-310 (1932)

[ShSh] Shiga, T., Shimizu, A.: Infinite dimensional stochastic differential equations and their applications. J. Math. Kyoto Univ. 20, 395-416 (1980)

[Scho] Schonmann, R.H.: Second order large deviation estimates for ferromagnetic systems in the pase coexistence region. Commun. Math. Phys. 112, 409-422 (1987)

[St] Stroock, D.W.: An introduction to the theory of large deviations. Springer (1984)

[Su] Sucheston, L.: On one-parameter proofs of almost sure convergence of multiparameter processes. Z. Wahrsch. verw. Gebiete 63, 43-50 (1982)

[Sug] Sugita, H.: Sobolev spaces of Wiener functionals and Malliavin's calculus. J. Math. Kyoto Univ. 25-1, 31-48 (1985)

[Te] Tempelman, A.A.: Specific characteristics and variational principle for homogeneous random fields. Z. Wahrscheinlichkeitstheorie verw. Geb. 65, 341-365 (1984)

[Th] Thouvenot, J.P.: Convergence en moyenne de l'information pour l'action de Z^2 . Z. Wahrscheinlichkeitstheorie verw. Geb. 24, 135-137 (1972)

[Va] Varadhan, S.R.S.: Large Deviations and Applications. SIAM Philadelphia (1984)

[Za] Zambrini, J.C.,: Schrödinger's stochastic variational dynamics. Preprint Univ. Bielefeld (1986)

STOCHASTIC MECHANICS AND RANDOM FIELDS

Edward NELSON

Originally published in: *Ecole d'Eté de Probabilités de Saint-Flour XV–XVII – 1985–87*, Lecture Notes in Mathematics, Vol. **1362**, 427–459, DOI: 10.1007/BFb0086184, © Springer-Verlag Berlin Heidelberg 1988, Reprint by Springer-Verlag Berlin Heidelberg 2012

E. NELSON : "STOCHASTIC MECHANICS AND RANDOM FIELDS"

This is intended as a guide to stochastic mechanics, and not as a systematic account. It is addressed to mathematicians with a background in probability theory; in particular, to those with some knowledge of Markov processes, martingales, and the Wiener process. It is by no means self-contained, for one of its purposes is to provide a guide to the literature.

Let us begin with some mathematical questions. There will be occasion later for historical remarks and a discussion of physical and philosophical issues.

I. Kinematics of Diffusion

1. The Wiener process and diffusions on $\mathbb{R}^n$

To fix some notation, recall that a *stochastic process indexed by* I is a function ξ from I to random variables on some probability space. Two stochastic processes ξ and η indexed by the same set I are *equivalent* in case they have the same finite joint distributions; that is, in case for all n and all $t_1, \ldots, t_n$ in I, the $\xi(t_i)$ and $\eta(t_i)$ are equidistributed. We use E to denote the expectation, and if $\mathcal{B}$ is a σ-algebra of measurable sets, we use $\mathrm{E}\{\ |\ \mathcal{B}\}$ to denote the conditional expectation with respect to $\mathcal{B}$.

A *difference process* is a stochastic process δ indexed by $\mathbb{R} \times \mathbb{R}$ such that for all t_1, t_2, and t_3 in $\mathbb{R}$ we have $\delta(t_1,t_2) + \delta(t_2,t_3) = \delta(t_1,t_3)$. Let δ be a difference process, choose t_0 in $\mathbb{R}$, choose a random variable $w(t_0)$ independent of the process δ, and define $w(t) = w(t_0)+\delta(t,t_0)$. Then w is a stochastic process indexed by $\mathbb{R}$ such that $\delta(t_1,t_2) = w(t_1)-w(t_2)$ for all t_1 and t_2 in $\mathbb{R}$. It is determined up to equivalence by specifying the *initial time* t_0 and the *initial distribution*, i.e., the probability distribution of the initial value $w(t_0)$. We shall write difference processes $\delta(t_1,t_2)$ simply as $w(t_1) - w(t_2)$, although this notation is abusive unless the initial time and initial distribution have been specified.

The one dimensional *Wiener difference process* w has independent increments that are Gaussian of mean 0 and variance the length of the interval; this description determines the process uniquely up to equivalence. The n dimensional Wiener difference process is the $\mathbb{R}^n$-valued process whose components w^i are independent one dimensional Wiener difference processes.

Let ξ be a stochastic process indexed by $\mathbb{R}$. For t in $\mathbb{R}$, let $\mathcal{P}_t$ be the σ-algebra generated by all $\xi(s)$ with $s \le t$ and let $\mathcal{F}_t$ be the σ-algebra generated by all $\xi(s)$ with $s \ge t$. Then $\mathcal{P}$ is a *forward filtration* (an increasing family of σ-algebras) and $\mathcal{F}$ is a *backward filtration* (a decreasing family of σ-algebras). Let $\mathcal{N}_t$ be the σ-algebra generated by $\xi(t)$. We use E_t as an abbreviation for $\mathrm{E}\{\ |\ \mathcal{N}_t\}$. We call $\mathcal{P}_t$ the *past*, $\mathcal{F}_t$ the *future*, and $\mathcal{N}_t$ the *present* at time t. Then ξ is a *Markov process* in case for all t the past and the future are conditionally independent given the present.

The theory of Markov processes is completely independent of a choice of the direction of time. This fact is of central importance in stochastic mechanics. The kinematics of diffusion is utterly symmetric with respect to the two directions of time (though it is not usually developed with this fact in mind), and the conservative dynamics that we shall develop likewise gives no preference to either direction of time.

We use dt as a strictly positive variable. If F is any function defined on $\mathbf{R}$, we define its *forward increment* by

$$d_+F(t) = F(t+dt) - F(t) \tag{1.1}$$

and its *backward increment* by

$$d_-F(t) = F(t) - F(t-dt). \tag{1.2}$$

An L^1 stochastic process ξ indexed by $\mathbf{R}$ is a *forward martingale* with respect to a forward filtration $\mathcal{P}$ in case

$$\mathrm{E}\{\xi(t) \mid \mathcal{P}_s\} = \xi(s), \qquad s \le t, \tag{1.3}$$

and a *backward martingale* with respect to a backward filtration $\mathcal{F}$ in case

$$\mathrm{E}\{\xi(t) \mid \mathcal{F}_s\} = \xi(s), \qquad s \ge t. \tag{1.4}$$

An L^1 difference process w is a *forward difference martingale* with respect to a forward filtration $\mathcal{P}$ in case

$$\mathrm{E}\{w(t) - w(s) \mid \mathcal{P}_s\} = 0, \qquad s \le t, \tag{1.5}$$

and a *backward difference martingale* with respect to a backward filtration $\mathcal{F}$ in case

$$\mathrm{E}\{w(t) - w(s) \mid \mathcal{F}_s\} = 0, \qquad s \ge t. \tag{1.6}$$

The Wiener process on $\mathbf{R}^n$ with an initial distribution at the initial time t_0 is a Markov process. Restricted to $t \ge t_0$ it is a forward martingale, and restricted to $t \le t_0$ it is a backward martingale.

Heuristically, if a particle moving according to the Wiener process is at x at the time $t \ge t_0$, then to find its position at an infinitesimal time dt later, choose a direction in $\mathbf{R}^n$ at random and move a distance $\sqrt{n\,dt}$ from x in that direction, so that each component $dw^i(t)$ of the Wiener increment is of mean 0 and variance dt; then repeat, choosing the directions independently each time. The paths will be very rough (non-differentiable) because of the large increments, but continuous due to cancellations among the different random directions. For times later than t_0, this process has no tendency to head in any particular direction; its forward drift is 0 at times $t \ge t_0$. A similar description holds for times $t \le t_0$ and its position an infinitesimal time dt earlier; the backward drift is 0 at times $t \le t_0$. The heuristic description of a general diffusion process is similar, except that there are time-dependent vector fields $b_\pm(x,t)$, the forward and backward drifts, and the particle is displaced by them in addition to the Wiener increments.

With this as motivation, we say that a (Markovian) *diffusion* on $\mathbf{R}^n$ is an $\mathbf{R}^n$-valued Markov process ξ indexed by $\mathbf{R}$ for which there exist Borel-measurable functions b_+ (the *forward drift*) and b_- (the *backward drift*) from $\mathbf{R}^n \times \mathbf{R}$ to $\mathbf{R}^n$ and Wiener difference processes w_+ and w_- such that for all $t_0 \le t_1$ the increment $w_+(t_1) - w_+(t_0)$ is independent of $\mathcal{P}_{t_0}$ and $w_-(t_1) - w_-(t_0)$ is independent of $\mathcal{F}_{t_1}$, and

$$\xi(t_1) - \xi(t_0) = \int_{t_0}^{t_1} b_+(\xi(t),t)\,dt + w_+(t_1) - w_+(t_0) \tag{1.7}$$

$$= \int_{t_0}^{t_1} b_-(\xi(t),t)\,dt + w_-(t_1) - w_-(t_0). \tag{1.8}$$

The processes w_+ and w_-, being Wiener difference processes, are equivalent, but they are not equal. We can express one process in terms of the other:

$$w_-(t_1) - w_-(t_0) = \int_{t_0}^{t_1} (b_+ - b_-)\big(\xi(t), t\big)dt + w_+(t_1) - w_+(t_0). \tag{1.9}$$

A diffusion is said to be a *finite energy diffusion* in case for all $t_0 \le t_1$

$$A = \int_{t_0}^{t_1} \mathrm{E}\tfrac{1}{4}(b_+^2 + b_-^2)\big(\xi(t), t\big)\,dt < \infty. \tag{1.10}$$

We call A the *kinetic energy integral*. It depends on the process and on the interval $[t_0, t_1]$.

By a *smooth diffusion* I shall mean a finite energy diffusion such that b_+ and b_- are smooth (i.e., C^∞). Let $\rho(x,t)$ be the probability density of the smooth diffusion $\xi(t)$, so that for any measurable set B in $\mathbb{R}^n$ we have

$$\Pr\{\,\xi(t) \in B\,\} = \int_B \rho(x,t)dx \tag{1.11}$$

where dx is Lebesgue measure on $\mathbb{R}^n$. The function ρ exists and is strictly positive and smooth on $\mathbb{R}^n \times \mathbb{R}$; this follows from the theory of partial differential equations because ρ is a solution of a parabolic equation with smooth coefficients, the *forward Fokker-Planck equation*

$$\frac{\partial \rho}{\partial t} = -\nabla\cdot(\rho b_+) + \tfrac{1}{2}\Delta\rho. \tag{1.12}$$

It also satisfies the anti-parabolic *backward Fokker-Planck equation*

$$\frac{\partial \rho}{\partial t} = -\nabla\cdot(\rho b_-) - \tfrac{1}{2}\Delta\rho. \tag{1.13}$$

There is a mechanical procedure for deriving backward equations from forward equations. Define the *time-reversed process* $\check{\xi}$ by $\check{\xi}(t) = \xi(-t)$, write the forward equation for $\check{\xi}$, and express the result in terms of ξ to obtain the backward equation for ξ.)

Although the paths of a smooth diffusion are non-differentiable, the forward drift and the backward drift express the *mean forward velocity* and the *mean backward velocity:*

$$\lim_{dt\downarrow 0} \mathrm{E}_t \frac{d_\pm \xi(t)}{dt} = b_\pm\big(\xi(t), t\big). \tag{1.14}$$

The *current velocity* v is the average of b_+ and b_- and the *osmotic velocity* u is one half their difference:

$$v = \tfrac{1}{2}(b_+ + b_-), \qquad u = \tfrac{1}{2}(b_+ - b_-). \tag{1.15}$$

The probability density ρ is related to the current velocity by the *current equation* (or equation of continuity)

$$\frac{\partial \rho}{\partial t} = -\nabla\cdot(\rho v), \tag{1.16}$$

and to the osmotic velocity by the *osmotic equation*

$$u = \tfrac{1}{2}\nabla \log \rho. \tag{1.17}$$

The current equation is obtained simply by averaging the forward and backward Fokker-Planck equations. If we take the difference of these two equations, we find that $-\rho u + \frac{1}{2}\nabla\rho$ is divergence-free, but to see that it is actually 0 (i.e., that (1.17) holds) takes a longer argument; see [N85]. We write

$$R = \tfrac{1}{2}\log \rho, \tag{1.18}$$

so that $\rho = e^{2R}$ and $\nabla R = u$.

2. An estimate for finite energy diffusions

Let ξ be a smooth diffusion on $\mathbf{R}^n$, and let f be a smooth function with compact support on $\mathbf{R}^n \times \mathbf{R}$. Then I claim that

$$
\begin{aligned}
&f(\xi(t_1),t_1) - f(\xi(t_0),t_0) \\
&= \int_{t_0}^{t_1} \left(\partial_t + b_+(\xi(t),t)\cdot\nabla + \tfrac{1}{2}\Delta\right) f(\xi(t),t)\,dt + \int_{t_0}^{t_1} \nabla f(\xi(t),t)\cdot d_+w_+(t) \qquad (2.1)\\
&= \int_{t_0}^{t_1} \left(\partial_t + b_-(\xi(t),t)\cdot\nabla - \tfrac{1}{2}\Delta\right) f(\xi(t),t)\,dt + \int_{t_0}^{t_1} \nabla f(\xi(t),t)\cdot d_-w_-(t). \qquad (2.2)
\end{aligned}
$$

These are the forward and backward *Itô equations*. The second integral in (2.1) is a *forward Itô stochastic integral*, and that in (2.2) is a *backward Itô stochastic integral*. They are defined as limits in L^2 of Riemann sums using forward and backward increments respectively, and will be discussed more in detail in §3. To establish (2.1), expand f in a Taylor polynomial of degree 2 and substitute into $d_+f(\xi(t),t)$, obtaining

$$d_+f = \partial_t f\,dt + \nabla f\cdot d_+\xi + \tfrac{1}{2}\partial_{x^i}\partial_{x^j} f\, d_+\xi^i d_+\xi^j + o(dt), \qquad (2.3)$$

where f is evaluated at $(\xi(t),t)$, the summation convention is used, and $o(dt)$ refers to the L^2 norm. Now $d_+\xi^i d_+\xi^j = d_+w_+^i d_+w_+^j + o(dt)$, and $d_+w_+^i d_+w_+^j$ can be replaced by $\delta^{ij}dt$, since for $i \neq j$ it is a random variable of mean 0 and variance $o(dt)$ (so that although its L^2 norm is not $o(dt)$, since these random variables for different values of t in the Riemann sum are orthogonal their sum is negligible), while for $i = j$ it differs from dt by such a random variable; this gives rise to the term $\frac{1}{2}\Delta$ in (2.1). These equations continue to hold for smooth f without compact support, since the trajectories of ξ are continuous with probability one. The Fokker-Planck equations are consequences of the Itô equations: the expectations of the stochastic integrals are 0; the other expectations can be expressed in terms of ρ; since the resulting equation holds for all test functions f, the ρ is by definition a weak solution of the Fokker-Planck equation.

In terms of v and u, we can write (1.10) as

$$
\begin{aligned}
A &= \int_{t_0}^{t_1} \mathrm{E}\tfrac{1}{2}(v^2+u^2)(\xi(t),t)\,dt \\
&= \int_{t_0}^{t_1}\int_{\mathbf{R}^n} \tfrac{1}{2}(v^2+u^2)(x,t)\rho(x,t)\,dx\,dt < \infty. \qquad (2.4)
\end{aligned}
$$

The current velocity v satisfies the current equation (1.16). This equation is still satisfied if we add to v any time-dependent vector field z such that for each t we have $\int z^2\rho\,dx < \infty$ and $\nabla\cdot(\rho z) = 0$ in the sense of distributions. Among all these vector fields $v + z$ there will be a unique one v_0 such that $\int_{\mathbf{R}^n} v_0^2\rho\,dx$ is a minimum. (A closed convex set in a Hilbert space has a unique element of smallest norm.) The condition on this v_0 is that

$$\frac{d}{d\lambda}\int_{\mathbf{R}^n}(v_0+\lambda z)^2\rho\,dx\,\bigg|_{\lambda=0} = 0; \qquad (2.5)$$

i.e., $\int_{\mathbf{R}^n} v_0\cdot z\rho\,dx = 0$. Notice that for a smooth diffusion ρ is strictly positive, since the osmotic velocity $u = \frac{1}{2}\nabla\log\rho$ is smooth. Since v_0 is orthogonal to the general divergence-free vector field $z\rho$, the field v_0 is a gradient. We set

$$\nabla S = v_0; \qquad (2.6)$$

then S is uniquely determined up to a function of t. By definition, S satisfies

$$-\nabla\cdot(\rho\nabla S) = \frac{\partial\rho}{\partial t}. \tag{2.7}$$

If we divide by 2ρ we find

$$-(\tfrac{1}{2}\Delta + u\cdot\nabla)S = \frac{\partial R}{\partial t}, \tag{2.8}$$

so that S is smooth for each t. By construction,

$$\int_{\mathbb{R}^n} (\nabla S)^2 \rho\, dx \leq \int_{\mathbb{R}^n} v^2 \rho\, dx. \tag{2.9}$$

The osmotic velocity is automatically a gradient for each t. We shall be interested primarily, but not exclusively, in *gradient diffusions*, for which v is also a gradient. (This is the same as saying that b_+, or equivalently b_-, is a gradient.) If v is a gradient, then $\nabla S = v$.

Let ξ be a smooth diffusion. Our goal is to estimate

$$\Pr\{\sup_{t_0\leq t\leq t_1} |R(\xi(t),t) - R(\xi(t_0),t_0)| \geq \lambda\} \tag{2.10}$$

purely in terms of λ and the energy integral A. (In [N85a] there is a similar estimate, but it involves more than just A.) I hope that this will be useful in constructing general finite energy diffusions whose drifts are not assumed to be smooth. The difficulty in doing this is that the drifts become exceedingly singular on the *nodes*, i.e., the region in $\mathbb{R}^n \times \mathbb{R}$ where $\rho = 0$ (and $R = -\infty$). But our estimate will show that it is impossible ever to reach the nodes; R cannot become $-\infty$.

For a smooth diffusion ξ, take the average of (2.1) and (2.2) where f is the smooth function R. This gives

$$\begin{aligned} R(\xi(t_1),t_1) - R(\xi(t_0),t_0) &= \int_{t_0}^{t_1} \Big(\partial_t + v(\xi(t),t)\cdot\nabla\Big) R(\xi(t),t)\,dt \\ &+ \tfrac{1}{2}\int_{t_0}^{t_1} \nabla R(\xi(t),t)\cdot d_+w_+(t) + \tfrac{1}{2}\int_{t_0}^{t_1} \nabla R(\xi(t),t)\cdot d_-w_-(t). \end{aligned} \tag{2.11}$$

Take one half the difference of (2.1) and (2.2) where now f is the smooth function S to find

$$\begin{aligned} &-\int_{t_0}^{t_1} \Big(\tfrac{1}{2}\Delta + u(\xi(t),t)\cdot\nabla\Big) S(\xi(t),t)\,dt \\ &= \int_{t_0}^{t_1} \nabla S(\xi(t),t)\cdot d_+w_+(t) - \int_{t_0}^{t_1} \nabla S(\xi(t),t)\cdot d_-w_-(t). \end{aligned} \tag{2.12}$$

But by (2.8), the integrand of the left hand side of this is $\partial_t R$, so we can substitute the right hand side for the integral of $\partial_t R$ in (2.11). Recalling that $\nabla R = u$, we find

$$\begin{aligned} R(\xi(t_1),t_1) - R(\xi(t_0),t_0) &= \int_{t_0}^{t_1} v(\xi(t),t)\cdot u(\xi(t),t)\,dt \\ &+ \tfrac{1}{2}\int_{t_0}^{t_1} (u+\nabla S)(\xi(t),t)\cdot d_+w_+(t) + \tfrac{1}{2}\int_{t_0}^{t_1} (u-\nabla S)(\xi(t),t)\cdot d_-w_-(t). \end{aligned} \tag{2.13}$$

This expresses the difference in R as the sum of three processes: an ordinary integral, a forward martingale, and a backward martingale. Replace t_1 by t, constrained to lie in the interval $[t_0, t_1]$, and let $X_1(t)$ be the ordinary integral, $X_2(t)$ the forward martingale, and $X_3(t)$ the backward martingale. Clearly

$$\sup_t |X_1(t)| \leq \int_{t_0}^{t_1} |v(\xi(t),t)\cdot u(\xi(t),t)|\, dt, \tag{2.14}$$

which has expectation $\leq A$. Hence by Chebyshev's inequality,

$$\Pr\{\sup_t |X_1(t)| \geq \lambda\} \leq \frac{A}{\lambda}. \tag{2.15}$$

By a martingale inequality,

$$\Pr\{\sup_t |X_2(t)| \geq \lambda\} \leq \frac{1}{\lambda}||X_2(t_1)||_1 \leq \frac{1}{\lambda}||X_2(t_1)||_2. \tag{2.16}$$

The estimate for X_3 is similar, but the roles of t_0 and t_1 are interchanged. By the triangle inequality, we have

$$\Pr\{\sup_t |X_3(t)| \geq \lambda\} \leq \frac{2}{\lambda}||X_3(t_1)||_2. \tag{2.17}$$

But

$$||X_2(t_1)||_2^2 + ||X_3(t_1)||_2^2 = \tfrac{1}{4}\mathrm{E}\int_{t_0}^{t_1} ((u+\nabla S)^2 + (u-\nabla S)^2)(\xi(t),t)dt \leq A. \tag{2.18}$$

If the event in (2.10) holds, then at least one of $\sup_t |X_\alpha(t)| \geq \lambda/3$, for $\alpha = 1$, 2, or 3, must hold. Consequently,

$$\Pr\{\sup_t |R(\xi(t),t) - R(\xi(t_0),t_0)| \geq \lambda\} \leq 6\frac{A+\sqrt{A}}{\lambda}. \tag{2.19}$$

We have proved the following slight extension of a theorem of Wallstrom [W88].

Theorem 1. *Let ξ be a smooth diffusion with kinetic energy integral A. Then for all $\lambda > 0$ and all $t_0 \leq t_1$ we have* (2.19).

3. Existence of finite energy diffusions

We must say more about how Itô stochastic integrals are defined.

If $t \mapsto \mathcal{B}_t$ is a family of σ-algebras, we say that the stochastic process ξ is *adapted* to the family in case each $\xi(t)$ is $\mathcal{B}_t$ measurable. Although we are interested only in stochastic integrals in which the integrand is adapted to the present, it is necessary to discuss integrands that are adapted to the past to define forward stochastic integrals. Let ξ be a diffusion. The set of all strongly measurable $\mathbf{R}^n$-valued functions η from $\mathbf{R}$ to L^2 of the underlying probability space such that

$$\int \mathrm{E}\eta^2(t)dt < \infty \tag{3.1}$$

and that are adapted to the past is a real Hilbert space $\mathcal{H}$. Let $\mathcal{H}_0$ be the subspace consisting of step functions. It is not difficult to see that $\mathcal{H}_0$ is a dense linear subspace of $\mathcal{H}$. (Partition a large compact interval with a small mesh and let η_0 be the step function that on each subinterval is the average of η over the previous subinterval.) For η in $\mathcal{H}_0$, define the forward stochastic integral $\int \eta(t) \cdot d_+ w_+(t)$ to be the obvious finite sum. This is a linear mapping from $\mathcal{H}_0$ to L^2 and a simple computation shows that it is isometric. Therefore it extends uniquely to an isometric linear map from $\mathcal{H}$ into L^2; this defines the forward stochastic integral when η is adapted to the past and satisfies (3.1). For η in $\mathcal{H}_0$ and $t_0 \leq t_1$ we have

$$\mathrm{E}_{t_0} \left(\int_{t_0}^{t_1} \eta(t) \cdot d_+ w_+(t) \right)^2 = \mathrm{E}_{t_0} \int_{t_0}^{t_1} \eta^2(t) dt, \tag{3.2}$$

so this continues to hold for all η in $\mathcal{H}$.

Suppose that η is adapted to the past and satisfies the weaker condition that

$$\int \eta^2(t) dt < \infty \tag{3.3}$$

with probability one. Let $\chi_m(t)$ be the indicator function of the event that $\int_{-\infty}^{t} \eta^2(s) ds \leq m$. Then $\chi_m \eta$ is in $\mathcal{H}$, so its forward stochastic integral is well defined. But with probability one, $\chi_m(t)\eta(t)$ is equal to $\eta(t)$ for all t for m sufficiently large, so we define its forward stochastic integral to be the common value for large m. In fact, the same argument defines the forward stochastic integral for any η adapted to the past on the set where (3.3) holds.

For η in $\mathcal{H}_0$ define ς by

$$\varsigma = \exp \left[\int \eta(t) \cdot d_+ w_+(t) - \tfrac{1}{2} \int \eta^2(t) dt \right]. \tag{3.4}$$

Then ς is a positive random variable, and $\mathrm{E}\varsigma = 1$. To see this, use the elementary fact that for a Gaussian random variable W of mean 0 we have

$$\mathrm{E} \exp \left[W - \tfrac{1}{2} \mathrm{E} W^2 \right] = 1, \tag{3.5}$$

write ς as a product, and take conditional expectations E_{t_i} where t_i is the left endpoint of an interval for the step function, working from right to left, to conclude the desired result. Any η in $\mathcal{H}$ is a limit of a sequence of elements of $\mathcal{H}_0$ and the corresponding stochastic integrals converge in L^2; by picking a subsequence, we can ensure that they and hence also the corresponding ς's converge with probability one. By Fatou's lemma, therefore, $\mathrm{E}\varsigma \leq 1$ for any η in $\mathcal{H}$. For ς adapted to the past and satisfying (3.3) with probability one, an entirely similar argument shows that $\mathrm{E}\varsigma \leq 1$. For a general η adapted to the past, we define ς to be 0 on the set where $\int \eta^2(t) dt = \infty$, so that we always have $\mathrm{E}\varsigma \leq 1$.

Let $\mathcal{M}_0$ be the set of all triples $D = \langle \rho, v, u \rangle$ of smooth functions mapping $\mathbf{R}^n \times \mathbf{R}$ into $\mathbf{R}$, $\mathbf{R}^n$, and $\mathbf{R}^n$, respectively, such that for each t the function $\rho(t) = \rho(\cdot, t)$ is a probability density, the current and osmotic equations are satisfied, and the kinetic energy integral is finite on every interval $[t_0, t_1]$.

There is a very simple construction of the diffusion associated to a triple D in $\mathcal{M}_0$, by means of a *Girsanov transformation* using the density ς. Let Ω be the set of all continuous functions from $\mathbf{R}$ to $\mathbf{R}^n$ and let $\mathcal{B}$ be the σ-algebra generated by the $\xi(t)$ given by $\xi(t)(\omega) = \omega(t)$ for t in $\mathbf{R}$, and similarly for $\mathcal{B}_{t_0,t_1}$ with $t \in [t_0, t_1]$. Let Pr_0 be the probability measure on $\langle \Omega, \mathcal{B} \rangle$ for the Wiener process with initial probability density $\rho(t_0)$ at the initial time t_0, and let $t_0 \leq t_1$. Let

$$\varsigma = \exp \left[\int_{t_0}^{t_1} b_+(\xi(t), t) \cdot d_+ \xi(t) - \tfrac{1}{2} \int_{t_0}^{t_1} b_+^2(\xi(t), t) dt \right], \tag{3.6}$$

where $b_+ = v + u$. Notice that this is a particular case of (3.4); here the diffusion ξ is the Wiener process, so for times after t_0 we have $d_+\xi = d_+w_+$. Let E_0 denote the expectation with respect to Pr_0. Then it can be shown that $\mathrm{E}_0\varsigma = 1$. Let Pr_{t_0,t_1} be the probability measure $\varsigma\,\mathrm{Pr}_0$. Then there is a unique probability measure Pr on $\langle\Omega, \mathcal{B}\rangle$ that agrees with each Pr_{t_0,t_1} on $\mathcal{B}_{t_0,t_1}$, and the process $\xi(t)$ is the diffusion associated to D. This is *Girsanov's formula.*

The heuristic reason for this is that

$$\mathrm{E}_t d_+\xi(t) = \mathrm{E}_t^0 \exp\left[b_+\big(\xi(t),t\big)\cdot d_+\xi(t) - \tfrac{1}{2}b_+^2\big(\xi(t),t\big)dt\right] d_+\xi(t) \tag{3.7}$$

where E_t^0 is the conditional expectation with respect to the present for the measure Pr_0. When the exponential is expanded to order dt we obtain

$$\begin{aligned}&1 + b_+\big(\xi(t),t\big)\cdot d_+\xi(t) + \tfrac{1}{2}\left[b_+\big(\xi(t),t\big)\cdot d_+\xi(t)\right]^2 - \tfrac{1}{2}b_+^2\big(\xi(t),t\big)dt = \\ &1 + b_+\big(\xi(t),t\big)\cdot d_+\xi(t) + o(dt),\end{aligned} \tag{3.8}$$

so that $\mathrm{E}_t d_+\xi(t) = b_+\big(\xi(t),t\big)dt + o(dt)$ as desired.

Let $\mathcal{M}$ be the set of all triples $D = \langle\rho, v, u\rangle$ of Borel measurable functions mapping $\mathrm{R}^n\times\mathrm{R}$ into R, R^n, and R^n, respectively, such that for all t the function $\rho(t) = \rho(\cdot,t)$ is a probability density and the current and osmotic equations are satisfied weakly, and the kinetic energy integral is finite on every interval $[t_0, t_1]$. Carlen [C84] constructed the diffusion associated to the general element of $\mathcal{M}$. His proof was by means of a partial differential equations approach, exploiting through intricate estimates the maximum principle and energy integral estimates. I conjecture that the Girsanov formula continues to be valid for all D in $\mathcal{M}$, in the following sense: There is a unique probability measure Pr on Ω such that for all $t_0 \le t_1$ it agrees on $\mathcal{B}_{t_0,t_1}$ with $\varsigma\,\mathrm{Pr}_0$ (where ς is put equal to 0 on the set where $\int_{t_0}^{t_1} b_+^2\big(\xi(t),t\big)dt = \infty$) such that with respect to it ξ is a finite energy diffusion with probability density ρ, current velocity v, and osmotic velocity u. This would strengthen Carlen's result. Notice that according to the conjecture, for each interval $[t_0, t_1]$ the diffusion is absolutely continuous with respect to the Wiener process, so that every property known to hold almost surely for the Wiener process on a finite time interval would hold almost surely for the general finite energy diffusion. The converse is false, since we can have $\varsigma = 0$ on a set of large Wiener measure.

Theorem 1 should play a role in the proof of this conjecture. But the situation is perhaps not as simple as it appears at first sight. Theorem 1 was established for diffusions corresponding to D in $\mathcal{M}_0$, but is $\mathcal{M}_0$ dense in any suitable sense in $\mathcal{M}$? Guerra [G85] introduced a metric on $\mathcal{M}$, which makes this question precise.

4. Action

For a particle of unit mass traveling along a smooth curve $t \mapsto \xi(t)$ during the time interval $[t_0, t_1]$, its kinetic action is

$$\int_{t_0}^{t_1} \frac{1}{2}\left(\frac{d\xi}{dt}\right)^2 dt. \tag{4.1}$$

For the trajectories of a diffusion, this makes no sense. Nevertheless, we can calculate $\big(d_+\xi(t)\big)^2$ to order dt^2 and see what it looks like.

Recall that $\big(d_+\xi(t)\big)^2$ is $\big(\xi(t_1) - \xi(t_0)\big)^2$ where $t_0 = t$ and $t_1 = t + dt$, and recall (1.7):

$$\xi(t_1) - \xi(t_0) = \int_{t_0}^{t_1} b_+\big(\xi(t),t\big)dt + w_+(t_1) - w_+(t_1).$$

We can estimate the integral by $(t_1 - t_0)b_+(\xi(t_0), t_0)$, but this does not give us the desired accuracy. To improve the accuracy, apply (1.7) itself to $\xi(t)$ in the integrand to obtain

$$\begin{aligned}&\xi(t_1) - \xi(t_0)\\&= \int_{t_0}^{t_1} b_+\Big(\xi(t_0) + \int_{t_0}^{t} b_+(\xi(s), s)\,ds + w_+(t) - w_+(t_0), t\Big)dt + w_+(t_1) - w_+(t_0).\end{aligned} \quad (4.2)$$

Now take a Taylor expansion to first order for b_+ at $(\xi(t_0), t_0)$ and let

$$W^k = \int_{t_0}^{t_1} (w_+^k(t) - w_+^k(t_0))\,dt. \quad (4.3)$$

Then

$$\begin{aligned}&\xi(t_1) - \xi(t_0)\\&= b_+(\xi(t_0), t_0)(t_1 - t_0) + \frac{\partial b_+}{\partial x^k}(\xi(t_0), t_0)W^k + w_+(t_1) - w_+(t_0) + o((t_1 - t_0)^{3/2}).\end{aligned} \quad (4.4)$$

Take the inner product of this with itself and revert to the t, $t + dt$ notation. We find

$$\tfrac{1}{2}\left(\frac{d_+\xi}{dt}\right)^2 = \tfrac{1}{2}b_+^2 + \frac{b_+ \cdot d_+ w_+}{dt} + \frac{\partial b_+}{\partial x^k}\frac{W^k \cdot d_+ w_+}{dt} + \tfrac{1}{2}\left(\frac{d_+ w_+}{dt}\right)^2 + o(1), \quad (4.5)$$

where everything is evaluated at $(\xi(t), t)$. Now take the conditional expectation with respect to the present. Notice that $\mathrm{E}_t b_+ \cdot d_+ w_+ = 0$ and that $\mathrm{E}_t(d_+ w_+)^2 = n\,dt$. Also,

$$\mathrm{E}_t \frac{\partial b_+}{\partial x^k} W^k \cdot d_+ w_+ = \tfrac{1}{2}\nabla \cdot b_+\,dt. \quad (4.6)$$

Therefore

$$\mathrm{E}_t \tfrac{1}{2}\left(\frac{d_+\xi}{dt}\right)^2 = \tfrac{1}{2}b_+^2 + \tfrac{1}{2}\nabla \cdot b_+ + \tfrac{1}{2}\frac{n}{dt} + o(1). \quad (4.7)$$

This contains the singular term $n/2dt$, but this term is a constant that is the same for all diffusions. When we study action, we are leaving kinematics and entering into dynamics, where action has played a fundamental role that has survived the revolutions of twentieth century physics. But action enters into *variational principles* that are not affected by an additive constant. We call

$$\mathrm{E}\int_{t_0}^{t_1} \Big(\tfrac{1}{2}b_+^2(\xi(t), t) + \tfrac{1}{2}\nabla \cdot b_+(\xi(t), t)\Big)dt \quad (4.8)$$

the *expected kinetic action*. If we express b_+ in terms of u and v, the expectation in terms of ρ integration, and integrate by parts, we find that the expected kinetic action is

$$\int_{t_0}^{t_1}\int_{\mathbf{R}^n} \tfrac{1}{2}(v^2 - u^2)(x, t)\rho(x, t)\,dx\,dt. \quad (4.9)$$

It differs from the kinetic energy integral (2.4) by having a minus sign instead of a plus sign. The derivation was for a smooth diffusion, but the result is meaningful for any finite energy diffusion.

A *potential* on $\mathbf{R}^n$ is a pair of Borel measurable functions $\phi: \mathbf{R}^n \to \mathbf{R}$ (the *scalar potential*) and $A: \mathbf{R}^n \to \mathbf{R}^n$ (the *vector potential*). Then the *expected potential action* is

$$\mathrm{E}\left[\int_{t_0}^{t_1} \phi(\xi(t),t)\,dt - \int_{t_0}^{t_1} A(\xi(t),t)\cdot\tfrac{1}{2}\big(d_+\xi(t) + d_-\xi(t)\big)\right]. \tag{4.10}$$

The minus sign is a matter of convention, but the use of

$$\tfrac{1}{2}\big(d_+\xi(t) + d_-\xi(t)\big) = v(\xi(t),t)\,dt + \tfrac{1}{2}\big(d_+w_+(t) + d_-w_-(t)\big) \tag{4.11}$$

is necessary to avoid asymmetry in the two directions of time. Of course, the expected potential action only exists if the expectation exists, in which case it can be written as

$$\int_{t_0}^{t_1}\int_{\mathbf{R}^n} [\phi(x,t) - A(x,t)\cdot v(x,t)]\,\rho(x,t)\,dx\,dt. \tag{4.12}$$

The *expected action* is the difference of the expected kinetic action and the expected potential action. Let $\chi: \mathbf{R}^n \times \mathbf{R} \to \mathbf{R}$ be smooth and of compact support (for simplicity of exposition). Then the transformation

$$\begin{cases} \phi \to \phi + \partial_t\chi \\ A \to A - \nabla\chi \end{cases} \tag{4.13}$$

is called a *gauge transformation*. The expected action is the same after a gauge transformation; this follows from the current equation after integration by parts.

I shall close this chapter with some brief comments on the extension of the theory of the kinematics of diffusion to Riemannian manifolds. The first observation to make is that this is not done simply for the sake of complicating the theory; the Riemannian metric is an intrinsic part of the probabilistic structure. Consider a diffusion ξ on a differentiable manifold M, and in local coordinates at a point x at which the diffusion starts, define σ^{ij} by

$$\sigma^{ij}dt = \mathrm{E}d\xi^i(t)d\xi^j(t) + o(dt).$$

Then the inverse matrix σ_{ij} is a Riemannian metric.

The notion of the Wiener difference process does not generalize to the context of a Riemannian manifold except in terms of its differential, but this is enough to enable us to define intrinsically the notions of mean forward and backward velocities. The current and osmotic equations hold in the more general context.

Dankel [D70] was the first to develop stochastic mechanics on a Riemannian manifold, with applications to spin. One problem was that to differentiate tensor fields along diffusion trajectories, one needs a notion of stochastic parallel transport. Itô's notion developed in [I62] was unsuitable, and a notion more adapted to the needs of stochastic mechanics was developed by Dohrn and Guerra [DG78, 79]. The Ricci curvature, but not the full uncontracted Riemannian curvature, plays a role in the Dohrn-Guerra notion of parallel transport.

When the expected kinetic action is computed, there is a term involving the scalar curvature, the Pauli-DeWitt term familiar to the physicists; see [D57]. All of these questions are discussed in [N85].

II. Conservative Dynamics of Diffusion

5. The variational principle

We seek a dynamics for diffusions that shall be analogous to conservative deterministic dynamics. Therefore we shall base it on a variational principle. Since we are seeking a dynamical law, in the beginning we make as many simplifying assumptions of smoothness as possible. Let the potential ϕ, A be smooth with compact support, and let $D = \langle \rho, v, u \rangle$ be in $\mathcal{M}_0$. What shall it mean for the diffusion D to be critical for the potential?

As we have seen, the expected action is

$$I = \int_{t_0}^{t_1} \int_{\mathbf{R}^n} \left(\tfrac{1}{2}v^2 - \tfrac{1}{2}u^2 - \phi + A \cdot v\right) \rho \, dx \, dt. \tag{5.1}$$

We shall say that D is critical in case this expected action is stationary, for every interval $[t_0, t_1]$, for variations in which $\rho(t_0)$ and $\rho(t_1)$ are held fixed. This is a direct analogue of Hamilton's principle of least action in deterministic mechanics.

More precisely, suppose that for all α in some neighborhood of 0 in $\mathbf{R}$ we have a $D(\alpha)$ in $\mathcal{M}_0$ such that $\rho(\alpha, x, t)$, $v(\alpha, x, t)$, and $u(\alpha, x, t)$ are smooth, that $D(0) = D$, and that $\rho(\alpha, t_0) = \rho(t_0)$ and $\rho(\alpha, t_1) = \rho(t_1)$ for all α. This is called a *variation* of D (for the interval $[t_0, t_1]$). We say that D is *critical* for the potential in case for all intervals $[t_0, t_1]$ and all variations of D,

$$\frac{d}{d\alpha} I(\alpha) \bigg|_{\alpha=0} = 0. \tag{5.2}$$

To distinguish this notion from other related notions, we also say that D is *critical in the sense of Lafferty*; see [L87] for this formulation and the result about to be derived.

Let D be critical and let $[t_0, t_1]$ be fixed. One possible type of variation is one in which ρ, and consequently also u, is fixed for all t in $[t_0, t_1]$, and only v varies with α, with $v(\alpha) = v + \alpha z$ where $\nabla \cdot (z\rho) = 0$ for each t (so that the current equation is satisfied for each α). Then

$$I'(0) = \int_{t_0}^{t_1} \int_{\mathbf{R}^n} (v + A) \cdot z \rho \, dx \, dt. \tag{5.3}$$

This must vanish for all choices of z with $\nabla \cdot (z\rho) = 0$, so a necessary condition for D to be critical is that for each t the vector field $v + A$ be a gradient vector field; i.e., that there exist a function S, uniquely defined up to a function of t, such that

$$v + A = \nabla S. \tag{5.4}$$

This is called the *stochastic Hamilton-Jacobi condition* because of its role in the generalization by Guerra and Morato of Hamilton-Jacobi theory to conservative diffusions; see [GM83].

Now we compute $I'(\alpha)$. In the following computation ρ, v, and u all depend on α, and for each α the current and osmotic equations hold. The idea of the computation is to bring all α-derivatives to ρ. Now $I'(\alpha)$ is the sum of three integrals:

$$\int_{t_0}^{t_1} \int_{\mathbf{R}^n} (v + A) \cdot \partial_\alpha (v\rho) \, dx \, dt, \tag{5.5}$$

$$\int_{t_0}^{t_1} \int_{\mathbf{R}^n} (-u) \cdot \partial_\alpha (u\rho) \, dx \, dt, \tag{5.6}$$

$$\int_{t_0}^{t_1} \int_{\mathbf{R}^n} \left(-\tfrac{1}{2}v^2 + \tfrac{1}{2}u^2 - \phi\right) \partial_\alpha \rho \, dx \, dt, \tag{5.7}$$

where we chose to include ρ with v and u in differentiating them and then to compensate by subtracting in the third integral.

The first integral at $\alpha = 0$ is

$$\begin{aligned}\int_{t_0}^{t_1}\int_{\mathbf{R}^n} \nabla S\cdot\partial_\alpha(v\rho)dx\,dt &= -\int_{t_0}^{t_1}\int_{\mathbf{R}^n} S\partial_\alpha\nabla\cdot(v\rho)dx\,dt \\ &= \int_{t_0}^{t_1}\int_{\mathbf{R}^n} S\partial_\alpha\partial_t\rho\,dx\,dt = -\int_{t_0}^{t_1}\int_{\mathbf{R}^n} \partial_t S\partial_\alpha\rho\,dx\,dt \end{aligned}\tag{5.8}$$

by the stochastic Hamilton-Jacobi condition, a spatial integration by parts, the current equation, and a temporal integration by parts. This last is justified by the fact that $\partial_\alpha\rho = 0$ at t_0 and t_1. Using the osmotic equation in the form $u = \nabla\rho/2\rho$, we see that the second integral is

$$\int_{t_0}^{t_1}\int_{\mathbf{R}^n} \tfrac{1}{2}\nabla\cdot u\partial_\alpha\rho\,dx\,dt \tag{5.9}$$

by a spatial integration by parts. Therefore at $\alpha = 0$ we have

$$I'(\alpha) = \int_{t_0}^{t_1}\int_{\mathbf{R}^n} \left[-\partial_t S + \tfrac{1}{2}\nabla\cdot u - \tfrac{1}{2}v^2 + \tfrac{1}{2}u^2 - \phi\right]\partial_\alpha\rho\,dx\,dt. \tag{5.10}$$

The only constraints on $\partial_\alpha\rho$ at $\alpha = 0$ are that its spatial integral be 0 (since each $\rho(\alpha)$ is a probability density for all t) and that it vanish at t_0 and t_1 (by the definition of a variation). Consequently, $I'(0) = 0$ if and only if the term in brackets is a function of t alone. We choose the additive function of t in S so that the term in brackets is 0; now S is uniquely determined up to a constant. Expressing u as ∇R and rearranging the terms in brackets, we find that D is critical if and only if

$$\partial_t S + \tfrac{1}{2}(\nabla S - A)^2 + \phi - \tfrac{1}{2}(\nabla R)^2 - \tfrac{1}{2}\Delta R = 0. \tag{5.11}$$

This is known as the *stochastic Hamilton-Jacobi equation.*

The triple $D = \langle\rho, v, u\rangle$ is determined by R and S (when the vector potential A is known and $v + A = \nabla S$). The current equation is

$$\partial_t R + \nabla R\cdot(\nabla S - A) + \tfrac{1}{2}\Delta S - \tfrac{1}{2}\nabla\cdot A = 0. \tag{5.12}$$

Together, (5.11) and (5.12) constitute necessary and sufficient conditions for a diffusion to be critical for a potential. They are a coupled system of nonlinear partial differential equations. But make the change of dependent variables

$$\psi = e^{R+iS}. \tag{5.13}$$

Then they are equivalent to the linear equation

$$\frac{\partial\psi}{\partial t} = -i\left[\tfrac{1}{2}(-i\nabla - A)^2 + \phi\right]\psi. \tag{5.14}$$

This is the *Schrödinger equation.* When the vector potential A is zero, it is customary to write V instead of ϕ for the scalar potential, so that the Schrödinger equation takes the form

$$\frac{\partial\psi}{\partial t} = -i[-\tfrac{1}{2}\Delta + V]\psi. \tag{5.15}$$

6. Stochastic mechanics

What is conserved by conservative diffusions? First it is necessary to say that the adjective "conservative" does not apply to the diffusion process itself—a diffusion is just a diffusion—but to the dynamics, i.e., the rules for associating a class of diffusions to a potential. The same comment applies to the assertion that stochastic mechanics is time-symmetric: it is not the diffusions themselves that are time-symmetric (whatever that might mean), but the dynamics.

The *current energy* is $\frac{1}{2}v^2$, the *osmotic energy* is $\frac{1}{2}u^2$, the *potential energy* is ϕ, and the *stochastic energy* E is their sum. Thus $t \mapsto E(\xi(t), t)$ is a stochastic process, and it certainly is not a constant. But if the potential ϕ, A is constant in time, a straightforward computation shows that the expected value $\mathrm{E}\, E(\xi(t), t)$ is constant in time.

From now on let us suppose that the vector potential A is zero, so that $\nabla S = v$ and we have a gradient diffusion. Then if one takes the gradient of the stochastic Hamilton-Jacobi equation (5.11) one finds

$$-\nabla V = (\partial_t + v\cdot\nabla)v - (\tfrac{1}{2}\Delta + u\cdot\nabla)u. \tag{6.1}$$

The left hand side is the familiar expression for the force in conservative dynamics. For any stochastic process η we define

$$D_\pm \eta(t) = \lim_{dt\downarrow 0} \mathrm{E}_t \frac{d_\pm \eta(t)}{dt} \tag{6.2}$$

when the limits exist. Thus $D_\pm \xi(t) = b_\pm(\xi(t), t)$. A simple computation shows that the right hand side of (6.1), evaluated at $(\xi(t), t)$, is

$$\tfrac{1}{2}(D_+D_- + D_-D_+)\,\xi(t), \tag{6.3}$$

which we call the *stochastic acceleration*. Hence (6.1) is called the *stochastic Newton equation*, because it is the analogue of the Newton equation $F = ma$. (What happened to the m? It is hidden away in the inner product $\cdot$ to simplify the formulas, and we can do this even when R^n is the configuration space of a system of particles with different masses. One other brief remark on physical dimensions: the inner product is also used to set the scale of the local fluctuations of our diffusions, via $\mathrm{E}(d_+w_+)^2 = dt$. But to make this dimensionally correct one must introduce a factor $\hbar$ with the dimensions of action. Throughout this account we have chosen units so that $\hbar = 1$.)

Here is an example of a critical diffusion, which I call the *one-slit process*. In this example $V = 0$, so the motion is said to be *free*. Let

$$v = \frac{t}{1+t^2}x, \tag{6.4}$$

$$u = \frac{-1}{1+t^2}x. \tag{6.5}$$

Then one can verify the stochastic Newton equation and the current equation, or equivalently one can verify that $\psi_0 = e^{R+iS}$ satisfies the free Schrödinger equation. (As always, R is determined by $\nabla R = u$ and the requirement that $\rho = e^{2R} = |\psi|^2$ be a probability density. The function S is determined up to an additive constant, which in this example is the constant value of S at $t = 0$.) The random variable $\xi(0)$ is Gaussian with mean 0 and variance $\frac{1}{2}$ times the identity matrix. The current velocity v is 0 at time 0; the particle is just resting near the origin. The random variable $\xi(t)$ is also Gaussian with mean 0, but its variance has grown to $\frac{1}{2}\sqrt{1+t^2}$ (whether t is positive or negative). The expected current and osmotic energies are

$$\mathrm{E}\tfrac{1}{2}v^2(\xi(t),t) = \frac{t^2}{4(1+t^2)}, \tag{6.6}$$

$$\mathrm{E}\tfrac{1}{2}u^2(\xi(t),t) = \frac{1}{4(1+t^2)}. \tag{6.7}$$

Their sum, the expected stochastic energy, is constantly $\frac{1}{4}$. Initially this is all osmotic energy, but as $t \to \pm\infty$ it changes into current energy.

So far in this example we have discussed only random variables at a single time. What about the stochastic process ξ itself? Let

$$f(t) = \frac{t-1}{1+t^2}, \tag{6.8}$$

so that $b_+(\xi(t),t) = (v+u)(\xi(t),t) = f(t)\xi(t)$. Then ξ satisfies the stochastic differential equation

$$d_+\xi(t) = f(t)\xi(t)dt + d_+w_+(t), \tag{6.9}$$

which is simply the differential form of (1.7). This is an inhomogeneous linear equation, whose solution is

$$\begin{aligned}\xi(t) &= e^{\int_0^t f(s)ds}\xi(0) + \int_0^t e^{\int_s^t f(r)dr} d_+w_+(s) \\ &= \sqrt{1+t^2}e^{-\arctan t}\left[\xi(0) + \int_0^t \frac{e^{\arctan s}}{\sqrt{1+s^2}} d_+w_+(s)\right].\end{aligned} \tag{6.10}$$

Notice that $f(t) \sim 1/t$ as $t \to \infty$. Consequently the forward drift $b_+(\xi(t),t) = f(t)\xi(t)$, evaluated at the position of the particle, converges in L^2 and with probability one to

$$\pi_+ = e^{-\pi/2}\left[\xi(0) + \int_0^\infty \frac{e^{\arctan s}}{\sqrt{1+s^2}} d_+w_+(s)\right]. \tag{6.11}$$

Asymptotically as $t \to \infty$, the particle travels with a constant drift (constant in time, but random). The asymptotic motion is that of a particle traveling in a straight line with constant velocity plus a Wiener process.

How does the process, which is Markovian and knows only its present position, remember to maintain the constant asymptotic average velocity? At time 0 the particle does not know what asymptotic regime it will eventually enter, but once it is there the drift, which is a function of position and time, happens to be just right to maintain the particle in the same asymptotic regime.

Now let us look at a more interesting free process, the *two-slit process*. Initially the particle is to be at rest ($v = 0$ for $t = 0$) and localized with equal probabilities near two points $\pm a$. We could take as the initial probability density

$$\tfrac{1}{2}\left(|\psi_0(x-a,0)|^2 + |\psi_0(x+a,0)|^2\right) \tag{6.12}$$

where ψ_0 is the wave function for the one-slit process. We assume that $|a|$ is several multiples of the standard deviation $1/\sqrt{2}$ of the Gaussian at time 0 for the one-slit process; then there is practically no overlap between the two probability densities at a and $-a$. But let us be a bit more clever: let $\psi_1(x,t) = \psi_0(x-a,t)$ and $\psi_2(x,t) = \psi_0(x+a,t)$. Then these describe the one-slit process but shifted by a and $-a$ respectively. Let $\psi = \gamma(\psi_1 + \psi_2)$, where γ is a normalization constant, very close to $1/\sqrt{2}$, to make ψ a unit vector in L^2. Then $|\psi(x,0)|^2$ is, as desired, a probability density concentrated equally near a and $-a$, and it is practically the same as (6.12).

But now we can exploit the fact that the Schrödinger equation is linear, so that if ψ_1 and ψ_2 are solutions, describing critical diffusions for a given potential, so is their sum $\psi_1 + \psi_2$ (multiplied by a normalization constant γ to make it a unit vector). We can ask for the current and osmotic velocities of the new process in terms of the old, and we find

$$v = \tfrac{1}{2}(v_1 + v_2) + \tfrac{1}{2}\frac{(v_1 - v_2)\sinh(R_1 - R_2) + (u_1 - u_2)\sin(S_1 - S_2)}{\cosh(R_1 - R_2) + \cos(S_1 - S_2)}, \tag{6.13}$$

$$u = \tfrac{1}{2}(u_1 + u_2) + \tfrac{1}{2}\frac{(u_1 - u_2)\sinh(R_1 - R_2) - (v_1 - v_2)\sin(S_1 - S_2)}{\cosh(R_1 - R_2) + \cos(S_1 - S_2)}. \tag{6.14}$$

Let us see how the particle in the two-slit process moves according to stochastic mechanics. For its forward drift we find

$$b_+(x,t) = \frac{1}{1+t^2}\left[(t-1)x + \frac{(1-t)\sinh\frac{a\cdot x}{1+t^2} - (1+t)\sin\frac{2ta\cdot x}{1+t^2}}{\cosh\frac{a\cdot x}{1+t^2} + \cos\frac{2ta\cdot x}{1+t^2}}a\right]. \tag{6.15}$$

Only the direction of a is interesting, so we take the dimension n to be 1. The denominator never vanishes; there are no nodes. But when t is reasonably large compared to a, it comes close to vanishing on the hyperbolas

$$\frac{2tax}{1+t^2} = (2m+1)\pi, \qquad m \in Z, \tag{6.16}$$

where the cos is -1 and the cosh is close to 1. But in this region the hyperbolas are practically equal to their asymptotes, the straight lines $2ax = (2m+1)\pi t$. For very small times t, the drift is practically $(t-1)x/(1+t^2)$, because of the large term $\cosh\bigl(ax/(1+t^2)\bigr)$. (Remember that the particle starts near $x = \pm a$.) But for larger times t, there is an enormous drift repelling the particles from the asymptotes; the particle finds itself trapped in one of the channels between these straight lines and its probability density has alternate peaks and troughs resembling those produced by interference phenomena in wave motion.

Let us discuss one other example of a critical diffusion. In this example $n = 1$ and the potential is $V(x) = \omega^2 x^2$, where ω is a constant. This is the *harmonic oscillator*. If we take $v = 0$ and $u = -\omega x$ we obtain a stationary Gaussian process of mean 0 and covariance

$$\mathrm{E}\xi(t)\xi(s) = \frac{1}{2\omega}e^{-\omega|t-s|}, \tag{6.17}$$

called the *ground state process* for the harmonic oscillator. For the harmonic oscillator, the stochastic Newton equation

$$\tfrac{1}{2}(D_+D_- + D_-D_+)\xi(t) = -\omega^2\xi(t) \tag{6.18}$$

is linear, so if we add to the ground state process a solution $\mu(t) = c_1\cos\omega t + c_2\sin\omega t$ of the deterministic harmonic oscillator equation

$$\frac{d^2}{dt^2}\mu(t) = -\omega^2\mu(t) \tag{6.19}$$

we again get a solution, called a *coherent state process*; see [GL81].

Since its beginning 35 years ago in the paper [F52] of Fényes, stochastic mechanics has been applied to a number of topics in quantum physics. There are references in [N85] and the recent book [BCZ87]. Here I shall just mention a few topics. Shucker [S80] showed that the asymptotic motion of a general free conservative diffusion is motion with a constant random velocity (whose probability density is given by the square modulus of the Fourier transform of the wave function) with the Wiener process superimposed, and this was extended to potential scattering by Carlen [C85, 86]. Stochastic mechanics allows one to investigate certain questions that cannot even be formulated in quantum mechanics; for example, one can ask for the probability law of the first time that a particle doing a critical diffusion enters a certain region. For results on this, and a discussion of possible physical tests, see [CT86] and the references cited there. Stochastic mechanics has also been applied to some macroscopic problems; see [DT87] and references in some of the articles appearing there.

7. Stochastic mechanics and nature

Recently I had lunch with a young mathematician who is familiar with stochastic mechanics but not working in the field. I told him about some of the exciting recent developments and he asked me what is the next step in stochastic mechanics. Without hesitation I said that the next step is to throw it away and start over. Let me try to explain why I feel that.

The predictions of stochastic mechanics are thoroughly confirmed by experiment. The outcome of any experiment can be described in terms of the positions of various objects (meters, pointers, marks on photographic plates, etc.) at a fixed time. According to stochastic mechanics, if it is assumed that all of the objects *including the measuring devices* perform a critical diffusion, then the probability density for finding the objects at certain positions x at times t is given by $|\psi(x,t)|^2$. Thus the only aspect of stochastic mechanics that is tested by experiment is the Schrödinger equation itself, and not the fascinating mathematical results on the behavior of the random trajectories.

Now the Schrödinger equation was discovered long before the advent of stochastic mechanics. There have been many attempts to understand its predictions, bringing in such concepts as complementarity, reduction of the wave packet by the consciousness of the observer, many worlds, etc. To these I much prefer the viewpoint so brilliantly expounded in [F85], in which Feynman says, "... we are not going to deal with *why* Nature behaves in the peculiar way that She does; there are no good theories to explain that."

The last phrase is a statement of fact. Nevertheless, one wants to understand why. Stochastic mechanics, as I see it, was an attempt to construct a naïvely realistic model of nature. To decide whether it was successful, I propose the following test. Suppose that we have measuring devices that are not themselves subject to the random fluctuations of stochastic mechanics, and that permit us to observe the random trajectories described by the theory. Is this consistent with what we know about nature?

This is a very stringent test. It goes far beyond showing that the predictions of stochastic mechanics cannot be falsified by experiment; as already remarked, that is not in doubt so long as we have only measuring devices that are themselves subject to the Schrödinger equation. There are two reasons for proposing this test. The first is one of intellectual coherence. It appears pointless to posit certain features of a physical theory and when they appear paradoxical to say that there is no problem because they are unobservable. The second reason is one of physics. So long as we do not understand why Nature behaves in the peculiar way that She does, we have no idea of the limitations of our theory. Quantum theorists are prone to insist that their rules are of universal validity, but that is challenged by many physicists who work on gravitation. I will not repeat here the analogy suggested by Kappler's experiment;

see pp. 117–118 of [N85] and also [BS34]. The point is that perhaps the randomness observed in phenomena on the microscopic scale has a physical cause, and there may be systems that are not subject to it. Alternatively, one may remark that stochastic mechanics is a dynamical theory of diffusion in a frictionless medium. When it is compared with quantum theory, the frictionless medium in question is the vacuum. But liquid helium and superconductors are also frictionless media; are there diffusion phenomena in these domains that can be observed without significantly disturbing the system under observation, and if so could we expect stochastic mechanics to describe them?

Stochastic mechanics fails the proposed test. I refer to the section entitled "The case against stochastic mechanics" of [N86].

III. Random Fields

8. The free Euclidean field

A field is a function defined on $\mathbf{R}^d$, so we would expect a random field to be a stochastic process indexed by $\mathbf{R}^d$. But it turns out that many interesting examples are too singular to be well-defined at points, so we define a *random field* to be a stochastic process ϕ indexed by the *test functions* on $\mathbf{R}^d$, i.e., the space $C_0^\infty(\mathbf{R}^d)$ of smooth functions with compact support, such that ϕ is linear. We shall only consider scalar (real-valued) fields.

Our first example is the *free Euclidean field* of Pitt [P71]. (It is also called the free Markov field.) It is the Gaussian stochastic process of mean 0 and covariance

$$\mathrm{E}\phi(f)\phi(g) = \langle f, (-\tfrac{1}{2}\Delta + m^2)^{-1} g\rangle. \tag{8.1}$$

Here $\langle , \rangle$ is the inner product in $L^2(\mathbf{R}^d)$ and m is a strictly positive constant, the *mass*. To verify that this process is linear, and hence a random field, it suffices to observe that the variance of $\phi(c_1 f_1 + c_2 f_2) - c_1\phi(f_1) - c_2\phi(f_2)$ is 0.

The *Sobolev space* $\mathcal{H}^{-1}$ is the real Hilbert space obtained by completing the test functions in the inner product

$$\langle f, g\rangle_{-1} = \langle f, (-\tfrac{1}{2}\Delta + m^2)^{-1} g\rangle = \int_{\mathbf{R}^d}\int_{\mathbf{R}^d} f(x)G(x-y)g(y)\,dx\,dy, \tag{8.2}$$

where G is the fundamental solution of $(-\frac{1}{2}\Delta + m^2)G = \delta$. This is a space of distributions containing many that are supported by hypersurfaces, but it does not contain delta functions at points if $d > 1$. The random field ϕ clearly extends to be a linear stochastic process indexed by $\mathcal{H}^{-1}$.

We have already seen the free Euclidean field for $d = 1$: it is the ground state process (6.17) for the harmonic oscillator. In fact, for $d = 1$ and $\omega = m$, the fundamental solution G is $e^{-\omega|t|}/2\omega$. This is a Markov process. What is the analogous property in higher dimensions?

Let ϕ be a random field indexed by $\mathcal{H}^{-1}$. For any subset B of $\mathbf{R}^d$, let $\mathcal{O}(B)$ be the σ-algebra generated by all $\phi(f)$ with $\operatorname{supp} f \subseteq B$. We denote the complement of B by B^c and the boundary of B by ∂B. Then we say that ϕ is a *Markov field* in case for all open sets U, the σ-algebras $\mathcal{O}(U)$ and $\mathcal{O}(U^c)$ are conditionally independent given $\mathcal{O}(\partial U)$.

I claim that the free Euclidean field is a Markov field. For any subset B of $\mathbf{R}^d$, let $\mathcal{M}(B)$ be the closed linear subspace of $\mathcal{H}^{-1}$ spanned by the f in $\mathcal{H}^{-1}$ with $\operatorname{supp} f \subseteq B$. Now for any two sets of Gaussian random variables of mean 0, the σ-algebras they generate are independent if and only if the sets are orthogonal. Also, the mapping $f \mapsto \phi(f)$ is an isometry of $\mathcal{H}^{-1}$ into L^2. Consequently, we need only show that for all open sets U, the spaces $\mathcal{M}(U)$ and $\mathcal{M}(U^c)$ are conditionally orthogonal in $\mathcal{H}^{-1}$ given $\mathcal{M}(\partial U)$; that is, we need only show that if $f \in \mathcal{M}(U)$ and g is its orthogonal projection onto $\mathcal{M}(U^c)$, so that

$$\langle h, (-\tfrac{1}{2}\Delta + m^2)^{-1} f \rangle = \langle h, (-\tfrac{1}{2}\Delta + m^2)^{-1} g \rangle, \qquad h \in \mathcal{M}(U^c), \tag{8.3}$$

then $g \in \mathcal{M}(\partial U)$. But let k be smooth with compact support in U^c, and let $h = (-\frac{1}{2}\Delta + m^2)k$. Then $h \in \mathcal{M}(U^c)$, so

$$\langle k, f \rangle = \langle k, g \rangle. \tag{8.4}$$

But the left hand side is 0 since $\operatorname{supp} f \subseteq U$. Hence the right hand side is 0, but since this holds for all such k, we have by definition that $\operatorname{supp} g \subseteq \partial U$, and the claim is established.

Notice that the Euclidean group (the group of all isometries of $\mathbb{R}^d$ as d-dimensional Euclidean space) acts by isometries on $\mathcal{H}^{-1}$ and hence as measure-preserving transformations on the underlying probability space of the free Euclidean field. The free Euclidean field has proved useful in constructive quantum field theory, in which one first replaces relativistic space-time (d-dimensional Minkowski space) by d-dimensional Euclidean space by means of an analytic continuation in time that replaces t by it; see [S74] and [GJ81]. But the free Euclidean field also arises in the extension of stochastic mechanics to the theory of random fields. This was first accomplished by Guerra and Ruggiero [GR73].

The study of a field as a mechanical system goes back to Jeans. But it is an abstract mechanical system; instead of material particles moving in time, Fourier coefficients change in time. Consider the deterministic scalar field satisfying the relativistic field equation

$$(\Box + m^2)\phi = 0. \tag{8.5}$$

Here ϕ is just a function on $\mathbb{R}^d$, and we have represented Minkowski space $\mathbb{R}^d$ as the Cartesian product of a space-like hyperplane $\mathbb{R}^{d-1}$ coordinatized by $\mathbf{x}$ and a time-like axis $\mathbb{R}$ coordinatized by t. The wave operator $\Box$ is $\partial_t^2 - \nabla_{\mathbf{x}}^2$. To simplify the discussion, put the system in a spatial box. That is, choose a number L and consider only functions ϕ that are periodic of period L in each of the $\mathbf{x}$-variables. Now for each t expand $\phi(\mathbf{x}, t)$ in a Fourier series

$$\phi(\mathbf{x}, t) = \sum q_{\mathbf{k}}(t) \cos \mathbf{k}\cdot\mathbf{x} + \sum q'_{\mathbf{k}}(t) \sin \mathbf{k}\cdot\mathbf{x}. \tag{8.6}$$

Then the field equation is satisfied if and only if each *oscillator* $q_{\mathbf{k}}$ and $q'_{\mathbf{k}}$ satisfies the deterministic oscillator equation (6.19) with $\omega^2 = \mathbf{k}^2 + m^2$. Guerra and Ruggiero replaced these deterministic oscillators by independent ground state processes for the harmonic oscillator, computed the covariance, and took the limit as $L \to \infty$ (in such a way that the sum in the Fourier series became a Lebesgue integral). The result was the covariance of the free Euclidean field. Thus stochastic mechanics applied to the free scalar field yields the free Euclidean field.

Their procedure breaks the relativistic symmetry of the problem. Starting with a problem that has relativistic (Poincaré) symmetry, the choice of hyperplane $\mathbb{R}^{d-1}$ leads to a solution that has Euclidean symmetry. Nevertheless, as discussed in §§2,4 of [N86], when one passes from the ground state of the free field to the field coupled to an external current or to the coherent state processes of the free field (which were discussed in the case $d = 1$ in §6), one finds that the expectation values satisfy relativistic equations. Roughly speaking, the means are relativistic and the fluctuations are Euclidean!

9. Bell's inequalities

The inequalities of Bell and their violation in the experiments of Aspect and collaborators are an important part of modern culture, essential to any attempt to understand the world in which we live. I refer to the discussions in [N84, 86]. Here I just want to take the occasion to correct a blunder in the second of these references, which was repeated in [N86a].

In the statement of the theorem on page 445 of [N86], and on pages 533–534 of [N86a], replace $\frac{1}{2}$ by $\frac{1}{3}$. (And delete the remark in the first reference to the effect that in view of Mermin's argument, it is surprising that the result holds with $\frac{1}{2}$ instead of $\frac{1}{3}$!) I am grateful to Lee Newberg for pointing out to me that the $\frac{1}{2}$ had to be wrong on conceptual grounds, precisely because the hypotheses of the theorem can be used to model the situation discussed by Mermin in [M81]. The mistake in the proof is that the center of the $p_1 p_2 p_3$-cube is a local minimum, where the value is $\frac{1}{2}$, but the absolute minimum occurs at a corner of the cube, where the value is $\frac{1}{3}$. The rest of the proof and the subsequent discussion are unaffected.

10. Speculations on a new starting point

Some of my friends who work on stochastic mechanics are unhappy when I express dissatisfaction with aspects of it. I think it quite likely that it is in some sense an approximation to a physically satisfactory theory, and as mathematics it is a lot of fun. But it is important to find a naïvely realistic model of nature that does not violate locality, or to show that this is impossible. I shall conclude by sketching some ideas that have not yet been explored. They come with no guarantee.

Consider first deterministic mechanics and deterministic field theory. What is the analogue in field theory of a time t? It is a space-like hypersurface Σ (a maximal one, that separates Minkowski space R^d into a past and a future). The analogue of a configuration at time t is a field configuration on Σ.

A stochastic process is a random configuration for each time t. Therefore a random field should be a random field configuration for each Σ, that is, a linear stochastic process indexed by test functions f each of which is supported on some Σ. Let us call this a *random space field* to distinguish it from a random field (a *random space-time field*) indexed by test functions F on space-time. Given such an F, it can be represented as a limit of linear combinations of f_i supported by Σ_i, but this can be done in many different ways depending on how the support of F is sliced up into space-like pieces. It is a kind of continuity assumption on a space field to require that for any such decomposition of F, the limit exists and is independent of the slicing. That is, a space field may not come from, or give rise to, a space-time field. If the space field ϕ is constructed as a limit in distribution of space-time fields ϕ_κ as a cut-off $\kappa \to \infty$, there is no reason to assume that it comes from a space-time field, especially as the notion of a momentum cut-off is relative to a space-like hypersurface.

Given an inertial frame, one can construct the free Euclidean field in the manner of Guerra and Ruggiero. It can then be restricted to test functions f supported on $t = constant$ hyperplanes. The same can be done for a different inertial frame, but the random variables are defined on a different probability space. Is there a way to knit all of these together as part of a single relativistically invariant random space field? I believe that the answer is yes.

A space-like hypersurface Σ has a space-like tangent plane at each of its points, which can be characterized by the forward unit time-like vector orthogonal to it (in the sense of Minkowski space). Let S_d be the set of all pairs $\langle x, u \rangle$ where $x \in \mathsf{R}^d$ and u is a forward unit time-like vector. This is a $2d-1$ dimensional manifold. The tangent space at each point of S_d splits naturally as the direct sum of three spaces: u can vary on the $d-1$ dimensional unit hyperboloid, keeping x fixed; x can move in the direction u with u carried along by parallel transport; or x can move in the $d-1$ directions orthogonal to u, again with u carried along by parallel transport. (This description is valid for any pseudo-Riemannian manifold with signature $(d-1,1)$.) But each of these three spaces has a natural Riemannian metric; taking their direct sum, we endow $\mathbf{S}_d$ in a natural way with a Riemannian (not pseudo-Riemannian) metric that is invariant under the action of the Poincaré group. Let $\tilde{\Delta}$ be the Laplace-Beltrami operator on $\mathbf{S}_d$ with respect to this metric, and let $\tilde{G}$ be the kernel of $(-\frac{1}{2}\tilde{\Delta}+m^2)^{-1}$, so that $\tilde{G}$ is a function of positive type on $\mathbf{S}_d \times \mathbf{S}_d$. Now let $\tilde{\phi}$ be the Gaussian stochastic process indexed by test functions on $\mathbf{S}_d$ of mean 0 and covariance given by $\tilde{G}$.

We want to construct a random space field from $\tilde{\phi}$. Let f be a test function on Σ, let $\tilde{\Sigma}$ be the hypersurface in $\mathbf{S}_d$ consisting of all points $\langle x,u\rangle$ in $\mathbf{S}_d$ with $x \in \Sigma$, and let $\tilde{f}$ be the function supported on $\tilde{\Sigma}$ such that $\tilde{f}(\langle x,u\rangle)=f(x)$. Now $\langle \tilde{f},\tilde{G}\tilde{f}\rangle=\infty$. But let $B_\kappa(u)$ be the ball of radius κ and center u in the unit hyperboloid and let C_κ be its volume (which does not depend on u); then it should be true that the following exists:

$$[f,g]_{-1}=\lim_{\kappa\to\infty} C_\kappa^{-2}\int_\Sigma\int_{B_\kappa(n_x)}\int_{\Sigma'}\int_{B_\kappa(n'_y)}\tilde{f}(\langle x,u\rangle)\tilde{G}(\langle x,u\rangle,\langle y,v\rangle)\tilde{g}(\langle y,v\rangle), \tag{10.1}$$

where f and g are test functions supported by Σ and Σ', and n_x is the forward unit normal at x in Σ and similarly for n'_y.

Notice that for fixed u (which amounts to fixing an inertial frame up to a spatial rotation), the set of all $\langle x,u\rangle$ is a submanifold of $\mathbf{S}_d$ whose induced metric is the *Euclidean* metric on $\mathbf{R}^d$, and $\mathbf{S}_d$ is isometric to the Cartesian product of Euclidean space and the unit hyperboloid. Therefore, for f and g supported by two $t=$ *constant* hyperplanes in this frame, I would expect that $[f,g]_{-1}=K_d\langle f\otimes\delta_1, g\otimes\delta_2\rangle_{-1}$ where δ_1 and δ_2 are delta functions in the variable orthogonal to the hyperplanes and K_d is a constant depending only on the dimension.

In this picture, the Guerra-Ruggiero field would be a random space field, equally valid in every inertial frame. It would serve to specify in a relativistically invariant way the kinematics of random space fields, and possibly open the way to construct a dynamics of critical random space fields that might avoid problems of non-locality and lack of relativistic covariance that have hitherto troubled the extension of the ideas of stochastic mechanics to field theory.

Some corrections to "Quantum Fluctuations" [N85]

p. 8, line 4: *For* mult *read* multi.

p. 8, line 6: *For* There *read* There-.

p. 77, line 5: The assertion that $\frac{1}{2}(D_+ + D_-)E=0$, or equivalently that

$$(\partial/\partial t + v\cdot\nabla)E=0,$$

where E is the stochastic energy $\frac{1}{2}(u^2+v^2)+\phi$, is in general false even for free diffusions. For otherwise we would have

$$(\partial/\partial t+\nabla S\cdot\nabla)\big((\nabla R)^2+(\nabla S)^2\big)=0.$$

Eliminate the time derivatives using the stochastic Hamilton-Jacobi equation for free motion and the current equation. What remains is a third order partial differential equation that R and S must satisfy at each time. But this is absurd, since the initial values can be chosen arbitrarily subject only to the normalization condition $\int e^{2R}=1$. I am grateful to R. Marra for help with this correction.

p. 87, (16.18): Replace $-$ by $+$.

p. 91, (17.5): The coefficient of the sinh term should have a minus sign (or replace the factor $t - 2\lambda^2$ next to it by $2\lambda^2 - t$). I am grateful to J. Fronteau for this correction.

p. 102, end of §20: There should be an acknowledgment of the priority of the work of Schulman [S68] and of Laidlaw and C. M. DeWitt [LD70] on the restrictions (expressed by these authors in terms of Feynman path integrals rather than stochastic mechanics) on the wave function when the configuration space is not simply connected.

References

[BCZ87] Ph. Blanchard, Ph. Combe, and W. Zheng, "Mathematical and Physical Aspects of Stochastic Mechanics," *Springer Lecture Notes in Phys.* **281**, 1987.

[BS34] R. Bowling Barnes and S. Silverman, "Brownian motion as a natural limit to all measuring processes," *Rev. Mod. Phys.* **6** (1934), 162–192.

[C84] Eric A. Carlen, "Conservative diffusions," *Commun. Math. Phys.* **94** (1984), 293–315.

[C85] —, "Potential scattering in stochastic mechanics," *Ann. Inst. H. Poincaré* **42** (1985), 407–428.

[C86] —, "The pathwise description of quantum scattering in stochastic mechanics," in *Stochastic Processes in Classical and Quantum Systems, Proc., Ascona, Switzerland 1985*, ed. S. Albeverio, G. Casati, and D. Merlini, *Springer Lecture Notes in Phys.* **262** (1986), 139–147.

[CT86] E. A. Carlen and A. Truman, "Sojourn times and first hitting times in stochastic mechanics," in *Fundamental Aspects of Quantum Theory*, ed. Vittorio Gorini and Alberto Frigerio, Plenum, New York (1986), 151–161.

[D57] Bryce S. DeWitt, "Dynamical theory in curved spaces I. A review of the classical and quantum action principles," *Rev. Mod. Phys.* **29** (1957), 377–397.

[D70] Thaddeus George Dankel, Jr., "Mechanics on manifolds and the incorporation of spin into Nelson's stochastic mechanics," *Arch. Rational Mech. Anal.* **37** (1970), 192–222.

[DG78] D. Dohrn and F. Guerra, "Nelson's stochastic mechanics on Riemannian manifolds," *Lettere al Nuovo Cimento* **22** (1978), 121–127.

[DG79] —, "Geodesic correction to stochastic parallel displacement of tensors," in *Stochastic Behavior in Classical and Quantum Physics*, ed. G. Casati and J. Ford, *Springer Lecture Notes in Phys.* **93** (1979), 241–249.

[DT87] Ian Davies and Aubrey Truman, eds., Proceedings of a conference held in Swansea 1986, to appear in *Springer Lecture Notes in Math.*

[F52] Imre Fényes, "Eine wahrscheinlichkeitstheoretische Begründung und Interpretation der Quantenmechanik," *Zeitschrift für Phys.* **132** (1952), 81–106.

[F85] Richard P. Feynman, "QED—The Strange Theory of Light and Matter," Princeton University Press, Princeton, NJ, 1985.

[G85] Francesco Guerra, "Carlen processes: a new class of diffusions with singular drifts," in *Quantum Probability and Applications II, Proc. Workshop Heidelberg 1984*, ed. L. Accardi and W. von Waldenfels, *Springer Lecture Notes in Math.* **1136** (1985), 259–267.

[GJ81] James Glimm and Arthur Jaffe, "Quantum Physics—A Functional Integral Point of View," Springer-Verlag, New York, 1981.

[GL81] Francesco Guerra and Maria I. Loffredo, "Thermal mixtures in stochastic mechanics," *Lettere al Nuovo Cimento* **30** (1981), 81–87.

[GM83] Francesco Guerra and Laura M. Morato, "Quantization of dynamical systems and stochastic control theory," *Phys. Rev. D* **27** (1983), 1774–1786.

[GR73] Francesco Guerra and Patrizia Ruggiero, "A new interpretation of the Euclidean-Markov field in the framework of physical Minkowski space-time," *Phys. Rev. Letters* **31** (1973), 1022–1025.

[I62] Kiyosi Itô, "The Brownian motion and tensor fields on a Riemannian manifold," *Proc. Int. Congress Math. (Stockholm)* (1962), 536–539.

[L87] John D. Lafferty, "The density manifold and configuration space quantization," to appear in *Trans. Am. Math. Soc.*

[LD70] Michael G. G. Laidlaw and Cécile Morette DeWitt, "Feynman functional integrals for systems of indistinguishable particles," *Phys. Rev. D* **3** (1971), 1375–1378.

[M81] N. D. Mermin, "Bringing home the quantum world: quantum mysteries for anybody," *Am. J. Phys.* **49** (1981), 940–943.

[N84] Edward Nelson, "Quantum fluctuations—an introduction," in *Mathematical Physics VII, Proc. VIIth Int. Congress on Math. Phys., Boulder, CO, 1983*, ed. W. E. Brittin, K. E. Gustafson, and W. Wyss, North-Holland, Amsterdam, (1984), 509–519.

[N85] —, "Quantum Fluctuations," Princeton University Press, Princeton, NJ, 1985.

[N85a] —, "Critical diffusions," in *Séminaire de Probabilités XIX 1983/84*, ed. J. Azéma and M. Yor, *Springer Lecture Notes in Math.* **1123** (1985), 1–11.

[N86] —, "Field theory and the future of stochastic mechanics," in *Stochastic Processes in Classical and Quantum Systems, Proc., Ascona, Switzerland 1985*, ed. S. Albeverio, G. Casati, and D. Merlini, *Springer Lecture Notes in Phys.* **262** (1986), 438–469.

[N86a] —, "The locality problem in stochastic mechanics," in *New Techniques and Ideas in Quantum Measurement Theory*, ed. Daniel M. Greenberger, *Ann. N. Y. Acad. Sci.* **480** (1986), 533–538.

[P71] L. Pitt, "A Markov property for Gaussian processes with a multidimensional parameter," *Arch. Rational Mech. Anal.* **43** (1971), 367–391.

[S68] Lawrence Schulman, "A path integral for spin," *Phys. Rev.* **176** (1968), 1558–1569.

[S74] Barry Simon, "The $P(\Phi)_2$ Euclidean (Quantum) Field Theory," Princeton University Press, Princeton, NJ, 1974.

[S80] D. S. Shucker, "Stochastic mechanics of systems with zero potential," *J. Functional Anal.* **38** (1980), 146–155.

[W88] Timothy C. Wallstrom, Thesis, Princeton University, to appear.

EXPOSES 1985

O. ADELMAN	Une curiosité géométrique concernant les trajectoires browniennes dans $\mathbb{R}^d$
P. ARTZNER	Les rôles de la loi des grands nombres en théorie des assurances
M. BABILLOT	Potentiel des chaînes semi-markoviennes : que se passe-t-il lorsque le point de départ de la chaîne s'en va à l'infini ?
P. BALDI	Optimisation globale et déstabilisation d'équilibres
L. BIRGE	Estimating decreasing densities. Asymptotics versus non-asymptotics
C. BOUTON	Approximation gaussienne pour des algorithmes stochastiques à dynamique markovienne
P. CHASSAING	Une loi forte des grands nombres pour un produit demi markovien de matrices
F. COMETS	Tunnelling and nucleation for a local mean-field magnetic model
A.R. DARWICH	Sur l'orthogonalité de deux probabilités correspondant à deux processus de Markov
L. GALLARDO	Critère de transience des marches aléatoires sur les hypergroupes commutatifs
P.L. HENNEQUIN	Les probabilités et les statistiques dans l'enseignement secondaire
C. KIPNIS	Un théorème central limite pour un système infini de particules
H. LAPEYRE	Grandes déviations pour certains systèmes différentiels aléatoires

R. LEANDRE	Application du calcul de Malliavin à certaines estimations de densité de processus
J.F. LE GALL	Un théorème central limite pour le nombre de points visités par une marche aléatoire plane récurrente
F. LE GLAND	Méthodes de Monte-Carlo en filtrage non-linéaire
G. LETAC	La loi de Cauchy-conforme et les fonctions qui préservent son type
P. MASSART	Majorations exponentielles pour les processus empiriques et principes d'invariance
F. MESSACI	Estimation de la densité spectrale d'un processus à temps continu par échantillonnage poissonnien
M. MIGUENS	Categorical data and small samples
J. PICARD	Discrétisation en temps du problème de Dobrushin
G. ROYER	Modèles de rotateurs et théorème de Dobrushin
F. RUSSO	Expression intégrale de certaines espérances conditionnelles relatives à des tribus engendrées par le drap brownien
E.H. SADI	Caractéristiques locales et processus à accroissements indépendants tangents
A. TOUATI	m-convergence en loi pour des suites de processus. Application aux semi-martingales
P. VALLOIS	Une extension d'un théorème de Ray-Knight concernant le temps local du mouvement brownien
S. WEINRYB	Etude asymptotique de l'image de la saucisse de Wiener par une mesure de $\mathbb{R}$. Applications à un problème d'homogénéisation et aux temps locaux d'intersection

EXPOSES 1986

P. BALDI	Sur le module de continuité des diffusions
J. BARTOSZEWICZ	Ordre de dispersion et transformation TTT
A. BENASSI	Ponts markoviens Problème de Dirichlet stochastique quasi-linéaire aux limites non homogènes
L. CANTO E CASTRO	Vitesse de convergence en théorie des valeurs extrêmes (modèles Normal et Gamma)
A. DERMOUNE	Minoration de l'état fondamental de l'équation de Schrödinger d'une particule de spin - $\frac{1}{2}$
N. EL KAROUI	Une méthode probabiliste de construction des réduites
J.P. FOUQUE	Hydrodynamique des systèmes de particules monotones et non symétriques : l'exclusion simple et le zero range
G. FOURT	Problèmes statistiques posés par l'estimation des paramètres dans des processus de cristallisation
C. GRAHAM	Système de particules réfléchies avec intéraction sur une frontière collante
R. HOPFNER	Asymptotic inference for continuous-time Markov chains
A. HUARD	Résolution par mixage d'opérateurs de l'équation de la cinétique des neutrons avec approximation probabiliste du terme de diffusion
A.T. LAWNICZAK	Gaussian Stochastic Processes with sample paths in Orlicz spaces
J.F. LE GALL	Quelques résultats de fluctuation pour la saucisse de Wiener

G. LETAC	Réciprocité des familles exponentielles sur $\mathbb{R}$
P. McGILL	Les sauts d'un processus de Lévy
M. MORA	Familles exponentielles naturelles sur $\mathbb{R}$ et leur fonction variance
D. NUALART	Equations d'onde stochastiques : Propagation des singularités
J. PICARD	Equation différentielle stochastique avec bruit coloré
R. RUSSO	Tribus séparantes et propriétés de Markov pour une classe de processus généralisés et ordinaires à n paramètres (gaussiens)
C. SAVONA	Approximation d'un problème de filtrage "linéaire par morceaux"
A.S. SZNITMAN	Propagation du chaos pour un système de splines browniennes se détruisant
P. VALLOIS	Intégration stochastique avec intégrant non adapté
S. WEINRYB	Application des résultats asymptotiques sur une ou plusieurs saucisses de Wiener localisées aux problèmes d'homogénéisation. Théorème central limite pour le potentiel associé.
L.M. WU	L'inégalité de Meyer sur l'espace de Poisson

EXPOSES 1987

L. ANDERSON	Prequantization of Infinite Dynamical Systems
J. ANGULO - R. GUTIERREZ	Sur des équations intégrales stochastiques de McShane dépendant d'un paramètre
M. BABILLOT	Fonctions harmoniques positives sur un espace riemannien symétrique
D. BOIVIN	Récurrence multiple et opérateurs linéaires
P. BOXLER	On a stochastic version of the center manifold theorem
W. CHOJNACKI	Non-trivial random cocycles
A. DERMOUNE	Décomposition en chaos et isométries définies par les intégrales stochastiques multiples
G. DEL GROSSO	Filtering for jump processes
G. DEWITT-MORETTE	Meiman's definition of "functional integrals as integrals on locally non compact groups with (rough) generalized measures"
W. DZIUBDZIELA	Limit laws for kth order statistics from conditionally mixing arrays of random variables
M. EMERY	Martingales dans les variétés
T. KOLSRUD	Dirichlet forms associated with boundary Dirichlet forms
T. KOLSRUD	Multiplicative random fields with applications in mathematical physics

P. KREE	Chaotic calculus, multiple stochastic integrals and Girsanov's formula
J. LAFFERTY	The index theorem and the action of a diffusion
J. LAFFERTY	Infinite dimensional geometry and stochastic mechanics
M. LAPIDUS	The Feynman-Kac formula with a Lebesgue-Stieltjes measure and Feynman's Operational Calculus
A.T. LAWNICZAK	RKHS for Gaussian measures on metric vector spaces
F. LE GALL	Mouvement brownien, cônes et processus stables
U. MANSMANN	About a random walk driven by a random media
M. PONTIER	Approximation d'un filtre avec observation sur une variété compacte riemannienne
S. ROSENBERG	An overview of the Atiyah-Singer Index Theorem
M. SANZ	Une application du calcul de variations stochastiques au problème du retournement du temps pour des diffusions
W. SZCZOTKA	Hereditary properties of queueing processes
A.S. USTUNEL	Quelques (petites) remarques sur les Diffusions Conditionnelles
J. VAN CASTEREN	Integral kernels and Schrödinger type equations : Pointwise inequalities for Schrödinger semigroups
W.D. WICK	Some recent results in the "hydrodynamic" theory of interacting particle systems
L.M. WU	Représentation de fonctionnelles et intégrale de Skorokhod sur l'espace de Poisson

LISTE DES AUDITEURS

Mle ALPIUM M.T.	(85)	Lisbonne (Portugal)
Mle AMINE S.	(87)	Paris VI
Mr. ANDERSSON L.	(87)	Stockholm (Suède)
Mr. ANGULO IBANEZ J.	(87)	Grenade (Espagne)
Mle ARENAS C.	(86)	Barcelone (Espagne)
Mr. ARTZNER P.	(85)	Strasbourg I
Mle ATHAYDE E.	(85)	Lisbonne (Portugal)
Mr. AZEMA J.	(87)	Paris VI
Mle BABILLOT M.	(85)	Paris VII
Mr. BADRIKIAN A.	(85-86-87)	Clermont II
Mr. BALDI P.	(86-87)	Pise (Italie)
Mr. BARTOSZEWICZ J.	(86)	Wroclaw (Pologne)
Mr. BENASSI A.	(86-87)	Paris VI
Mr. BERNARD P.	(85-86-87)	Clermont II
Mr. BERTHUET R.	(85)	Clermont II
Mr. BOIVIN D.	(87)	Brest
Mr. BOUAZIZ M.	(87)	Clermont II
Mle BOUTON C.	(86-87)	Palaiseau
Mle BOXLER P.	(86)	Bremen (R.F.A.)
Mr. CANDELPERGHER B.	(87)	Nice
Mr. CANELA M.	(85)	Barcelone (Espagne)
Mme CANTO E CASTRO L.	(86)	Lisbonne (Portugal)
Mme CHALEYAT-MAUREL M.	(85-86)	Paris VI
Mr. CHASSAING P.	(86)	Nancy I
Mle CHEVET S.	(85-87)	Clermont II
Mr. CHOJNACKI W.	(87)	Varsovie (Pologne)
Mle CHOU S.	(87)	Paris
Mr. COMETS F.	(85)	Paris XI
Mr. DARWICH A.	(85)	Paris VI
Mr. DELL'ANTONIO G.	(87)	Rome (Italie)
Mr. DERMOUNE A.	(86-87)	Paris VI
Mme DEWITT-MORETTE	(87)	Austin(U.S.A.)
Mr. DOSS H.	(85)	Paris VI
Mr. DZIUBDZIELLA W.	(87)	Wroclaw (Pologne)
Mme ELIE L.	(85)	Paris VII
Mme EL KAROUI N.	(86)	Paris VI
Mr. EMERY	(87)	Strasbourg I

Mr. FEUERVERGER A.	(86)	Toronto (Canada)
Mr. FOUQUE J.P.	(86)	Paris VI
Mr. FOURT G.	(85-86-87)	Clermont II
Mme GABRIELLA DEL GROSSO	(87)	Rome (Italie)
Mr. GALLARDO L.	(85-86)	Nancy II
Mr. GOLDBERG J.	(86-87)	Lyon
Mr. GRAHAM C.	(86)	Palaiseau
Mr. GRORUD A.	(87)	Marseille
Mr. GUISSE M.	(86)	Abidjan (Cote d'Ivoire)
Mr. GUTIERREZ J.	(87)	Grenade (Espagne)
Mr. HENNEQUIN P.L.	(85-86-87)	Clermont II
Mr. HU Y.	(87)	Strasbourg I
Mr. HOPFNER R.	(86)	Ludwigs (R.F.A.)
Mr. HUARD A.	(86)	Besançon
Mme JOLIS M.	(86)	Barcelone (Espagne)
Mr. KERKYACHARIAN G.	(87)	Nancy I
Mr. KIPNIS C.	(85)	Paris VII
Mr. KOLSRUD T.	(87)	Stockholm (Suède)
Mle KOUKIOU F.	(86)	Lausanne (Suisse)
Mr. KREE P.	(87)	Paris VI
Mr. LAFFERTY J.	(87)	Cambridge (U.S.A.)
Mme LAPEYRE H.	(85)	Paris VI
Mr. LAPEYRE B.	(85)	Paris
Mr. LAPIDUS M.	(87)	Georgia (U.S.A.)
Mle LAWNICZAK A.	(85-86-87)	Toronto (Canada)
Mr. LEANDRE R.	(85)	Besançon
Mr. LE GALL J.F.	(85-86-87)	Paris VI
Mr. LE GLAND F.	(85)	Sophia Antipolis
Mr. LOBRY C.	(87)	Nice
Mr. LOTI-VIAUD	(85)	Paris VI
Mr. MANSMANN U.	(87)	Berlin (R.F.A.)
Mme MASSAM H.	(86)	Ontario (Canada)
Mr. MASSART P.	(85)	Paris XI
Mr. McGILL	(86)	Dublin (Irlande)
Mle MESSACI F.	(85)	Rouen
Mle MIGUENS M.	(85)	Lisbonne (Portugal)
Mle MILLET A.	(85)	Angers
Mr. MOGHA G.	(85-86)	Clermont II
Mle MORA M.	(86)	Toulouse

Mr. NUALART D.	(86)	Barcelone (Espagne)
Mme PARK S.	(87)	Austin (U.S.A.)
Mr. PEREZ	(87)	Grenade (Espagne)
Mme PERRIN Y.	(85-87)	Clermont II
Mr. PETRITIS D.	(86)	Lausanne (Suisse)
Mme PICARD D.	(86-87)	Paris VI
Mr. PICARD J.	(85-86-87)	Sophia Antipolis
Mr. PISTONE G.	(86)	Gênes (Italie)
Mme PONTIER M.	(87)	Orléans
Mr. ROSENBERG S.	(87)	Boston (U.S.A.)
Mr. ROUX D.	(85-86-87)	Clermont II
Mr. ROYER G.	(85)	Clermont II
Mr. RUSSO F.	(85-86)	Lausanne (Suisse)
Mle SAADA E.	(85)	Paris VI
Mr. SADI E. H.	(85-86)	Paris VI
Mme SANZ M.	(87)	Barcelone (Espagne)
Mle SAVONA C.	(86-87)	Lyon
Mr. SONG S.	(87)	Paris VI
Mr. SZCZOTKA W.	(87)	Worclaw (Pologne)
Mr. SZNITMAN A.	(86-87)	Paris VI
Mr. TERRANOVA D.	(87)	Milan (Italie)
Mr. TOUATI A.	(85)	Bizerte (Tunisie)
Mr. UGO	(87)	Italie
Mr. USTUNEL	(87)	Issy les Moulineaux
Mr. VALLOIS P.	(85-86)	Paris VI
Mle WEINRYB S.	(85-86)	Palaiseau
Mr. WICK D.	(87)	Colorado (U.S.A.)
Mr. WU L.	(86-87)	Paris VI
Mr. YCART B.	(87)	Pau

Part I

Sergio Albeverio: Theory of Dirichlet forms and applications

Originally published in: *Ecole d'Eté de Probabilités de Saint-Flour XXX – 2000*, Lecture Notes in Mathematics, Vol. **1816**, 1–6, DOI: 10.1007/3-540-44922-1, © Springer-Verlag Berlin Heidelberg 2003, Reprint by Springer-Verlag Berlin Heidelberg 2012

Table of Contents

Summary. The theory of Dirichlet forms, Markov semigroups and associated processes on finite and infinite dimensional spaces is reviewed in an unified way. Applications are given including stochastic (partial) differential equations, stochastic dynamics of lattice or continuous classical and quantum systems, quantum fields and the geometry of loop spaces.

0 Introduction

The theory of Dirichlet forms is situated in a vast interdisciplinary area which includes analysis, probability theory and geometry.
Historically its roots are in the interplay between ideas of analysis (calculus of variations, boundary value problems, potential theory) and probability theory (Brownian motion, stochastic processes, martingale theory).
First, let us shortly mention the connection between the "phenomenon" of Brownian motion, and the probability and analysis which goes with it. As well known the phenomenon of Brownian motion has been described by a botanist, R. Brown (1827), as well as by a statistician, in connection with astronomical observations, T.N. Thiele (1870), by an economist, L. Bachelier (1900), (cf. [455]), and by physicists, A. Einstein (1905) and M. Smoluchowski (1906), before N. Wiener gave a precise mathematical framework for its description (1921-1923), inventing the prototype of interesting probability measures on infinite dimensional spaces (Wiener measure). See, e.g., [394] for the fascinating history of the discovery of Brownian motion (see also [241], [16] for subsequent developments).
This went parallel to the development of infinite dimensional analysis (calculus of variation, differential calculus in infinite dimensions, functional analysis, Lebesgue, Fréchet, Gâteaux, P. Lévy...) and of potential theory.

Although some intimate connections between the heat equation and Brownian motion were already implicit in the work of Bachelier, Einstein and Smoluchowski, it was only in the 30's (Kolmogorov, Schrödinger) and the 40's that the strong connection between analytic problems of potential theory and fine properties of Brownian motion (and more generally stochastic processes) became clear, by the work of Kakutani. The connection between analysis and probability (involving the use of Wiener measure to solve certain analytic problems) as further developed in the late 40's and the 50's, together with the application of methods of semigroup theory in the study of partial differential equations (Cameron, Doob, Dynkin, Feller, Hille, Hunt, Martin, ...).

The theory of stochastic differential equations has its origins already in work by P. Langevin (1911), N. Bernstein (30's), I. Gikhman and K. Ito (in the 40's), but further great developments were achieved in connection with the above mentioned advances in analysis, on one hand, and martingale theory, on the other hand.

By this the well known relations between Markov semigroups, their generators and Markov processes were developed, see, e.g. [162], [160], [207], [208], [209], [276], [463].

This theory is largely concerned with processes with "relatively nice characteristics" and with "finite dimensional state space" E (in fact locally compact state spaces are usually assumed). From many areas, however, there is a demand of extending the theory in two directions:

1) "more general characteristics", e.g. allowing for singular terms in the generators
2) infinite dimensional (and nonlinear) state spaces.

As far as 1) is concerned let us mention the needs of handling Schrödinger operators and associated processes in the case of non smooth potentials, see [70].

As far as 2) is concerned let us mention the theory of partial differential equations with stochastic terms (e.g. "noises"), see, e.g. [201], [28], [37], [38], [129], [127] the description of processes arising in quantum field theory (work by Friedrichs, Gelfand, Gross, Minlos, Nelson, Segal...) or in statistical mechanics, see, e.g. [16], [15], [344], [242]. Other areas which require infinite dimensional processes are the study of variational problems (e.g. Dirichlet problem in infinite dimensions) [278], the study of certain infinite dimensional stochastic equations of biology, e.g. [474], the representation theory of infinite dimensional groups, e.g. [68], the study of loop groups, e.g. [30], [12], the study of the development of interest rates in mathematical finance, e.g. [416], [337], [502].

The theory of Dirichlet forms is an appropriate tool for these extensions. In fact it is central for it to work with reference measures μ which are neither necessarily "flat" nor smooth and in replacing the Markov semigroups on continuous functions of the "classical theory" by Markov semigroups on

$L^2(\mu)$-spaces (thus making extensive use of "Hilbert space methods" [211]). The theory of Dirichlet forms was first developed by Feller in the 1-dimensional case, then extended to the locally compact case with symmetric generators by Beurling and Deny (1958-1959), Silverstein (1974), Ancona (1976), Fukushima (1971-1980) and others (see, e.g., [244], [258]).(Extensions to non symmetric generators were given by J. Elliott, S. Carrillo-Menendez (1975), Y. Lejan (1977-1982), a.a., see, e.g. [367]).
The case of infinite dimensional state spaces has been investigated by S. Albeverio and R. Høegh-Krohn (1975-1977), who were stimulated by previous analytic work by L. Gross (1974) and used the framework of rigged Hilbert spaces (along similar lines is also the work of P. Paclet (1978)). These studies were successively considerably extended by Yu. Kondratiev (1982-1987), S. Kusuoka (1984), E. Dynkin (1982), S.Albeverio and M.Röckner (1989-1991), N. Bouleau and F. Hirsch (1986-1991), see [39], [147], [278], [367], [230], [172], [465], [234], [235], [236], [237], [238], [239], [256].
An important tool to unify the finite and infinite dimensional theory was provided by a theory developed in 1991, by S. Albeverio, Z.M. Ma and M. Röckner, by which the analytic property of quasi regularity for Dirichlet forms has been shown in "maximal generality" to be equivalent with nice properties of the corresponding processes.
The main aim of these lectures is to present some of the basic tools to understand the theory of Dirichlet forms, including the forefront of the present research. Some parts of the theory are developed in more details, some are only sketched, but we made an effort to provide suitable references for further study.
The references should also be understood as suggestions in the latter sense, in particular, with a few exceptions, whenever a review paper or book is available we would quote it rather than an original reference. We apologize for this "distortion", which corresponds to an attempt of keeping the reference list into some reasonable bounds - we hope however the references we give will also help the interested reader to reconstruct historical developments.
For the same reason, all references of the form "see [X]" should be understood as "see [X] and references therein".

1 Functional analytic background: semigroups, generators, resolvents

1.1 Semigroups, Generators

The natural setting used in these lectures is the one of normed linear spaces B over the closed algebraic field $\mathbb{K} = \mathbb{R}$ or $\mathbb{C}$. Some of the results are however depending on the additional structure of completeness, therefore we shall assume most of the time that B is a Banach space.
We are interested in describing operators like the Laplacian Δ and the associated semigroup (heat semigroup), and vast generalizations of them.
Let $L \equiv (L, D(L))$ be a linear operator on a normed space B over $\mathbb{K}$, defined on a linear subset $D(L)$ of B, the definition domain of L.
We say that two such operators $L_i, i = 1, 2$ are equal if $D(L_1) = D(L_2)$ and $L_1 u = L_2 u, \forall u \in D(L_1)$.
L is said to be bounded if $\exists C \geq 0$ s.t. $\|Lu\| \leq C\|u\|, \forall u \in D(L) = B$.
We then have, setting $\|L\| \equiv \sup_{u \in B, \|u\| \leq 1} \|Lu\| \in [0, +\infty]$

$$L \text{ bounded } \Leftrightarrow \|L\| < +\infty.$$

L is said to be continuous at 0 ($\in D(L)$!) if $u_n \to 0, u_n \in D(L)$ implies $Lu_n \to 0, n \to \infty$.
L is said to be continuous if $u_n \to u, u_n \in D(L)$ implies $u \in D(L)$ and $Lu_n \to Lu, n \to \infty$.
One easily shows

$$L \text{ bounded } \Leftrightarrow L \text{ continuous at } 0 \Leftrightarrow L \text{ continuous.}$$

We define $L = \alpha_1 L_1 + \alpha_2 L_2, \alpha_i \in \mathbb{K}, i = 1, 2$, by $D(L) = D(L_1) \cap D(L_2), Lu = \alpha_1 L_1 u + \alpha_2 L_2 u, \forall u \in D(L)$.
Moreover we define for L_1, L_2
$L_1 L_2 u \equiv L_1(L_2 u), \forall u \in D(L_1 L_2) \equiv L_1 D(L_2) \equiv \{u \in B | L_2 u \in D(L_1)\}$

Definition 1. *A linear bounded operator A on a normed linear space B is a contraction if $\|A\| \leq 1$. A family $T = (T_t)_{t \geq 0}$ of linear bounded operators on B is said to be a strongly continuous semigroup or C_0-semigroup if*

i) $T_0 = 1$ *(the identity on B)*
ii) $\lim_{t \downarrow 0} T_t u = u, \forall u \in B$ *(strong continuity)*
iii) $(T_t)_{t \geq 0}$ *is a semigroup i.e.*
$T_t T_s = T_s T_t = T_{s+t}, \forall t, s > 0.$
$(T_t)_{t \geq 0}$ *is said to be a C_0-semigroup of contractions or a C_0-contraction semigroup if, in addition,*
iv) T_t *is a contraction for all $t \geq 0$.*

Originally published in: *Ecole d'Eté de Probabilités de Saint-Flour XXX – 2000*, Lecture Notes in Mathematics, Vol. **1816**, 7–17, DOI: 10.1007/3-540-44922-1_1, © Springer-Verlag Berlin Heidelberg 2003, Reprint by Springer-Verlag Berlin Heidelberg 2012

Exercise 1. Show that i),ii),iv) imply that $t \to T_t u$ is continuous, for all $t \geq 0, \forall u \in B$.

Definition 2. *Let $T \equiv (T_t)_{t \geq 0}$ be a C_0-contraction semigroup on B. The linear operator L is said to be generator of T if:*

i) $D(L) \equiv \left\{ u \in B | \lim\limits_{t \downarrow 0} \frac{1}{t}(T_t u - u) \text{ exists in } B \right\}$
ii) $Lu = \lim\limits_{t \downarrow 0} \frac{1}{t}(T_t u - u) \, \forall u \in D(L)$

Exercise 2. Show that the "strong derivative" $\frac{d}{dt} T_t u \equiv \lim\limits_{h \downarrow 0} \frac{(T_{t+h} - T_t)u}{h}$ exists in B, for all $u \in D(L)$ and $\frac{d}{dt} T_t u = L T_t u = T_t L u \, \forall t \geq 0, \forall u \in D(L)$.
In particular $Lu = \frac{d}{dt} T_t u|_{t=0}, \forall u \in D(L)$.

It is easy to convince oneself that even simple operators like the Laplacian Δ are not bounded, e.g. in $B = L^2(\mathbb{R}^d)$. For this reason it is useful to introduce the concept of a closed operator.

Definition 3. *A linear operator L in B is called <u>closed</u> if $u_n \in D(L)$, $u_n \to u$ as $n \to \infty$, Lu_n convergent as $n \to \infty$, in B, imply that $u \in D(L)$, and $Lu_n \to Lu$.*

Exercise 3. Show that L closed $\Leftrightarrow$ $G(L)$ closed in $B \times B$, where $G(L) \equiv \{\{u, Lu\}, u \in D(L)\}$ is the graph of L.

Proposition 1. *Let $T = (T_t)_{t \geq 0}$ be a C_0-contraction semigroup on a Banach space B, with generator L. Then $T_t u = u + \int\limits_0^t T_s L u \, ds, u \in D(L)$ where the integral on the r.h.s is to be understood in the natural sense of strong integrals on Banach spaces (Bochner integral [1]).*

Proof. This follows immediately from Exercise 2, via integration. □

Proposition 2. *The generator L of a C_0-contraction semigroup $T = (T_t)_{t \geq 0}$ on a Banach space is a closed operator.*

Proof. This easily follows from Proposition 1, the strong continuity (Exercise 1), the fact that for $u_n \to u$, Lu_n convergent to v, $\|T_s L u_n\| \leq \|L u_n\| \leq C$, for some $C \geq 0$, independent of n, as Lu_n converges, and dominated convergence. □

Proposition 3. *The generator L of a C_0-contraction semigroup $T = (T_t)_{t \geq 0}$ on a Banach space is densely defined.*

[1] See, e.g. [506], p.132

Proof. One easily shows that for any $u \in B$, with $v_t \equiv \int_0^t T_s u ds$:

$$\frac{1}{r}\left[v_{t+r} - v_t\right] = \frac{1}{r}\left[T_r v_t - v_t\right] \to T_t u - u, \text{ as } r \downarrow 0$$

hence $v_t \in D(L)$.
On the other hand
$\frac{v_t}{t} \to u, t \downarrow 0$, yielding an approximation of an arbitrary $u \in B$ by elements $\frac{v_t}{t}$ in $D(L)$. □

Corollary 1. *If $T = (T_t)_{t \geq 0}, S = (S_t)_{t \geq 0}$ are two C_0-contraction semigroups on a Banach space with the same generator L, then $T_t = S_t \quad \forall t \geq 0$.*

Proof. From Exercise 2 we have easily $\frac{d}{ds} T_{t-s} S_s u = 0, \forall 0 \leq s \leq t, \forall u \in D(L)$ from which $T_t u = S_t u \forall u \in D(L)$ follows, hence $T_t = S_t$, these being bounded and $D(L)$ being dense. □

The above corollary implies that the usual notation $T_t = e^{tL}, t \geq 0$ for the semigroup with generator L is justified.
The question when a given densely defined linear operator L is the generator of a C_0-contraction semigroup is answered by the theory of Hille-Yosida. For this we recall some basic definitions.
If L is a linear injection (1-1 map), then L^{-1} is defined on $D(L^{-1}) = LD(L)$, by $L^{-1}u = v, u \in D(L^{-1})$, with v s.t. $Lv = u$.
For a linear operator L the resolvent set is defined by:
$\rho(L) \equiv \{\alpha \in \mathbb{K} | \alpha - L : D(L) \to B$ is an injection onto B i.e.
$D((\alpha - L)^{-1}) = B$. Moreover $(\alpha - L)^{-1}$ is bounded.$\}$

Exercise 4. Show that if $\rho(L) \neq 0$ then $\rho(L)$ is closed (use that $(\alpha - L)^{-1}$ for $\alpha \in \rho(L)$ is bounded).

The spectrum $\sigma(L)$ of L is by definition the complement in $\mathbb{K}$ of $\rho(L)$. For $\alpha \in \rho(L), G_\alpha \equiv (\alpha - L)^{-1}$ (which exists as a bounded operator on B) is called the resolvent of L at α.
$(G_\alpha)_{\alpha \in \rho(L)}$ is called the <u>resolvent family</u> associated to L.

Exercise 5. Show that $(G_\alpha)_{\alpha \in \rho(L)}$ satisfies the resolvent identity
$G_\alpha - G_\beta = (\beta - \alpha) G_\alpha G_\beta = (\beta - \alpha) G_\beta G_\alpha, \forall \alpha, \beta \in \rho(L)$.

Proposition 4. *Let L be the generator of a C_0-contraction semigroup on a Banach space. Then $(0, \infty) \subset \rho(L)$ and for any*
$Re\alpha > 0 : (\alpha - L)^{-1} u = G_\alpha u = \int_0^{+\infty} e^{-\alpha t} T_t u dt$
(where the integral is in Bochner's sense) and $\|G_\alpha\| \leq \frac{1}{Re\alpha}$.

Proof. Set $R_\alpha \equiv \int_0^{+\infty} e^{-\alpha t} T_t dt$.
It is easily seen that $(\alpha - L)R_\alpha u = u, \forall u \in B, Re\alpha > 0$. Since L is closed for all $u \in D(L) : LR_\alpha u = R_\alpha Lu$, from which one deduces that $\alpha - L$ is injective for $Re\alpha > 0$ (in particular for $\alpha > 0$) and $R_\alpha = G_\alpha$. The bound in Proposition 4 then follows from the definition of R_α. □

Remark 1. G_α is the Laplace transform of T_t (in the sense given by Proposition 4).

Theorem 1. *(Hille-Yosida, for C_0-contraction semigroups):*
Let L be a linear operator in a Banach space B. The following are equivalent:

i) L is the generator of a C_0-contraction semigroup $T = (T_t)_{t \geq 0}$ on B.
ii) L is densely defined and
α) $(0, \infty) \subset \rho(L)$
β) $\|\alpha(\alpha - L)^{-1}\| \leq 1 \quad \forall \alpha > 0$

Corollary 2. *If ii) is fullfilled then L is closed and uniquely determined.*

Proof. ii) implies i) by Theorem 1 and hence that L is closed by Proposition 2. The rest follows from Corollary 1. □

Proof. (of Theorem 1)
i) ⇒ ii): From i) we have L closed, densely defined (Propositions 2,3). That $(0, \infty) \subset \rho(L)$ and ii) holds follows from Proposition 4.
ii) ⇒ i): For details we refer to, e.g.[413]. In the proof the following Proposition is useful.

Proposition 5. *Let L satisfy the conditions ii) of Theorem 1. Set $G_\alpha = (\alpha - L)^{-1}, \alpha > 0$. Then*

i) $\alpha G_\alpha u \to u$ in B, as $\alpha \to +\infty$
ii) Define $L^{(\alpha)} \equiv -\alpha + \alpha^2 G_\alpha, \alpha > 0$ ("Yosida approximation of L"). Then $L^{(\alpha)}$ is bounded, $D(L^{(\alpha)}) = B, L^{(\alpha)}u \to Lu, \alpha \uparrow +\infty, u \in D(L)$, and $e^{tL^{(\alpha)}}u$ converges as $\alpha \uparrow +\infty$ for all $u \in D(L)$ to $\tilde{T}_t u$, where $\tilde{T}_t$ is a C_0-contraction semigroup, with generator L. Moreover $\tilde{T}_t$ coincides with the semigroup T_t generated by L mentioned in i).

Proof. For $u \in D(L)$ we have

$$\begin{aligned} \|\alpha G_\alpha u - u\| &= \|\alpha(\alpha - L)^{-1}u - (\alpha - L)(\alpha - L)^{-1}u\| \\ &= \|L(\alpha - L)^{-1}u\| \\ &= \|(\alpha - L)^{-1}Lu\| \\ &\leq \frac{1}{\alpha}\|Lu\| \to 0, \alpha \uparrow +\infty \end{aligned}$$

(where we used Proposition 4). But αG_α is a contraction by Proposition 4 and $D(L)$ is dense by assumption, hence $\alpha G_\alpha u \to u$ as $\alpha \uparrow +\infty$, for all $u \in B$.

From this it is easy to see that $\alpha G_\alpha Lu \to Lu, u \in D(L)$, as $\alpha \uparrow +\infty$, and thus $L^{(\alpha)}u = -\alpha u + \alpha^2 G_\alpha u = \alpha G_\alpha Lu \to Lu$ as $\alpha \uparrow +\infty$.
The rest follows by realizing that

$$e^{tL^{(\alpha)}}u = \sum_{n=0}^{\infty} \frac{t^n}{n!} L^{(\alpha)^n} u = e^{\alpha t} e^{-\alpha^2 G_\alpha} u$$

Remark 2. Another useful "approximation formula" for T_t in terms of the resolvent is the following one:

$$T_t u = \lim_{n\to\infty} \left(\frac{n}{t}\right)^n (G_{\frac{n}{t}} u)^n, \forall u \in B$$

(see, e.g., [413], p. 33).

Remark 3. In the formulation of Hille-Yosida's theorem i) can be replaced by a statement involving the generator of a C_0-contraction resolvent family according to the following definition.

Definition 4. *A C_0-contraction resolvent family is a family $(G_\alpha)_{\alpha>0}$ such that*

$$\alpha G_\alpha u \to u, \alpha \uparrow +\infty, \|\alpha G_\alpha\| \le 1, \alpha > 0$$

and the resolvent identity in Exercise 5 holds.
Hille-Yosida's theorem holds then with i) replaced by:

i') L is the generator of a C_0-contraction resolvent family $(G_\alpha)_{\alpha>0}$ in the sense that $G_\alpha = (\alpha - L)^{-1}$ on B. There is a one-to-one correspondence between C_0-contraction semigroups $(T_t)_{t\ge 0}$ and C_0-contraction resolvent families $(G_\alpha)_{\alpha>0}$ given by the Laplace-transform formula in Proposition 4 (and Remark 1) resp. Proposition 5 or Remark 2 after Proposition 5.

Hille-Yosida's characterization of generators L involves the resolvent G_α. A pure characterization of L, under some "direct restrictions" on L is given by the Lumer-Phillips theorem, for which we need a definition.

Definition 5. *The duality set $F(u)$ for any element u in a Banach space B is defined by*

$$F(u) \equiv \{u^* \in B^* | \langle u^*, u\rangle = \|u\|^2 = \|u^*\|^2\},$$

where B^* is the dual of B (the space of continuous linear functionals on B) and $\langle , \rangle$ is the dualization between B and B^*.
An operator L is dissipative on B if for any $u \in D(L)$ there exists some $u^* \in F(u)$ such that $Re\langle u^*, Lu\rangle \le 0$.
($-L$ is then said to be accretive).

Proposition 6. *L is dissipative iff*

$$\|(\alpha - L)u\| \geq \alpha\|u\|, \forall u \in D(L) \forall \alpha > 0$$

Proof. See, e.g. [413] (Theorem 4.2). □

Proposition 7. *Let L be dissipative. Then L is closed iff Range* $(\alpha - L)$ *is closed, for all* $\alpha > 0$.

Proof. The proof is left as an exercise (cf,e.g., [413]). □

We recall that an operator L_0 in a Banach space is said to be closable if there exists at least one closed extension $\tilde{L}_0$ of it, i.e. $\tilde{L}_0$ closed and $\tilde{L}_0 u = L_0 u, \forall u \in D(L_0) \subset D(\tilde{L}_0)$. One calls closure $\overline{L}_0$ of L_0 the minimal closed extension of L_0.

Theorem 2. *(Lumer-Phillips)*
Let L be a linear closable operator in a Banach space. Then the closure $\overline{L}$ *of L generates a* C_0*-contraction semigroup on B iff*

a) $D(L)$ *is dense in B*
b) L is dissipative
c) The range of $\alpha_0 - L$ *is dense in B, for some* $\alpha_0 > 0$.

Proof. See, e.g., [413] Theorem 4.3 □

Remark 4. If L is the generator of a C_0-contraction semigroup on B then a) holds, c) holds for all $\alpha > 0$ and b) holds, see, e.g. [413], [424].

Remark 5. If L is a linear operator satisfying a),b) then L is closable. This, together with c) gives that $\overline{L}$ generates a C_0-contraction semigroup.
See [424],(p.240 and p.345).

1.2 The case of a Hilbert space

We shall consider here the special case where the Banach space B of section 1.1 is a Hilbert space $\mathcal{H}$, with scalar product $(,)$.
We first observe that if R is a contraction then

$$|(Ru,u)| \leq \|Ru\| \|u\| \leq \|u\|^2.$$

Hence $Re(Ru,u)$ and $Im(Ru,u)$ are bounded absolutely by $\|u\|^2$.
If $(T_t)_{t\geq 0}$ is self-adjoint, i.e. $T_t^* = T_t$ (where R^* means the adjoint to R) and T_t is a C_0-contraction semigroup on $\mathcal{H}$ with generator L, then for all $u,v \in D(L)$, using the self-adjointness of T_t :

$$\begin{aligned}(-Lu,v) &= \lim_{t\downarrow 0} \frac{1}{t}(u - T_t u, v) \\ &= (u, -Lv)\end{aligned}$$

i.e. L is symmetric in H (in the sense that L^* is an extension of L or, equivalently, $(u,Lv) = (Lu,v), \forall u,v \in D(L)$).

Remark 6. If A is a symmetric operator in $\mathcal{H}$ we have $(u,Au) = (Au,u), \forall u \in D(A)$. On the other hand $\overline{(u,Au)} = (Au,u)$ (by the properties of the scalar product), hence $(u,Au) = \overline{(u,Au)}$ for symmetric operators, i.e. (u,Au) is real.
For A bounded with $D(A) = B$ we have A symmetric iff A is self-adjoint (but this is not so in general for A unbounded!).
In particular a C_0-contraction semigroup is symmetric iff it is self-adjoint. It is easily seen that the following are equivalent:

i) $(T_t)_{t\geq 0}$ is a symmetric C_0-contraction semigroup
ii) $(G_\alpha)_{\alpha>0}$ is a symmetric C_0-contraction resolvent family

(use, e.g., the Laplace transformation Proposition 4, resp. Proposition 5).
We also see that if (T_t) is a symmetric C_0-contraction semigroup then

$$|(u,T_t u)| = |(T_t u,u)| \leq \|u\|^2, \text{ for all } u \in \mathcal{H}. \tag{1}$$

On the other hand $\lim_{t\downarrow 0} \left(\frac{T_t - 1}{t}u, u\right) = (Lu,u), \forall u \in D(L)$.
But $\left(\frac{(T_t-1)u}{t}, u\right)$ is real (by the symmetry property) and negative, by (1), hence $(Lu,u) \leq 0$.
One calls a densely defined operator A in a Hilbert space positive if $(u,Au) \geq 0, \forall u \in D(A)$.

Remark 7. A positive implies $-A$ dissipative. The above says that $(-L)$ is positive, or equivalently, that L is negative.
By Lumer-Phillips theorem the range of $\alpha_0 - L$ is dense in H, for some $\alpha_0 > 0$. Hence we have proven:

Proposition 8. *The generator of a symmetric C_0-contraction semigroup in a Hilbert space is a negative densely defined closed symmetric operator L s.t. the range of $\alpha_0 - L$ is dense, for some $\alpha_0 > 0$.*

Remark 8. One easily shows that the fact that the range of $\alpha_0 - L$ is dense for some $\alpha_0 > 0$ implies that L is self-adjoint (see, e.g. [424]).
Viceversa, if L is linear, symmetric (hence closable) densely defined on $\mathcal{H}$, negative and such that the range of $\alpha_0 - L$ is dense in $\mathcal{H}$ for some $\alpha_0 > 0$ then, by Lumer-Phillips theorem, its closure $\overline{L}$ (which is self-adjoint by the above remark) generates a symmetric C_0-contraction semigroup (symmetry can be seen, e.g., by the symmetry of $G_\alpha = (\alpha - \overline{L})^{-1}$ and the above considerations on the symmetry properties of G_α resp. T_t).

Remark 9. $\overline{L}$ in Remark 8 can be easily replaced by any self-adjoint negative extension $\tilde{L}$ of L. In fact then both $\tilde{L}$ and its adjoint $\tilde{L}^* = \tilde{L}$ are negative hence dissipative and then they generate a C_0-contraction semigroup, see [424],p.248.

Remark 10. Spectral theory also gives a direct relation between self-adjoint properties of generators L and corresponding semigroups, recalling that $L = \int_{\sigma(L)} \lambda dE(\lambda)$, $T_t = \int_{\sigma(L)} e^{t\lambda} dE(\lambda)$, $E(\lambda)$ being the spectral family associated with L. Here $\sigma(L) \subset (-\infty, 0]$.

1.3 Examples

We shall concentrate, in this section, on:
Semigroups in Banach or Hilbert spaces associated with differential operators over finite dimensional spaces.
The typical situation is given by the finite dimensional space $\mathbb{R}^d$ and the ("finite dimensional") differential operator Δ (the Laplacian) acting, e.g. in the Hilbert space $\mathcal{H} = L^2(\mathbb{R}^d)$ resp. on the Banach space $B = C_b(\mathbb{R}^d)$.
Let us first consider the case $\mathcal{H} = L^2(\mathbb{R}^d)$.
We see that $(\Delta, C_0^\infty(\mathbb{R}^d))$ (or, e.g., $(\Delta, \mathcal{S}(\mathbb{R}^d))$) is densely defined and symmetric in $\mathcal{H}$ (as a consequence of an integration by parts).
Let U be the map from $L^2(\mathbb{R}^d)$ into $L^2(\hat{\mathbb{R}}^d)$ defined by L^2-Fourier transform i.e.

$$(Uf)(k) \equiv (2\pi)^{-\frac{d}{2}} \int_{\mathbb{R}^d} e^{ik\cdot x} f(x)dx, k \in \hat{\mathbb{R}}^d$$

($\hat{\mathbb{R}}^d$ a copy of $\mathbb{R}^d$, for the Fourier transform variables). Then U is unitary (by Parseval's theorem), i.e. $U^*U = UU^* = 1$.
Let M be the multiplication operator given by $M\hat{u}(k) \equiv |k|^2\hat{u}(k), k \in \hat{\mathbb{R}}^d, \hat{u} \in L^2(\hat{\mathbb{R}}^d)$, on its natural domain $D(M) \equiv \left\{\hat{u} \in L^2(\hat{\mathbb{R}}^d) | M\hat{u} \in L^2(\hat{\mathbb{R}}^d)\right\}$.
M is self-adjoint positive (since $(M + \alpha)$, has dense range for all $\alpha > 0$).
Let us set

$$H_0 = U^*MU$$

with

$$\begin{aligned} D(H_0) &= \{u \in L^2(\mathbb{R}^d) | Uu \in D(M)\} \\ &= \{U^*D(M)\} \end{aligned}$$

(i.e. $u \in D(H_0) \leftrightarrow \hat{u} \in D(M)$).

Remark 11. One easily shows that $D(H_0) = H^{2,2}(\mathbb{R}^d)$ is the Sobolev space obtained by closing $C_0^\infty(\mathbb{R}^d)$ in the norm given by the scalar product

$$(u, v)_2 \equiv \sum_{|\alpha|\le 2} \int \overline{D^\alpha u} D^\alpha v \, dx.$$

H_0 is self-adjoint positive in $L^2(\mathbb{R}^d)$, being unitary equivalent to the self-adjoint positive operator M (positivity is immediate; self-adjointness follows e.g. by spectral theory, the spectrum of H_0 being the same as the one of M and the spectral family of H_0 being $U^*E_\lambda U$, where E_λ is the spectral family to M).
By Lumer-Phillips theorem (or spectral theory) we have that $e^{-tM}, t \ge 0$, is a symmetric C_0-contraction semigroup on $L^2(\mathbb{R}^d)$, hence

$$e^{-tH_0} = U^*e^{-tM}U, t \ge 0$$

is also a symmetric C_0-contraction semigroup on $L^2(\mathbb{R}^d)$.
Its spectral representation can be obtained by the one of M, in fact since $e^{-tM}\hat{u}(k) = e^{-t|k|^2}\hat{u}(k)$, we have for all $u \in L^2(\mathbb{R}^d)$.

$$e^{-tH_0}u(x) = \int_{\mathbb{R}^d} \pi_t(x,y)u(y)\,dy, \tag{2}$$

where $\pi_t(x,y) \equiv (4\pi t)^{\frac{-d}{2}} e^{-\frac{|x-y|^2}{4t}}, t > 0$ is the heat kernel density.
(2) holds for $t = 0$ with $\pi_t(x,y)dy$ replaced by the Dirac measure $\delta_x(dy)$ (since $e^{-tH_0}|_{t=0}$ is the unity operator in $L^2(\mathbb{R}^d)$).

Remark 12. Formula (2) easily extends to $t \in \mathbb{C}$ with $Re(t) > 0$.
In particular we have a representation for the unitary group $e^{itH_0}, t \in \mathbb{R}$. This unitary group (uniquely associated to H_0 by Stone's theorem) gives the time evolution in the quantum mechanics of one (non relativistic) particle, see, e.g. [423],[424], [425],[426].

One can ask the question:
do there possibly exist other semigroups $e^{t\tilde{L}}, t \geq 0$ (unitary groups $e^{it\tilde{L}}, t \in \mathbb{R}$) generated by self-adjoint extensions $\tilde{L}$, different from the closure $\overline{L}$ of Δ from $C_0^\infty(\mathbb{R}^d)$ in B?

That the answer is no, for $B = L^2(\mathbb{R}^d)$ (or $C_b(\mathbb{R}^d)$) , can be seen using the following important Theorem, for which we need a definition.

Definition 6. *Let L be a closed linear operator on a Banach space B. A linear subset D in $D(L)$ is called a core for L if $\overline{L \restriction D} = L$ (i.e. the closure of the restriction $L \restriction D$ of L to D is precisely L).*

Theorem 3. *(Nelson)*
Let L be the generator of a C_0-contraction semigroup on a Banach space B. Let $D_0 \subset D_1 \subset D(L), \overline{D_0} = B$, such that e^{tL} maps D_0 into D_1. Then D_1 is a core for L.

Proof. See, e.g., [393], [227] p.17, [424]. For extensions see [501]. □

For the application of the theorem to our situation, let us take $e^{tL} = e^{-tH_0}$, with $H_0 = U^*MU$ as above. To see that Nelson's theorem can be applied with $D_0 = D_1 = \mathcal{S}(\mathbb{R}^d)$ we observe that $D(L)$ contains $\mathcal{S}(\mathbb{R}^d)$ (as seen from the fact that $U\mathcal{S}(\mathbb{R}^d) = \mathcal{S}(\hat{\mathbb{R}}^d)$, and M maps $\mathcal{S}(\hat{\mathbb{R}}^d)$ into itself, and $U^*\mathcal{S}(\hat{\mathbb{R}}^d) = \mathcal{S}(\mathbb{R}^d)$) and by (2) we have $e^{-tH_0}\mathcal{S}(\mathbb{R}^d) \subset \mathcal{S}(\mathbb{R}^d)$ (the smoothness of the elements of $e^{-tH_0}\mathcal{S}(\mathbb{R}^d)$ can be checked directly, using, e.g., dominated convergence). Thus we have shown that $\mathcal{S}(\hat{\mathbb{R}}^d)$ is a core for e^{-tH_0}.
To see that also $C_0^\infty(\mathbb{R}^d)$ is a core in $L^2(\mathbb{R}^d)$, let us set $A \equiv -\Delta$ on $C_0^\infty(\mathbb{R}^d)$. Let $v \in D(A^*)$, then

$$(-\Delta u, v) = (Au, v) = (u, A^*v), \forall u \in C_0^\infty(\mathbb{R}^d).$$

Hence, $-\Delta v$,defined by looking at $v \in L^2(\mathbb{R}^d)$ as a distribution, is equal to $A^*v \in L^2(\mathbb{R}^d)$.
Thus $v \in H^{2,2}(\mathbb{R}^d)$ and $A^*v = H_0 v$ (by the fact that $D(H_0) = H^{2,2}(\mathbb{R}^d)$).
This shows that $D(A^*) \subset D(H_0)$ and H_0 is an extension of A^*. Conversely, for $v \in D(H_0)$ we have $H_0 v \in L^2(\mathbb{R}^d)$, hence $(H_0 u, v) = (u, H_0 v) \forall u \in C_0^\infty(\mathbb{R}^d)$, thus $v \in D(A^*), A^*v = H_0 v$, i.e. A^* is an extension of H_0. Thus H_0 must coincide with A^*, and then $A^* = A^{**}$ (since $H_0 = H_0^*$ by self-adjointness), which shows that the closure of A is self-adjoint and coincides with H_0, thus $C_0^\infty(\mathbb{R}^d)$ is a core for H_0, in $L^2(\mathbb{R}^d)$.

Remark 13. From the explicit formula (2) we see that the r.h.s. of (2) also maps the Banach space $B = C_\infty(\mathbb{R}^d)$ (the continuous functions on $\mathbb{R}^d$ vanishing at infinity with supremum norm), into itself, and is a C_0-contraction semigroup $\tilde{P}_t$.
Let us call $\tilde{L}$ the generator of $\tilde{P}_t$.
$D(\tilde{L}) \supset \mathcal{S}(\hat{\mathbb{R}}^d)$ as easily verified by the definition of the generator and (2).
In fact $\tilde{L} = -\Delta$ on $\mathcal{S}(\hat{\mathbb{R}}^d)$ and by Nelson's theorem applied to $D_0 = D_1 = \mathcal{S}(\hat{\mathbb{R}}^d), B = C_\infty(\mathbb{R}^d)$ we have that $\mathcal{S}(\hat{\mathbb{R}}^d)$ is a core for $\tilde{P}_t$ in $C_\infty(\mathbb{R}^d)$.

Remark 14. P_t and $\tilde{P}_t$ can be identified in the following sense.
P_t and $\tilde{P}_t$ on $C_\infty(\mathbb{R}^d) \cap L^2(\mathbb{R}^d)$, as C_0-contraction semigroups, coincide, hence by the density of $C_\infty(\mathbb{R}^d) \cap L^2(\mathbb{R}^d)$ in $L^2(\mathbb{R}^d), P_t = \tilde{P}_t$ on $L^2(\mathbb{R}^d)$.
Similarly one can show $P_t = \tilde{P}_t$ in $C_\infty(\mathbb{R}^d)$, by exploiting the boundedness of $P_t, \tilde{P}_t$ in $C_\infty(\mathbb{R}^d)$ and their equality on the dense subset $C_\infty(\mathbb{R}^d) \cap L^2(\mathbb{R}^d)$ of $C_\infty(\mathbb{R}^d)$.
In this sense then the heat semigroup e^{-tH_0} can be identified in $C_\infty(\mathbb{R}^d)$ and $L^2(\mathbb{R}^d)$ with the semigroup with generator Δ having $\mathcal{S}(\hat{\mathbb{R}}^d)$ (or $C_0^\infty(\mathbb{R}^d)$) as core, both in $C_\infty(\mathbb{R}^d)$ and $L^2(\mathbb{R}^d)$.

2 Closed symmetric coercive forms associated with C_0-contraction semigroups

2.1 Sesquilinear forms and associated operators

Sesquilinear forms Let $\mathcal{H}$ be a Hilbert space over $\mathbb{K} = \mathbb{R}$ or $\mathbb{C}$, with scalar product $(\cdot,\cdot)$ (conjugate linear in the first argument, linear in the second argument), and corresponding norm $\|\cdot\|^2 = (\cdot,\cdot)$.

Let D be a linear subspace of $\mathcal{H}$.

Definition 7. *A map $\mathcal{E} : D \times D \to \mathbb{K}$, conjugate linear in the first argument, linear in the second argument is called a sesquilinear form (on D, in $\mathcal{H}$).*

D is called the domain of $\mathcal{E}$. One writes $(\mathcal{E}, D)$ whenever it is important to specify the domain.

$\mathcal{E}[u] \equiv \mathcal{E}(u,u)$, $u \in D$ is called the associated quadratic form.

Remark 15. For $\mathcal{K} = \mathbb{C}$, $(\mathcal{E}[u], u \in D)$ uniquely determines $(\mathcal{E}, D)$ by the polarization formula

$$\mathcal{E}(u,v) = \tfrac{1}{4}(\mathcal{E}[u+v] - \mathcal{E}[u-v] + i\mathcal{E}[u+iv] - i\mathcal{E}[u-iv]).$$

This is not so, in general, for $\mathbb{K} = \mathbb{R}$ (see, e.g., [495])

Definition 8. *A sesquilinear form $\mathcal{E}$ is said to be symmetric if $\forall u, v \in D$:*

$$\mathcal{E}(u,v) = \overline{\mathcal{E}(v,u)}$$

(where $-$ stands for complex conjugation).

Remark 16. The quadratic form associated with a symmetric sesquilinear form is real-valued.

Definition 9. *A sesquilinear form $\mathcal{E}$ is said to be lower bounded if there exists $\gamma \in \mathbb{R}$ such that:*

$$\mathcal{E}[u] \geq \gamma \|u\|^2, \quad \forall u \in D(\mathcal{E})$$

One writes then $\mathcal{E} \geq \gamma$. γ is said to be the lower bound for $\mathcal{E}$.

$\mathcal{E}$ is called positive if $\gamma = 0$.

Remark 17. If $\mathcal{E}$ is positive then

$$|\mathcal{E}(u,v)| \leq (\mathcal{E}[u])^{1/2}(\mathcal{E}[v])^{1/2}$$

Proof. This is Cauchy-Schwarz' inequality.

Originally published in: *Ecole d'Eté de Probabilités de Saint-Flour XXX – 2000*, Lecture Notes in Mathematics, Vol. **1816**, 18–32, DOI: 10.1007/3-540-44922-1_2, © Springer-Verlag Berlin Heidelberg 2003, Reprint by Springer-Verlag Berlin Heidelberg 2012

Example 1. Let A be a linear operator with domain $D(A)$ in $\mathcal{H}$. Define for $u, v \in D(A)$:

$$\mathcal{E}(u,v) = (u, Av).$$

Then $\mathcal{E}$ is a sesquilinear form with domain $D(\mathcal{E}) = D(A)$. The following equivalences follow immediately from the definitions.

$\mathcal{E}$ is symmetric iff A is symmetric.

$\mathcal{E} \geq \gamma$ iff $A \geq \gamma$ (in the sense that $(u, Au) \geq \gamma \|u\|^2$ for some $\gamma \in \mathbb{R}$, $\forall u \in D(A)$; in which case one says that A is lower bounded with lower bound γ).

$\mathcal{E} \geq 0$ iff $A \geq 0$ (in which case one says that A is positive).

Closed forms Let $\mathcal{E}$ be a sesquilinear, lower bounded form on $\mathcal{H}$.

Definition 10. *A sequence $(u_n)_{n \in \mathbb{N}}$ is said to be $\mathcal{E}$-convergent to $u \in \mathcal{H}$, for $n \to \infty$, and one writes $u_n \stackrel{\mathcal{E}}{\to} u$, $n \to \infty$, if $u_n \in D(\mathcal{E})$, $u_n \to u$ (i.e. (u_n) converges to u in $\mathcal{H}$) and $\mathcal{E}[u_n - u_m] \to 0$, $n, m \to \infty$ (i.e. u_n is an "$\mathcal{E}$-Cauchy sequence").*

N.B. u is not required to be in $D(\mathcal{E})$.

Definition 11. *$\mathcal{E}$ is said to be closed if $u_n \stackrel{\mathcal{E}}{\to} u$, $n \to \infty$, implies $u \in D(\mathcal{E})$ and $\mathcal{E}[u_n - u] \to 0$, as $n \to \infty$.*

Let $\mathcal{E}$ be a symmetric, positive sesquilinear form. Define for any $\alpha > 0$:

$$\mathcal{E}_\alpha(u,v) \equiv \mathcal{E}(u,v) + \alpha(u,v), \quad \forall u, v \in D(\mathcal{E}).$$

Then $D(\mathcal{E})$ taken with the norm given by

$$\|u\|_1 \equiv (\mathcal{E}_1[u])^{\frac{1}{2}} \quad , u \in D(\mathcal{E})$$

is a pre Hilbert space, in the sense that $(D(\mathcal{E}), \|\cdot\|_1)$ has all properties of a Hilbert space, except for completeness. We call $D(\mathcal{E})_1$ this space.

Remark 18. a) $\underline{u} \equiv (u_n)_{n \in \mathbb{N}}$ is $\mathcal{E}$-convergent iff $\underline{u}$ is Cauchy in $D(\mathcal{E})_1$.

b) $u_n \stackrel{\mathcal{E}}{\to} u$, $n \to \infty$, $u \in D(\mathcal{E})$ iff $\|u_n - u\|_1 \to 0$, $n \to \infty$.

Proposition 9. *A lower bounded form $\mathcal{E}$ is closed iff $D(\mathcal{E})_1$ is complete.*

Proof: This is left as an exercise (cf., e.g., [312], p. 314).

Example 2. Let S be a linear operator with domain $D(S) \subset \mathcal{H}$. Define $\mathcal{E}(u,v) \equiv (Su, Sv)$, $D(\mathcal{E}) = D(S)$. Then $\mathcal{E}$ is a positive, symmetric sesquilinear form. $\mathcal{E}$ is closed iff S is closed (the proof of the latter statement is left as an exercise).

Closed forms

Definition 12. *A sesquilinear lower bounded form* $\overset{\circ}{\mathcal{E}}$ *is said to be closable if it has a closed extension* $\mathcal{E}$, *i.e.,* $\mathcal{E}$ *is closed,* $D(\mathcal{E}) \supset D(\overset{\circ}{\mathcal{E}})$ *and* $\mathcal{E} = \overset{\circ}{\mathcal{E}}$ *on* $D(\overset{\circ}{\mathcal{E}})$.

Proposition 10. *A sesquilinear lower bounded form* $\overset{\circ}{\mathcal{E}}$ *is closable iff* $u_n \overset{\mathcal{E}}{\to} 0$, $n \to \infty$ *implies* $\mathcal{E}[u_n] \to 0$, $n \to \infty$.

Proof. This is left as an exercise (cf., e.g., [312], p. 315).

Definition 13. *The smallest closed extension of a sesquilinear lower bounded form* $\mathcal{E}$ *is by definition* <u>*the closure*</u> $\bar{\mathcal{E}}$ *of* $\mathcal{E}$.

Example 3. Let $\mathcal{E}$ be as in Example 2, i.e. $\mathcal{E}(u,v) = (Su, Sv)$, $\forall u, v \in D(\mathcal{E}) = D(S)$, S a linear operator on $\mathcal{H}$. Then $\mathcal{E}$ is closable iff S is closable. In the latter case one has $\bar{\mathcal{E}}(u,v) = (\bar{S}u, \bar{S}v)$, where $\bar{S}$ is the closure of the operator S (a linear operator A is said to be closable if it has a closed extension, cf. Definition 3 in Chapter 1 for the concept of closed operators). Moreover one has $\mathcal{E}$ closed iff S is closed.

The proofs are left as execises.

Remark 19. Not every sesquilinear symmetric positive form is closable. Consider, e.g., $\mathcal{H} = L^2(\mathbb{R})$, $\mathcal{E}(u,v) \equiv \bar{u}(0)v(0)$, $u, v \in D(\mathcal{E}) = C_o^\infty(\mathbb{R})$. Then $\mathcal{E}$ is sesquilinear, symmetric, and positive but not closable. In fact take a sequence $u_n \in C_o^\infty(\mathbb{R})$, with $u_n(x) = 0$ for $|x| \geq \frac{c}{n}$, $u_n(0) = 1$, $u_n(x) \leq 1$, $\forall x \in \mathbb{R}$, then we have, (by the mean-value theorem) $\|u_n\| \leq \frac{2c}{n} \to 0$, hence $u_n \to 0$, $n \to \infty$, moreover

$$\mathcal{E}[u_m - u_n] = (\bar{u}_m(0) - \bar{u}_n(0)) \cdot (u_m(0) - u_n(0)) = 0$$

hence $u_n \overset{\mathcal{E}}{\to} 0$, $n \to \infty$. On the other hand $\mathcal{E}[u_n] = \bar{u}_n(0)u_n(0) = 1$ does not converge to 0 as $n \to \infty$, which shows by Proposition 10 that $\mathcal{E}$ is not closable.

N.B. Concerning closability the situation with forms and densely defined operators is thus very different: every symmetric densely defined operator A is namely closable! (since A symmetric means by definition that the adjoint A^*, which exists uniquely since A is densely defined, is an extension of A, but every adjoint operator is closed, see, e.g. [312], p. 168).

Forms constructed from positive operators

Proposition 11. *Let* A *be a positive symmetric operator. Then*

$$\overset{\circ}{\mathcal{E}}_A(u,v) \equiv (u, Av), \quad u, v \in D(\overset{\circ}{\mathcal{E}}_A) = D(A)$$

is a sesquilinear, symmetric, positive, closable form.

Proof. $\overset{\circ}{\mathcal{E}}_A$ is clearly sesquilinear, symmetric, positive. To prove the closability, let $u_n \overset{\mathcal{E}}{\to} 0$, $n \to \infty$. We have to show $\overset{\circ}{\mathcal{E}}_A[u_n] \to 0$, $n \to \infty$. But by the triangle inequality resp. Cauchy-Schwarz inequality:

$$\begin{aligned}\overset{\circ}{\mathcal{E}}_A[u_n] &\le |\overset{\circ}{\mathcal{E}}_A(u_n, u_n - u_m)| + |\overset{\circ}{\mathcal{E}}_A(u_n, u_m)| \\ &\le \overset{\circ}{\mathcal{E}}_A[u_n]^{1/2}[\overset{\circ}{\mathcal{E}}_A[u_n - u_m]]^{1/2} + |(u_n, Au_m)|\end{aligned} \tag{3}$$

where for the latter term we have used the definition of $\overset{\circ}{\mathcal{E}}_A$.

But from the assumption $u_n \overset{\mathcal{E}}{\to} 0$, $n \to \infty$, we have for any given $\epsilon > 0$, that there exists $N(\epsilon)$ s.t. for $n, m > N(\epsilon)$:

$$\overset{\circ}{\mathcal{E}}_A[u_n - u_m] \le \epsilon^2. \tag{4}$$

Moreover, by the symmetry of A

$$|(u_n, Au_m)| = |(Au_n, u_m)| \le \|Au_n\|\|u_m\| \overset{m\to\infty}{\to} 0 \tag{5}$$

for any fixed $n \in \mathbb{N}$ since $u_m \overset{\mathcal{E}}{\to} 0$, $m \to \infty$ implies $\|u_m\| \to 0$, $m \to \infty$.

Hence from (3)–(5), for any given $\epsilon > 0$, for some $N(\epsilon)$ large enough,

$$\overset{\circ}{\mathcal{E}}_A[u_n] \le \overset{\circ}{\mathcal{E}}_A[u_n]^{1/2}\epsilon, \quad n > N(\epsilon). \tag{6}$$

For given $n > N(\epsilon)$, either $\overset{\circ}{\mathcal{E}}_A[u_n] = 0$, or $\overset{\circ}{\mathcal{E}}_A[u_n] > 0$, in which case from (6) we deduce $\overset{\circ}{\mathcal{E}}_A[u_n]^{1/2} \le \epsilon$. In both cases $\overset{\circ}{\mathcal{E}}_A[u_n] \le \epsilon$, $n > N(\epsilon)$, which shows that $\overset{\circ}{\mathcal{E}}_A[u_n] \to 0$, $n \to \infty$. □

Positive closed operators from positive symmetric closed forms

Theorem 4 (Friedrichs representation theorem). *Let $\mathcal{E}$ be a densely defined sesquilinear, symmetric, positive, closed form. Then there exists a unique self-adjoint positive operator $A_\mathcal{E}$ s.t.*

i) $D(A_\mathcal{E}) \subset D(\mathcal{E})$, $\mathcal{E}(u, v) = (u, A_\mathcal{E} v)$, $\forall u \in D(\mathcal{E}), v \in D(A_\mathcal{E})$.

ii) $D(A_\mathcal{E})$ is a core for $\mathcal{E}$ (in the sense that the closure of the restriction of $\mathcal{E}$ to $D(A_\mathcal{E})$ coincides with $\mathcal{E}$, i.e. $\overline{\mathcal{E}_{|D(A_\mathcal{E})}} = \mathcal{E}$).

iii) $D(\mathcal{E}) = D(A_\mathcal{E}^{1/2})$ (where $A_\mathcal{E}^{1/2}$ is the unique square root of the positive self-adjoint operator $A_\mathcal{E}$, defined, e.g., by the spectral theorem), and:

$$\mathcal{E}(u, v) = (A_\mathcal{E}^{1/2} u, A_\mathcal{E}^{1/2} v), \quad \forall u, v, \in D(\mathcal{E}).$$

And viceversa: if A is a self-adjoint positive operator, then $\mathcal{E}$ defined by $\mathcal{E}(u,v) \equiv (A^{1/2}u, A^{1/2}v)$ with $D(\mathcal{E}) = D(A^{1/2})$ is a densely defined sesquilinear form. $\mathcal{E}$ is the closure $\overline{\mathring{\mathcal{E}}}$ with

$$\mathring{\mathcal{E}}(u,v) = (u, Av), \quad v \in D(A), \quad u \in D(\mathring{\mathcal{E}}) = D(A).$$

Remark 20. One says $A_{\mathcal{E}}$ (in the first part of the theorem) is the self-adjoint operator associated with the form $\mathcal{E}$. Viceversa, in the second part of the theorem, $\mathcal{E}$ is the form associated with the operator A.

One often writes $-L_{\mathcal{E}}$ instead of $A_{\mathcal{E}}$

The proof of the first part relies on following

Lemma 1. *Let $\mathcal{H}_1$ be a dense subspace of a Hilbert space $\mathcal{H}$. Let a scalar product $(\cdot,\cdot)_1$ (in general different from the scalar product $(\cdot,\cdot)$ in $\mathcal{H}$) be defined on $\mathcal{H}_1$, so that $(\mathcal{H}_1, (\cdot,\cdot)_1)$ is a Hilbert space. Suppose that there exists a constant $\kappa > 0$ s.t. $\kappa\|u\|^2 \leq \|u\|_1^2$ for all $u \in \mathcal{H}$. Then there exists uniquely a self adjoint operator A in $\mathcal{H}$ s.t. $D(A) \subset \mathcal{H}_1$, $(Au, v) = (u,v)_1, \forall u \in D(A), v \in \mathcal{H}_1$, and, moreover, $A \geq \kappa$.*

A is described by

$$D(A) = \{u \in \mathcal{H}_1 \mid \exists \hat{u} \in \mathcal{H} \mid (u,v)_1 = (\hat{u}, v) \forall v \in \mathcal{H}_1\}, \quad Au = \hat{u}.$$

$D(A)$ is both dense in $\mathcal{H}_1$ with respect to the $\|\cdot\|_1$-norm and in $\mathcal{H}$ with respect to the $\|\cdot\|$-norm.

Proof. (cf. e.g., [495], [427]): We first remark that $\hat{u}$ in the definition of $D(A)$ is uniquely defined, since $\mathcal{H}_1$ is dense in $\mathcal{H}$ by assumption. Moreover, $u \mapsto \hat{u}$ is linear, from the definition, thus A is linear.

Let $J : \mathcal{H} \to \mathcal{H}_1$ with $D(J) = \mathcal{H}_1 \subset \mathcal{H}$, $Jf = f$, $\forall f \in D(J)$. Then J is closed from $D(J) = \mathcal{H}_1 \subset \mathcal{H}$ to $\mathcal{H}_1$ (in the sense that $f_n \in D(J)$, $f_n \to f$, $n \to \infty$, in $\mathcal{H}$, $Jf_n \to h$ in $\mathcal{H}_1$ implies $f \in D(J)$ and $Jf = h$:

in fact $Jf_n = f_n$ and $Jf_n \to h$ in $\mathcal{H}_1$ implies $f_n \to h$ in $\mathcal{H}$ by $\|f_n - h\|^2 \leq \frac{1}{\kappa}\|f_n - h\|_1^2$. But then $Jf_n = f_n \to f$ in $\mathcal{H}$, by assumption, and $f_n \to h$ in $\mathcal{H}_1$, again by assumption, imply $f = h$ in $\mathcal{H}_1 = D(J)$ hence $f \in D(J)$, $Jf = f = h$ by the definition of J and the fact that $f = h$ as elements of $\mathcal{H}_1$.

J is densely defined from $\mathcal{H}$ into $\mathcal{H}_1$, with $D(J) = \mathcal{H}_1$ and closed (a fortiori closable), then J^* is uniquely and densely defined, closed from $\mathcal{H}_1$ into $\mathcal{H}$ (by Th. 5.29 in [312], p. 168).

Set $A_0 = J^*$. Then we have $\forall u \in D(J^*)$, $v \in \mathcal{H}_1$:

$$(A_0 u, v) = (J^* u, v) = (u, Jv)_1 = (u,v)_1.$$

Set $A = A_0$, looked upon as an operator from $\mathcal{H}$ into $\mathcal{H}$. It is then clear that $D(A) \subset \mathcal{H}_1 \subset \mathcal{H}$,

$$(Au, v) = (A_0 u, v) = (u, v)_1 \quad \forall u \in D(A), v \in \mathcal{H}_1. \qquad (*)$$

That $A \geq \kappa$ follows from the fact that $(Au, u) = (u, u)_1 \geq \kappa(u, u)$, $\forall u \in D(A)$, by the definition of $(\cdot, \cdot)_1$. That A is symmetric in $\mathcal{H}$ follows from $(Au, v) = (u, v)_1$, $\forall u \in D(A), v \in \mathcal{H}_1$ and, for $v \in D(A)$:

$$(u, Av) = (u, A_0 v) = (u, J^* v) = (Ju, v)_1 = (u, v)_1.$$

Also the description of $D(A)$ given in the lemma is proven, since $D(A)$ is characterized by the definition of A_0 and J^* as the set of all $u \in \mathcal{H}_1$ s.t.

$$(Au, v) = (A_0 u, v) = (u, v)_1 \quad \forall v \in \mathcal{H}_1.$$

That $D(A)$ is $(\cdot, \cdot)_1$-dense in $\mathcal{H}_1$ is clear from the fact that $D(J^*)$ is $(\cdot, \cdot)_1$-dense in $\mathcal{H}_1$.

That $D(A)$ is $(\cdot, \cdot)$-dense is also clear from the relation between the $\|\cdot\|_1$ and $\|\cdot\|$-norms.

It remains to show that A is self-adjoint. For this it is enough to prove that the range of A is $\mathcal{H}$ (cf., e.g., [495]). Let us consider $v \in \mathcal{H}, w \in \mathcal{H}_1$:

$$|(v, w)| \leq \|v\| \|w\| \leq \tfrac{1}{\sqrt{\kappa}} \|v\| \|w\|_1$$

where in the latter inequality we used the relation between $\|\cdot\|$ and $\|\cdot\|_1$. This shows that, $\forall v \in \mathcal{H}$, $w \mapsto (v, w)$ is a continuous linear functional on $\mathcal{H}_1$, hence there exists, by Riesz' theorem (see, e.g., [423]) a $\tilde{v} \in \mathcal{H}_1$ s.t.

$$(\tilde{v}, w)_1 = (v, w) \quad \forall w, v \in \mathcal{H}.$$

By the definition (*) of A (used with w replacing v and $\tilde{v}$ replacing u) we have then $(v, w) = (Au, w)$ for any $v \in \mathcal{H}$, $\forall w \in \mathcal{H}$, which shows that any $v \in \mathcal{H}$ can be written as Au for some $u \in D(A)$, hence the range of A is the whole of $\mathcal{H}$.

The uniqueness of A in the lemma is proven as follows: Let B be self-adjoint in $\mathcal{H}$ s.t.

$$(Bu, v) = (u, v)_1 \quad \forall u \in D(B), v \in \mathcal{H}_1.$$

Then by definition of A, A is an extension of B (i.e. $B \subset A$). But A is self-adjoint so

$$B \subset A = A^* \subset B^*$$

B being itself self-adjoint, this implies $B = A$. This finishes the proof of the lemma and of the theorem. □

2.2 The relation between closed positive symmetric forms and C_0-contraction semigroups and resolvents

The basic relations

Theorem 5. *Le $\mathcal{E}$ be a densely defined positive symmetric sesquilinear form which is closed, in a Hilbert space $\mathcal{H}$. Let $-L_{\mathcal{E}}$ be the associated self-adjoint positive operator given by Theorem 4 (in 2.1) so that*

$$\mathcal{E}(u,v) = ((-L_{\mathcal{E}})^{1/2}u, (-L_{\mathcal{E}})^{1/2}v) \quad \forall u, v \in D(\mathcal{E}).$$

Then $L_{\mathcal{E}}$ generates a C_0-contraction semigroup $T_t = e^{tL_{\mathcal{E}}}$, $t \geq 0$, in $\mathcal{H}$.

And viceversa, if T_t is a symmetric C_0-contraction semigroup, then its generator L is self-adjoint, negative (i.e., $-L$ is positive) and the associated form given by Theorem 4 in 2.1 is positive, symmetric, closed.

One has

$$\lim_{t\searrow 0} \tfrac{1}{t}(u - T_t u, v) = \mathcal{E}(u,v), \quad \forall u, v \in D(\mathcal{E})$$

Proof. The direct way follows from the Theorem 4 given in Chapter 2, 2.1. The viceversa part follows from the fact that L is self-adjoint, negative and the same Theorem 4. □

Theorem 6. *All statements in Theorem 5 hold with the semigroup $(T_t)_{t\geq 0}$ replaced by the symmetric resolvent family $(G_\alpha)_{\alpha>0}$, $G_\alpha \equiv (\alpha - L_{\mathcal{E}})^{-1}$, corresponding to $(T_t)_{t\geq 0}$.*

One has for all $u \in \mathcal{H}, v \in D(\mathcal{E})$:

$$\mathcal{E}_\alpha(G_\alpha u, v) = (u,v)$$

(where we recall the definition $\mathcal{E}_\alpha(u,v) \equiv \mathcal{E}(u,v) + \alpha(u,v)$).

Moreover,

$$\mathcal{E}(u,v) = \lim_{\alpha\to+\infty} \alpha(u - \alpha G_\alpha u, v), \quad \forall u, v \in D(\mathcal{E}).$$

Proof. $(G_\alpha)_{\alpha>0}$ is self-adjoint, by the spectral theorem. The relation for $\mathcal{E}_\alpha$ holds because of

$$\mathcal{E}_\alpha(G_\alpha u, v) = \mathcal{E}(G_\alpha u, v) + \alpha(G_\alpha u, v) \tag{7}$$

(as seen using the definition of $\mathcal{E}_\alpha$, noting the fact that $G_\alpha u \in D(L) \subset D(\mathcal{E})$, for L the operator associated to $\mathcal{E}$ in the sense of Theorem 4 in Chapter 2,2.1). But

$$\mathcal{E}(G_\alpha u, v) = (-LG_\alpha u, v) \tag{8}$$

by the relation between $\mathcal{E}$ and L. The r.h.s. of latter relation can be written as

$$((-L + \alpha - \alpha)G_\alpha u, v) = (u,v) - (\alpha G_\alpha u, v), \tag{9}$$

where we used $G_\alpha = (\alpha - L)^{-1}$. The relation involving $\mathcal{E}_\alpha$ then follows from (7)-(9).

For the limit relation we use (7), the relation just shown for $\mathcal{E}_\alpha$ to get

$$(u, v) = \mathcal{E}(G_\alpha u, v) + \alpha(G_\alpha u, v)$$

hence

$$\alpha(u, v) = \mathcal{E}(\alpha G_\alpha u, v) + \alpha^2(G_\alpha u, v),$$

and the fact that $\alpha G_\alpha \to 1$ as $\alpha \to +\infty$. □

Remark 21. The "relations $\mathcal{E} \leftrightarrow L \leftrightarrow T \leftrightarrow G$" as described in Theorems 4, 5, 6 can be summarized in the following two tables:

Table 1

$B =$ Banach space over $\mathbb{K} = \mathbb{R}, \mathbb{C}$

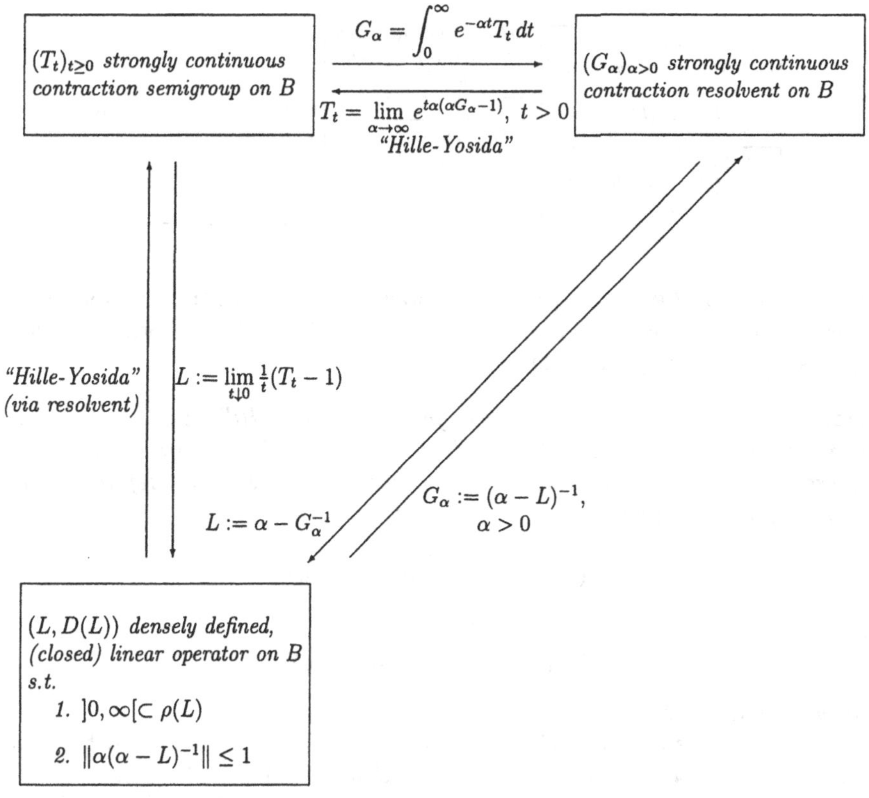

Table 2

$\mathcal{H}$ = Hilbert space over $\mathbb{R}$ with inner product $(\ ,\ )$ and norm $\|\ \| := (\ ,\ )^{1/2}$.

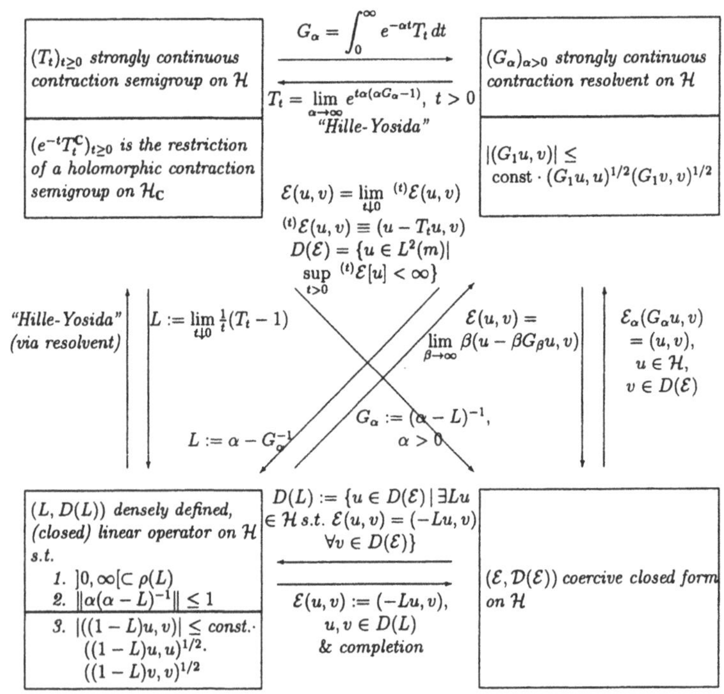

Extension to the case of coercive forms in a real Hilbert space In this section we consider a real Hilbert space $\mathcal{H}$. Sesquilinear forms on such spaces will be simply called bilinear.

Definition 14. *Let $\mathcal{E}$ be a bilinear form on a real Hilbert space $\mathcal{H}$, with dense domain $D(\mathcal{E})$ (i.e. both $u \to \mathcal{E}(u,v)$ and $v \to \mathcal{E}(u,v)$ are linear). The symmetric (resp. antisymmetric) part $\tilde{\mathcal{E}}$ (resp. $\check{\mathcal{E}}$) of $\mathcal{E}$ is by definition the bilinear form given by:*

$$\tilde{\mathcal{E}}(u,v) \equiv \tfrac{1}{2}[\mathcal{E}(u,v) + \mathcal{E}(v,u)]$$

resp.

$$\check{\mathcal{E}}(u,v) \equiv \tfrac{1}{2}[\mathcal{E}(u,v) - \mathcal{E}(v,u)],$$

for all $u, v \in D(\mathcal{E})$.

One then has $\mathcal{E} = \tilde{\mathcal{E}} + \check{\mathcal{E}}$. $\tilde{\mathcal{E}}$ is a symmetric bilinear form and $\check{\mathcal{E}}$ is an antisymmetric bilinear form (in the sense that $\check{\mathcal{E}}(u,v) = -\check{\mathcal{E}}(v,u)$).

Suppose $\mathcal{E}$ is positive definite (i.e. $\mathcal{E}(n,n) \geq 0$ for all $u \in D(\mathcal{E})$). Then one says that $\mathcal{E}$ satisfies the weak sector condition with constant $k \geq 0$ if

$$|\mathcal{E}_1(u,v)| \le k\mathcal{E}_1[u]^{1/2}\mathcal{E}_1[v]^{1/2}.$$

$(\mathcal{E}, D(\mathcal{E}))$ is called a coercive closed form if $\mathcal{E}$ satisfies the weak sector condition and $(\mathcal{E}, D(\mathcal{E}))$ is closed.

The relations $\mathcal{E} \leftrightarrow L \leftrightarrow T \leftrightarrow G$ discussed in Theorems 4, 5, 6 (and in Tables 2.2, 2.2 extend to the case of a real Hilbert space, with the symmetry and positivity in $\mathcal{E}$, $-L$, T, G replaced respectively by:

a) Coerciveness for $\mathcal{E}$.
b) L is closed operator with $\rho(L) \subset (0,\infty)$ s.t. $\|\alpha(\alpha - L)^{-1}\| \le 1$, for $\alpha > 0$,

$$|((1-L)u,v)| \le C((1-L)u,u)^{1/2}((1-L)v,v)^{1/2}.$$

c) G is a C_0-contraction resolvent family with

$$|(G_1u,v)| \le C(G_1u,u)^{1/2}(G_1v,v)^{1/2} \quad u,v \in \mathcal{H}.$$

d) T is a C_0-contraction semigroup s.t. its natural linear extension to the complexification $\mathcal{H}_{\mathbb{C}} = \mathcal{H} + i\mathcal{H}$ of $\mathcal{H}$ satisfies the following condition: the operator $e^{-t}T_t^{\mathbb{C}}$ is the restriction of a holomorphic contraction semigroup on the sector

$$\{z \in \mathbb{C} \mid |Im(z)| \le \tfrac{1}{k}Re(z)\} \text{ (with } k \text{ as in Definition 14).}$$

Moreover, G, T are accompanied by dual semigroups $\hat{G}$, $\hat{T}$ (that only in the symmetric case coincide with G, T), see, e.g., [312], [427], [367].

Remark 22. The direct relation between $\mathcal{E}$ and T has been discussed in [407] and [136]. E.g. one has the result, relating $\mathcal{E}$ and T:

$$\mathcal{E}(u,v) = \lim_{t\searrow 0} {}^{(t)}\mathcal{E}(u,v), \quad \forall u,v \in D(\mathcal{E}),$$

with ${}^{(t)}\mathcal{E}(u,v) \equiv \frac{1}{t}(u - T_tu, v)$, $D(\mathcal{E}) = \{u \in \mathcal{H} \mid \sup_{t>0} {}^{(t)}\mathcal{E}[u] < \infty\}$.

An example In the whole course, we shall have two basic examples, one in finite dimensions and one in infinite dimensions. Here is the first basic example, the second one will be introduced in Chapter 3, 4.2.

Let μ be a positive Borel measure on $\mathbb{R}^d$ with $\operatorname{supp}\mu = \mathbb{R}^d$. Let us consider the bilinear form in $\mathcal{H} = L^2(\mathbb{R}^d, \mu)$:

$$\overset{\circ}{\mathcal{E}}_\mu(u,v) = \int_{\mathbb{R}^d} < \nabla u, \nabla v > d\mu, \quad u,v \in C_0^\infty(\mathbb{R}^d),$$

with $\nabla u(x) = (\frac{\partial u}{\partial x_1}, \dots \frac{\partial u}{\partial x_d})$, $x = (x_1, \dots, x_d) \in \mathbb{R}^d$, $< \nabla u, \nabla v >= \sum_{i=1}^d \frac{\partial u}{\partial x_i}\frac{\partial v}{\partial x_i}$.

We call $\overset{\circ}{\mathcal{E}}_\mu$ with domain $D(\overset{\circ}{\mathcal{E}}_\mu) = C_0^\infty(\mathbb{R}^d)$ the classical pre-Dirichlet form given by μ. $\overset{\circ}{\mathcal{E}}_\mu$ is symmetric and positive. The basic question is: For which μ is $\overset{\circ}{\mathcal{E}}_\mu$ closable?

In Proposition 11 in 2.1, we have already indicated a condition for a positive symmetric sesquilinear form to be closable, namely that it can be expressed by a symmetric operator B s.t.

$$\overset{\circ}{\mathcal{E}}_\mu(u,v) = (u, Bv)_\mu, \quad u, v \in D(B) = D(\overset{\circ}{\mathcal{E}}_\mu)$$

(where $(u,v)_\mu$ stands for the $L^2(\mathbb{R}^d, \mu)$-scalar product). When can we find such a B? The problem is solved when we can derive an "integration by parts formula" ("IP"), writing

$$\overset{\circ}{\mathcal{E}}_\mu(u,v) = -\int_{\mathbb{R}^d} u \Delta v d\mu - \int_{\mathbb{R}^d} u < \beta_\mu, \nabla v > d\mu, \tag{10}$$

with $< \beta_\mu(x), y > d\mu(x) =$ "$< \nabla_x, y > d\mu(x)$", $y \in \mathbb{R}^d$, whenever μ is differentiable in a suitable sense. For this it suffices, e.g., that μ is absolutely continuous with respect to Lebesgue measure, with density ρ s.t.

$$\frac{1}{\rho}\frac{\partial \rho}{\partial x_i} \in L^2(\mathbb{R}^d, \mu), \quad i = 1, \dots, d$$

because then

$$\beta_\mu = \frac{\nabla \rho}{\rho} = \nabla \ln \rho. \tag{11}$$

In this case

$$\overset{\circ}{\mathcal{E}}_\mu(u,v) = (u, A_\mu v)_\mu \tag{12}$$

with

$$A_\mu = -\Delta - < \beta_\mu(x), \nabla_x >$$

on $C_0^\infty(\mathbb{R}^d)$ or shortly $A_\mu = -\Delta - \beta_\mu \cdot \nabla$ (thus the operator B we were seeking is this A_μ). A_μ is called the classical pre-Dirichlet operator given by μ. The quantity β_μ in (11) is the logarithmic derivative of ρ. More generally: whenever there is a measurable vector field $\beta_\mu = (\beta_\mu^1, \dots, \beta_\mu^d)$ s.t.

$$\int \frac{\partial u}{\partial x_i} d\mu = -\int u \beta_\mu^i d\mu, \quad \forall u \in C_0^\infty(\mathbb{R}^d) \tag{13}$$

one says that β_μ is the logarithmic derivative of μ (in $L^2(\mathbb{R}^d, \mu)$) whenever $\beta_\mu^i \in L^2(\mathbb{R}^d, \mu), \forall i = 1, \dots, d$.

In this case one easily sees that (10) holds, hence $\overset{\circ}{\mathcal{E}}_\mu$ is closable. Detailed conditions for closability to hold have been worked out, based on the above idea. E.g. for $d = 1$, $\overset{\circ}{\mathcal{E}}_\mu$ is closable iff μ has a density $\rho \in L^1_{loc}(\mathbb{R})$ with respect to the Lebesgue measure s.t. $\rho = 0$ a. e. on $S(\rho) \equiv \mathbb{R} - R(\rho)$, where

$$R(\rho) \equiv \{x \in \mathbb{R} \mid \int_{x-\epsilon}^{x+\epsilon} \frac{dy}{\rho(y)} < +\infty, \exists \epsilon > 0\}$$

is the regular set for ρ ("Hamza's condition") (cf. [244],[119]).

Remark 23. A condition of this type is also necessary and sufficient for the "partial classical pre-Dirichlet form" $\int_{\mathbb{R}^d} \frac{\partial u}{\partial x_k} \frac{\partial v}{\partial x_k} d\mu$, $u, v \in C_0^\infty(\mathbb{R}^d)$ to be closable, for all $k = 1, \dots, d$, see [119]. For an example where the above condition is not satisfied, yet the total classical pre-Dirichlet form is closable, see [258].

Definition 15. *If the classical pre-Dirichlet form $\overset{\circ}{\mathcal{E}}_\mu$ given by μ is closable, the closure $\mathcal{E}_\mu$ is called classical Dirichlet form associated with μ. The corresponding self-adjoint negative operator L_μ s.t.*

$$((-L_\mu)^{1/2}u, (-L_\mu)^{1/2}v)_\mu = \mathcal{E}_\mu(u, v)$$

is called the classical Dirichlet operator associated with μ.

The corresponding classical Dirichlet form shares with other forms an essential "contraction property", which shall be discussed in Chapter 3 to which we refer also for other comments on classical Dirichlet forms.

Let us however discuss already at this stage briefly why classical Dirichlet forms are important, relating them to (generalized) Schrödinger operators.

Let μ be absolutely continuous with respect to the Lebesgue measure on $\mathbb{R}^d$ with density ρ. To start with we assume $\rho(x) > 0$ for all $x \in \mathbb{R}^d$ and ρ smooth.

Let us consider the map $W : L^2(\mathbb{R}^d) \to L^2(\mathbb{R}^d, \mu)$ given by

$$Wu \equiv \frac{u}{\sqrt{\rho}}, \quad u \in L^2(\mathbb{R}^d)$$

(where $L^2(\mathbb{R}^d)$ denotes the space of square summable functions with respect to the Lebesgue measure). W is unitary (and $W^*v = \sqrt{\rho}v$, $v \in L^2(\mathbb{R}^d, \mu)$, as seen from the construction). Let A_μ be given by (12) and consider on the domain $W^*D(A_\mu)$, the operator A:

$$A \equiv W^* A_\mu W.$$

If $D(A_\mu)$ is dense in $L^2(\mathbb{R}^d)$ (which is the case discussed above, where $D(A_\mu) \supset C_0^\infty(\mathbb{R}^d)$) then, by unitarity, $W^*D(A_\mu)$ is dense in $L^2(\mathbb{R}^d, \mu)$, hence A is densely defined in $L^2(\mathbb{R}^d, \mu)$. We have, for $u, v \in W^*D(A_\mu)$,

$$(Wu, A_\mu Wv) = (u, W^* A_\mu Wv) = (u, Av)$$

where $(\cdot, \cdot)$ is the scalar product on $L^2(\mathbb{R}^d)$.

Set $\sqrt{\rho} = \varphi$, assuming $\varphi(x) > 0$, $\forall x \in \mathbb{R}^d$, $\varphi \in C^\infty(\mathbb{R}^d)$. We compute the l.h.s of the above equality for $u, v \in C_0^\infty(\mathbb{R}^d)$ (observing that then also

$\frac{u}{\varphi}, \frac{v}{\varphi} \in C_0^\infty(\mathbb{R}^d)$ by our assumption on φ) using the relations between A_μ and $\overset{\circ}{\mathcal{E}}_\mu$ given by (12) and the definition of $\overset{\circ}{\mathcal{E}}_\mu$:

$$(Wu, A_\mu Wv)_{L^2(\mu)} = \overset{\circ}{\mathcal{E}}_\mu(Wu, Wv) = \int_{\mathbb{R}^d} \nabla(\frac{u}{\varphi})\nabla(\frac{v}{\varphi})\varphi^2 dx. \tag{14}$$

But $\nabla(\frac{u}{\varphi}) = \frac{1}{\varphi}\nabla u - \frac{1}{\varphi^2}(\nabla\varphi)\, u$ and correspondingly with u replaced by v. Inserting these equalities into (14) we get the following four terms:

a)

$$\int_{\mathbb{R}^d} \frac{1}{\varphi}(\nabla u)\frac{1}{\varphi}(\nabla v)\varphi^2 dx = -\int_{\mathbb{R}^d} u(\Delta v)dx$$

(where we first observed that the φ-terms cancel, and then we integrated by parts, using $u, v \in C_0^\infty(\mathbb{R}^d)$),

b)

$$\int_{\mathbb{R}^d} \frac{1}{\varphi^2}(\nabla\varphi)u\frac{1}{\varphi^2}(\nabla\varphi)v\varphi^2 dx = \int_{\mathbb{R}^d} u(\tfrac{1}{4}\beta_\mu^2)v dx$$

(where we used that by (11): $\beta_\mu = \nabla \ln \varphi^2 = 2\frac{\nabla\varphi}{\varphi}$),

c) Two mixed terms:

$$-\int_{\mathbb{R}^d} \frac{\nabla\varphi}{\varphi}u(\nabla v)dx - \int_{\mathbb{R}^d} \frac{\nabla\varphi}{\varphi}v(\nabla u)dx$$

$$= -\int_{\mathbb{R}^d} \frac{\nabla\varphi}{\varphi}\nabla(uv)dx = \int_{\mathbb{R}^d} \nabla(\frac{\nabla\varphi}{\varphi})\, u\, v\, dx = \frac{1}{2}\int_{\mathbb{R}^d} (div\, \beta_\mu)\, u\, v\, dx$$

(where we have used again integration by parts and $\beta_\mu = 2\frac{\nabla\varphi}{\varphi}$).

In total we then get:

$$(Wu, A_\mu Wv)_\mu = (u, Av), \tag{15}$$

with $A = -\Delta + V(x)$, with

$$V(x) \equiv \frac{1}{4}\beta_\mu(x)^2 + \frac{1}{2}div\beta_\mu(x).$$

Remark 24. The operator A is a Schrödinger operator with potential term V, acting in $L^2(\mathbb{R}^d)$ (in this interpretation it is appropriate to take $L^2(\mathbb{R}^d)$ as the complex Hilbert space of square integrable functions). A is densely defined (on $C_0^\infty(\mathbb{R}^d)$) in $L^2(\mathbb{R}^d)$, positive and symmetric as seen from (15) and the positivity and symmetry of A_μ (coming from corresponding properties of $\overset{\circ}{\mathcal{E}}_\mu$, see (14)). It is well known that Schrödinger operators describe the dynamics of quantum systems. More precisely: any self-adjoint extension H of A can be taken to define the dynamics of a quantum mechanical particle moving in the force field given by the potential V (having made the inessential convention $\frac{\hbar^2}{2m} = 1$, where $\hbar$ is the reduced Planck's constant and m the mass of the particle; obviously a "scaling" of variables will remove this restriction). The corresponding Schrödinger equation is then

$$i\frac{\partial}{\partial t}\psi = H\psi, \quad t \geq 0,$$

with $\psi_{|t_0} = \psi_0 \in D(H) \subset L^2(\mathbb{R}^d)$; it is solved by $e^{-itH}\psi_0$ (the unitary group, when we let $t \in \mathbb{R}$, generated by H via Stone's theorem).

Assumptions on V are known s.t. H is uniquely determined by its restriction A to $C_0^\infty(\mathbb{R}^d)$ (i.e. A is essentially self-adjoint on $C_0^\infty(\mathbb{R}^d)$; see, e.g. [423] for this concept), e.g. V bounded is enough (but in fact H is uniquely determined in much more general situations, see, e.g. [426]).

Remark 25. We can write, using $\beta_\mu = 2\frac{\nabla\varphi}{\varphi}$:

$$V = \left(\frac{\nabla\varphi}{\varphi}\right)^2 + div\left(\frac{\nabla\varphi}{\varphi}\right) = \frac{(\nabla\varphi)^2}{\varphi^2} + \frac{\varphi\Delta\varphi - (\nabla\varphi)^2}{\varphi^2} = \frac{\Delta\varphi}{\varphi}.$$

This shows

$$(-\Delta + V)\varphi = 0$$

i.e. φ is a "$(-\Delta + V)$-harmonic function" ".

The passage $A_\mu \to A$ can thus be seen as a particular case of "Doob's transform" technique, going from an operator A_μ in $L^2(\mathbb{R}^d, \mu)$, with 1 as harmonic function (1 is in the domain, if μ is finite) to an operator A in $L^2(\mathbb{R}^d)$, here $-\Delta + V$, with φ as harmonic function (see, e.g. [217] for the discussion of Doob's transform). Doob's transform is also called, in the context of the above operator, "ground state transformation" (see, e.g. [426], [467]).

Viceversa: Let us consider a "stationary Schrödinger equation" of the form

$$(-\Delta + \tilde{V})\varphi = E\varphi$$

for some $\tilde{V} : \mathbb{R}^d \to \mathbb{R}$, $E \in \mathbb{R}$. If there is a solution $\varphi \in L^2(\mathbb{R}^d)$ s.t. $\varphi(x) > 0$ $\forall x \in \mathbb{R}^d$, then setting $V \equiv \tilde{V} - E$ we get $V = \frac{\Delta\varphi}{\varphi}$. If, e.g., $D(V) \supset C_0^\infty(\mathbb{R}^d)$, we can define $A = -\Delta + V$ on $C_0^\infty(\mathbb{R}^d)$, as a linear densely defined symmetric operator. The associated sesquilinear form

$$\tilde{\overset{\circ}{\mathcal{E}}}(u, v) = (u, Av)$$

defined for $u, v \in C_0^\infty(\mathbb{R}^d)$ is then symmetric and closable in $L^2(\mathbb{R}^d)$. Moreover, if

$$\|\nabla\varphi\|_2^2 + (\varphi, V\varphi) \geq 0$$

(which is, e.g. the case whenever $V \geq 0$), then $\tilde{\overset{\circ}{\mathcal{E}}}$ is positive. We also observe that, defining $A_\mu = -\Delta - \beta_\mu \cdot \nabla$ with $\beta_\mu = \frac{\nabla \rho}{\rho} = 2\nabla ln(\varphi)$ with $\rho = \varphi^2$, $d\mu = \rho\, dx$, we have

$$\tilde{\overset{\circ}{\mathcal{E}}}(u,v) = (Wu, A_\mu W v)_\mu,$$

with $A_\mu \equiv WAW^*$, $u, v \in C_0^\infty(\mathbb{R}^d)$.

Defining on the other hand

$$\tilde{\overset{\circ}{\mathcal{E}}}_\mu(u,v) = \int < \nabla u, \nabla v > d\mu$$

we see, as in the beginning of this section, that

$$\tilde{\overset{\circ}{\mathcal{E}}}_\mu(u,v) = (u, A_\mu v)_\mu = (u, Av) = \tilde{\overset{\circ}{\mathcal{E}}}(u,v).$$

Note that $\tilde{\overset{\circ}{\mathcal{E}}}_\mu$ is a form in $L^2(\mathbb{R}^d, \mu)$, whereas $\tilde{\overset{\circ}{\mathcal{E}}}$ is a form in $L^2(\mathbb{R}^d)$.

Thus to a generalized Schrödinger operator of the form $-\Delta + \tilde{V}$ in $L^2(\mathbb{R}^d)$ we have associated a classical pre-Dirichlet form $\tilde{\overset{\circ}{\mathcal{E}}}_\mu$ and its closure $\tilde{\mathcal{E}}_\mu$ and corresponding associated densely defined operators L_μ° resp. L_μ. Even though $\tilde{\overset{\circ}{\mathcal{E}}}$ and A in $L^2(\mathbb{R}^d)$ have a more direct physical interpretation, rather than their corresponding objects $\mathcal{E}_\mu$ resp. L_μ in $L^2(\mathbb{R}^d, \mu)$, the latter are more appropriate whenever discussing "singular interactions" (see, e.g. [44], [104], [20], [106], [40], [152], [157], [164], [176], [193], [183], [266], [503] or the case where $\mathbb{R}^d$ is replaced by an infinite dimensional space E (like in quantum field theory), see, below and, e.g., [278], [15], [41] (in fact in the latter case there are interesting probability measures μ on E, whereas no good analogue of Lebesgue measure on E exists). In this sense operating with $\mathcal{E}_\mu$, L_μ is more natural and general than operating with Schrödinger operators.

Remark 26. In the above discussion of closability and Doob's transformation we have assumed φ, V, $\tilde{V}$ to be smooth and $\varphi > 0$. These assumptions can be strongly relaxed. Moreover, the considerations extend to cover the cases where, instead of A resp. A_μ, general elliptic symmetric operators with L^p_{loc}-coefficients are handled, see e.g. [367], [499], [104], [106], [20], [22], [174], [364], [375] where various questions (including, e.g. closability) are discussed (we shall give more references in Chapter 4).

3 Contraction properties of forms, positivity preserving and submarkovian semigroups

3.1 Positivity preserving semigroups and contraction properties of forms- Beurling-Deny formula.

In this whole section the Banach respectively Hilbert space of Chapter 2 will be the functional space $L^2(m) \equiv L^2(E, \mathcal{B}, m; \mathbb{K})$, where $(E, \mathcal{B}, m)$ is a measure space and $L^2(E, \mathcal{B}, m; \mathbb{K})$ stands for the $\mathbb{K}$-valued (equivalence class of) functions over E which are square integrable with respect to m. Let $(,)$ be the scalar product in $L^2(m)$, $\| - \|$ the corresponding norm.

Definition 16. *A linear operator A in $L^2(m)$ is said to be positivity preserving (p.p.) iff $u \geq 0$ m-a.e. implies $Au \geq 0$ m-a.e., for any $u \in L^2(m)$. A semigroup $T = (T_t)_{t \geq 0}$ is said to be positivity preserving if T_t is p.p. for all $t \geq 0$.*

Proposition 12. *Let L be a self-adjoint operator in $L^2(m)$ s.t. $L \leq 0$ (i.e. $-L \geq 0$). The corresponding symmetric C_0-contraction semigroups $(e^{tL})_{t>0}$ is positivity preserving iff the corresponding resolvent family $(\alpha - L)^{-1} = G_\alpha$, $\alpha > 0$, is positivity preserving.*

This is an immediate consequence of the Laplace-transform formula, see chapter 1:
$G_\alpha u = \int_0^\infty e^{-\alpha t} e^{tL} u \, dt$, $u \in L^2(m)$ resp. the Yosida approximation formula

$$e^{tL} = \lim_{\alpha \to +\infty} e^{tL^{(\alpha)}}$$

with $L^\alpha \equiv -\alpha + \alpha^2 G_\alpha$.

Theorem 7. *(Beurling-Deny representation for positivity preserving semigroups)*
Let $\mathbb{K} = \mathbb{R}$ and let L be a self-adjoint operator in $L^2(m)$, with $L \leq 0$. Then the following statements are equivalent:

i) $(e^{tL})_{t \geq 0}$ *is positivity preserving*
ii) $\mathcal{E}(|u|) \leq \mathcal{E}[u]$ *for all* $u \in D(\mathcal{E})$
(where $\mathcal{E}$ is the positive, symmetric, closed form associated with L, according to Friedrichs representation theorem of Chapter 2)

Remark 27. This is an important theorem since it expresses the positivity preserving property of semigroups (and in probability theory, as we shall see, one is primarily interested in such semigroups, e.g., transition semigroups) through corresponding properties of associated forms, which are often easier to verify directly.

Originally published in: *Ecole d'Eté de Probabilités de Saint-Flour XXX – 2000*, Lecture Notes in Mathematics, Vol. **1816**, 33–42, DOI: 10.1007/3-540-44922-1_3, © Springer-Verlag Berlin Heidelberg 2003, Reprint by Springer-Verlag Berlin Heidelberg 2012

Proof. *i)* → *ii)*
One first remarks that $|e^{tL}u| \leq e^{tL}|u|$ (which follows from $e^{tL}(|u| \pm u) \geq 0$, by the assumption in i)).
But for all $u \in L^2(m)$, using the semigroup property: $(u, e^{tL}u) = \|e^{\frac{t}{2}L}u\|^2$, which is then bounded from above by $\|e^{\frac{t}{2}L}|u|\|^2 = \left(|u|, e^{tL}|u|\right)$, where in the latter equality we again used the semigroup property.
From this we deduce

$$\left(u, \left(e^{tL} - 1\right) u\right) \leq \left(|u|, \left(e^{tL} - 1\right) |u|\right). \qquad \forall t \geq 0$$

For $u \in D(L)$ the left hand side divided by $t > 0$ by Theorem 5 in chapter 2 section 2.2 converges for $t \downarrow 0$ to $(u, Lu) = -\mathcal{E}[u]$.
The right hand side is non positive, since e^{tL} is positivity preserving, thus:

$$-\mathcal{E}[u] \leq \frac{1}{t}\left(|u|, \left(e^{tL} - 1\right) |u|\right) \leq 0 \qquad \forall t \geq 0.$$

Hence the limit for $t \downarrow 0$ of $\frac{1}{t}\left(|u|, \left(e^{tL} - 1\right) |u|\right)$ exists by subsequences, and by the definition of $D(L)$ we must have then that this is equal to $\left(|u|, L|u|\right)$.
Thus $-\mathcal{E}[u] \leq \left(|u|, L|u|\right) = -\mathcal{E}\left[|u|\right] \leq 0$ for all $u \in D(L)$. Since $D(L)$ is a core for $D(\mathcal{E})$, and L is closed, this extends by continuity to all $u \in D(\mathcal{E}) = D((-L)^{1/2})$, showing that i) → ii).

ii) → *i)*:
We have, for $u \in D(L), v \in D(L)$:

$$\mathcal{E}_\alpha[u + v] = \mathcal{E}_\alpha[u] + \mathcal{E}_\alpha[v] + 2((\alpha - L)u, v).$$

Replace now u by $w = (\alpha - L)^{-1}u$ ($\in D(L)!$), and take $v \geq 0$, $v \in D(\mathcal{E})$, then $\mathcal{E}_\alpha[w + v] = \mathcal{E}_\alpha[w] + \mathcal{E}_\alpha[v] + 2(u, v)$. Taking $u \geq 0$, v as before, we get:

$$\mathcal{E}_\alpha[w + v] \geq \mathcal{E}_\alpha[w] + \mathcal{E}_\alpha[v]. \tag{16}$$

But

$$\mathcal{E}_\alpha[|w| + v] = \mathcal{E}_\alpha[w + (|w| - w) + v] = \mathcal{E}_\alpha[w + v'], \tag{17}$$

with $v' \equiv |w| - w + v$. Applying (16) on the right hand side, with v replaced by v' we get:

$$\mathcal{E}_\alpha[w + v'] \geq \mathcal{E}_\alpha[w] + \mathcal{E}_\alpha[v']. \tag{18}$$

Hence from (17), (18):

$$\mathcal{E}_\alpha[|w| + v] \geq \mathcal{E}_\alpha[w] + \mathcal{E}_\alpha[v']. \tag{19}$$

By assumption ii) the r.h.s. is bounded from below by

$$\mathcal{E}_\alpha[|w|] + \mathcal{E}_\alpha[v']. \quad (20)$$

Taking now $v = 0$, we get from (19):

$$\mathcal{E}_\alpha[|w|] \geq \mathcal{E}_\alpha[|w|] + \mathcal{E}_\alpha[|w| - w] \quad (21)$$

(where we inserted the definition of v'). On the other hand, by $\mathcal{E}_\alpha \geq 0$:

$$\mathcal{E}_\alpha[|w| - w] \geq 0. \quad (22)$$

Hence, from (21), (22):

$$\mathcal{E}_\alpha[|w| - w] = 0$$

thus $|w| - w{=}0$ i.e. $|w| = w$, in particular $w \geq 0$. From the definition of w, then:

$$(\alpha - L)^{-1}u \geq 0 \ \ for\ u \geq 0,$$

which proves i), using Proposition 12.

Remark 28. 1) There are several other statements which are equivalent to the ones of Theorem 7, see e.g., [367],[368], [136].
2) In a similar way one proves the Theorem for the complex Hilbert space $L^2_{\mathbb{C}}(m)$. It is also easy to show that i),ii) are equivalent with iii), e^{tL} is reality preserving and $\mathcal{E}[u^+] \leq \mathcal{E}[u]$ for all real $u \in D(\mathcal{E})$, with $u^+ \equiv \sup(u, 0)$. This is left as an exercise (cf. [423]).

3.2 Beurling-Deny criterium for submarkovian contraction semigroups

Theorem 8. *Let $\mathbb{K} = \mathbb{R}$ and let L be a self-adjoint operator in $L^2(m)$, with $L \leq 0$, s.t. $T_t u \geq 0$ for $u \geq 0$, with $T_t = e^{tL}$, $t \geq 0$. Then the following are equivalent:*

a) e^{tL} is a contraction on $L^\infty(m) \cap L^2(m)$, $\forall t \geq 0$.
b) For $u \geq 0$, $u \in D(\mathcal{E})$, then $\mathcal{E}[u \wedge 1] \leq \mathcal{E}[u]$ (with $u \wedge 1 \equiv inf(u, 1)$).

Proof. The proof of *b)* $\rightarrow$ *a)* is similar to the one of *i)* $\rightarrow$ *ii)* in Theorem 7. For *a)* $\Rightarrow$ *b)* the idea is to show

$$(u \wedge 1, (1 - T_t)(u \wedge 1)) \leq (u, (1 - T_t)u) \ \ \forall t \geq 0. \quad (23)$$

Once this is shown a) follows easily by dividing the inequality by $t > 0$ and going to the limit $t \downarrow 0$, which yields, for $u \in D(L)$

$$\mathcal{E}[u \wedge 1] \leq (u, (1 - L)u) = \mathcal{E}[u].$$

b) then follows by the fact that $D(L)$ is a core for $\mathcal{E}$. The proof of (23) is obtained by proving it first for special step functions and then going to the limit, exploiting the fact that $F(u) \equiv u \wedge 1$ satisfies $|F(u)| \leq |u|$, $F(u) - F(v) \leq |u - v|$, see, e.g., [258], [367], [423] for details.

3.3 Dirichlet forms

Let (E, B, M) be a σ-finite measure space, and let $\mathbb{K} = \mathbb{R}$.

Definition 17. *Let S be a linear operator in $L^2(m)$ with $D(S) = L^2(m)$, S bounded. S is called* <u>*submarkovian*</u> *if $0 \leq Su \leq 1$ m-a.e., whenever $0 \leq u \leq 1$ m-a.e., for all $u \in L^2(m)$*

Remark 29. If S is positivity preserving in $L^2(m)$ and $1 \in L^2(m)$, then $Su \leq 1$ for $u \leq 1$ m-a.e., $u \in L^2(m)$ follows from $S(u-1) \leq 0$ and linearity, since $u - 1 \in L^2(m)$.

Definition 18. *Let $\mathcal{E}$ be a positive, symmetric, bilinear closely defined form in $L^2(m)$ (not necessarily closed). Given $\Phi : \mathbb{R} \to \mathbb{R}$, we say that $\mathcal{E}$ contracts under Φ if $u \in D(\mathcal{E})$ implies $\Phi(u) \in D(\mathcal{E})$ and $\mathcal{E}[\Phi(u)] \leq \mathcal{E}[u]$. If $\Phi(t) = (0 \vee t) \wedge 1$, $t \in \mathbb{R}$, then we call Φ the "unit contraction". Let, for all $\epsilon > 0$, Φ_ϵ be such that $\Phi_\epsilon(t) = t$, $t \in [0,1]$*

$$-\epsilon \leq \Phi_\epsilon(t) \leq 1 + \epsilon \ \ \forall t \in \mathbb{R},$$

$$0 \leq \Phi_\epsilon(t_2) - \Phi_\epsilon(t_1) \leq t_2 - t_1, \ t_1 \leq t_2,$$

$$\Phi_\epsilon(u) \in D(\mathcal{E}),$$

for $v \in D(\mathcal{E})$ and $\liminf_{\epsilon \downarrow 0} \mathcal{E}[\Phi_\epsilon(u)] \leq \mathcal{E}[u]$, then we call Φ_ϵ "ϵ-approximation" of the unit contraction and we say that $\mathcal{E}$ is <u>*submarkovian*</u>.

Theorem 9. *Let $\mathcal{E}$ be a positive, symmetric, bilinear, closed densely defined form and let T_t, G_α be the associated C_0-contraction semigroup resp. contraction resolvent. Then the following assertions are equivalent:*

a) T_t is submarkovian for all $t \geq 0$
b) αG_α is submarkovian for all $\alpha > 0$
c) $\mathcal{E}$ is submarkovian
d) $\mathcal{E}$ contracts under the unit contraction
e) The infinitesimal generator L of T_t has the "Dirichlet contraction property" $(u, L(u-1)^+) \leq 0 \ \forall u \in D(L)$.

Proof. $a) \to b)$: Proposition 12 in Chapter 3.3.1
$d) \to c)$: take $\Phi_\epsilon = unit$
$b) \to d)$: this is similar to the proof of $i) \to ii)$ Theorem 7 in Chapter 3.3.1
$c) \to b)$: this is similar to the proof of $b) \to a)$ in Theorem 8 in Chapter 3.3.2.
For the remaining parts see [244], [258], [367], [172]. □

For the applications the following "Addendum" is useful.

Proposition 13. *Let $\overset{\circ}{\mathcal{E}}$ be a symmetric, positive, closable, densely defined bilinear form. Assume moreover that $\overset{\circ}{\mathcal{E}}$ is submarkovian on $D(\overset{\circ}{\mathcal{E}})$, then the closure $\mathcal{E}$ of $\overset{\circ}{\mathcal{E}}$ satisfies a)-e) of Theorem 9.*

Proof. (Sketch): Let $u \in D(\mathcal{E})$, since $\overset{\circ}{\mathcal{E}}$ is closable by assumption, there exists a sequence $u_n \in D(\overset{\circ}{\mathcal{E}})$, s.t. $u_n \to u$ in $L^2(m)$, for $n \to \infty$, and $\overset{\circ}{\mathcal{E}}[u_n] \to \mathcal{E}[u]$. Choose $\Phi_{\frac{1}{n}}$ as $\epsilon = \frac{1}{n}$- approximation of the unit contraction, for $n \in \mathbb{N}$. Then $\sup \overset{\circ}{\mathcal{E}}_1[\Phi_{\frac{1}{n}}(u_n)] \leq sup\overset{\circ}{\mathcal{E}}_1[u_n]$ and this is finite, by the convergence of $\overset{\circ}{\mathcal{E}}_1[u_n]$. $(D(\mathcal{E}), \mathcal{E}_1)$ is a Hilbert space and by the Banach-Saks theorem (cf. e.g. Th 2.2. in App.2 in [367]), $g_n \equiv \frac{1}{n}\sum_1^n \Phi_{\frac{1}{j}}(u_j)$ converges strongly for $n \to \infty$ to $g \in D(\mathcal{E})$ and $\overset{\circ}{\mathcal{E}}[g_n] \to \mathcal{E}[g]$. On the other hand: $g_n \to \Phi_0[u]$ as $n \to \infty$
$\Phi_\epsilon \to \Phi_0$ m-a.e. as $\epsilon \downarrow 0$.
So by these convergences we get $\Phi_0[u] = g$ m-a.e., $\Phi_0 \in D(\mathcal{E})$ and $\mathcal{E}(\Phi_0[u])^{\frac{1}{2}} = \lim_{n\to\infty} \overset{\circ}{\mathcal{E}}[g_n]^{\frac{1}{2}}$. By the triangle inequality and the definition of g_n the r.h.s. is bounded from above by $\limsup \frac{1}{n}\sum_{j=1}^n \overset{\circ}{\mathcal{E}}[\varphi_{\frac{1}{j}}(u_j)]^{\frac{1}{2}}$. Since the Cesaro-limit and the ordinary limits coincide whenever they exist, we get that this is $\mathcal{E}[u]^{\frac{1}{2}}$ and thus $\mathcal{E}$ contracts under the unit contraction. □

Definition 19. *A form $\mathcal{E}$ with the properties as in Theorem 9 (i.e. bilinear, positive, symmetric, closed, densely defined and submarkovian) is called a (symmetric) Dirichlet form.*

Remark 30. Proposition 13 says that the closure $\overline{\overset{\circ}{\mathcal{E}}} = \mathcal{E}$ of the form $\overset{\circ}{\mathcal{E}}$ in Proposition 13 is a Dirichlet form. In applications the forms $\overset{\circ}{\mathcal{E}}$ are usually given on a nice subset D of $L^2(m)$ where they are closable. Proposition 13 permits then to pass from the contraction properties of $\overset{\circ}{\mathcal{E}}$ on D to corresponding ones for its closure $\mathcal{E}$.

Remark 31. Dirichlet forms are studied extensively in [150], [244], [258], [468], [469], [367], [309] (and references therein).

3.4 Examples of Dirichlet forms

Standard finite dimensional example Let $E = \mathbb{R}^d$, $B = \mathcal{B}(\mathbb{R}^d)$, $m = \mu$, $\mu(dx) = \rho(x)\,dx$, $r \geq 0$, (where dx is Lebesgue measure on $\mathbb{R}^d$).
Consider $\overset{\circ}{\mathcal{E}}_\mu(u,v) \equiv \int \nabla u \cdot \nabla v\, d\mu$ in $L^2_{\mathbb{R}}(\mathbb{R}^d,\mu) \equiv L^2(\mathbb{R}^d,\mu;\mathbb{R}) \equiv L^2(\mu)$, $u, v \in C_0^\infty(\mathbb{R}^d)$ (with $\nabla u(x) \equiv (\frac{\partial}{\partial x_1}u, \cdots, \frac{\partial}{\partial x_d}u)(x)$, $x \in \mathbb{R}^d$).

The basic question to be answered is whether $\overset{\circ}{\mathcal{E}}_\mu$ is closable. In Chapter 1 we saw that for this to be the case it is enough to find a (symmetric, positive) operator $\overset{\circ}{A}_\mu$ in $L^2(\mu)$ s.t. $\overset{\circ}{\mathcal{E}}_\mu(u,v) = (u, \overset{\circ}{A}_\mu v)$.
By "integration by parts" we see that $\overset{\circ}{A}_\mu = -A - \beta_\mu \cdot \nabla$ on $C_0^\infty(\mathbb{R}^d)$. This is well defined whenever

$$\beta_{\mu,i}(x) \equiv \frac{1}{\rho}\partial_i \rho(x) \in L^2(\mu), \qquad \forall i \in \{1, ..., d\}$$

with $\partial_i \equiv \frac{\partial}{\partial x_i}$ (taken in the distributional sense).

"Optimal" conditions for closability of $\overset{\circ}{\mathcal{E}}$ have been given in [119], see also [223], [474].

The second basic question to be answered is whether the closure $\mathcal{E}_\mu \equiv \overline{\overset{\circ}{\mathcal{E}}}_\mu$ of $\overset{\circ}{\mathcal{E}}_\mu$ is a Dirichlet form.

The answer is yes, since by using Proposition 13 it is enough to verify that $\overset{\circ}{\mathcal{E}}_\mu$ is submarkovian. Let us take $\epsilon > 0$, $\Phi_\epsilon \in C^\infty$, Φ_ϵ-contraction, an ϵ-approximation of the unit contraction $|\Phi'_\epsilon(\cdot)| \leq 1$. Then:
$\overset{\circ}{\mathcal{E}}[\Phi_\epsilon(u)] = \int \nabla\Phi_\epsilon(u)\nabla\Phi_\epsilon(u)d\mu = \int \Phi'_\epsilon(u)\nabla u\cdot\Phi'_\epsilon(u)\nabla u d\mu = \int |\Phi'_\epsilon(u)|^2|\nabla u|^2 d\mu \leq \int |\nabla u|^2 d\mu = \overset{\circ}{\mathcal{E}}[u]$, $\forall u \in C_0^\infty(\mathbb{R}^d)$, where we used the definition of $\overset{\circ}{\mathcal{E}}$, Leibniz differentiation formula in the second equality, and $|\Phi'_\epsilon| \leq 1$ for the latter inequality.

This shows that $\overset{\circ}{\mathcal{E}}$ contracts under Φ_ϵ and by Proposition 13 we then have that $\mathcal{E}_\mu$ is a Dirichlet form. □

The standard infinite dimensional example Let E be a separable real Banach space, E' its topological dual. Let μ be the probability measure on the Borel subsets B in E generated by all open subsets of E, and suppose $supp\mu = E$ (i.e. $\mu(U) > 0$ for all $U \subset E$, U open, $U \neq \emptyset$).
Let $FC_b^\infty \equiv FC_b^\infty(E) = \{u : E \to \mathbb{R} \mid \exists m, \exists f \in C_b^\infty(\mathbb{R}^m), \exists l_1, \cdots, l_m \in E' : u(z) = f(< z, l_1 >, \cdots, < z, l_m >)\}$. Then FC_b^∞ is dense in $L^2(\mu) \equiv L^2_{\mathbb{R}}(E, \mu)$ (this is essentially a form of the Stone-Weierstrass theorem, see, e.g. [396], [277], [290]).
For $u \in FC_b^\infty(E)$, $z \in E$, $k \in K \subset E$ (K a linear subspace of E) we define the (Gâteaux-)derivative in the direction k by:

$$\frac{\partial u}{\partial k}(z) \equiv \frac{d}{ds}u(z+sk)|_{s=0} \left(= \sum_{i=1}^m \partial_i f(< z, l_1 >, \cdots, < z, l_m >) < k, l_i > \right).$$

We define

$$\overset{\circ}{\mathcal{E}}_{\mu,k}(u,v) \equiv \int \frac{\partial u}{\partial k} \cdot \frac{\partial v}{\partial k} d\mu,$$

for $u, v \in FC_b^\infty$.

We easily see that $\overset{\circ}{\mathcal{E}}_{\mu,k}$ is a symmetric, positive, bilinear form on $L^2(\mu)$, with domain $D(\tilde{\mathcal{E}}_{\mu,k}) = FC_b^\infty$ (dense in $L^2(\mu)$!). As before the first question to be answered is: for which μ is $\overset{\circ}{\mathcal{E}}_{\mu,k}$ defined on FC_b^∞ closable? A sufficient condition for this is (analogously as in the finite dimensional case) the existence of $\beta_{\mu,k} \in L^2(\mu)$ s.t.

$$\int \frac{\partial u}{\partial k} v d\mu = -\int u \frac{\partial v}{\partial k} d\mu - \int uv\beta_{\mu,k} d\mu, \ \forall u, v \in FC_b^\infty. \qquad (24)$$

Remark 32. This is an "integration by parts" formula, in the spirit of those in the theory of smooth measures [205], [59], [61], [62], [65], [119] and in "Malliavin calculus" (originally on Wiener space, see [372], [396]).
Assuming (24) we have then that

$$\overset{\circ}{L}_{\mu,k} \equiv \frac{\partial^2}{\partial k^2} + \beta_{\mu,k}\frac{\partial}{\partial k}$$

is a well defined linear operator with $D(\overset{\circ}{L}_{\mu,k}) = FC_b^\infty$, symmetric and such that

$$\overset{\circ}{\mathcal{E}}_{\mu,k}(u,v) = (u, (-\overset{\circ}{L}_{\mu,k})v)$$

(with (,) the $L^2(\mu)$-scalar product). From Proposition 11 in Chapter 2 we see that $\overset{\circ}{\mathcal{E}}_{\mu,k}$ is closable in $L^2(\mu)$.

The next question is whether the closure $\mathcal{E}_{\mu,k} = \bar{\overset{\circ}{\mathcal{E}}}_{\mu,k}$ of $\overset{\circ}{\mathcal{E}}_{\mu,k}$ in $L^2(\mu)$ is a Dirichlet form. This is proven similarly as in the finite dimensional case. In fact let Φ_ε be an ε-approximation of the unit contraction which is C^∞ on $\mathbb{R}$. Then, by Leibniz formula, $u \to \Phi_\varepsilon(u) \in FC_b^\infty$ if $u \in FC_b^\infty$ and

$$\tilde{\mathcal{E}}_{\mu,k}[\Phi_\varepsilon(u)] \leq \int |\Phi_\varepsilon'(u)|^2 |\nabla u|^2 d\mu \leq \int |\nabla u|^2 d\mu = \overset{\circ}{\mathcal{E}}_{\mu,k}[u].$$

By using Proposition 13 this yields the proof that $\overset{\circ}{\mathcal{E}}_{\mu,k}$ is a Dirichlet form.
We assume that E and its dual E' are s.t.

$$E' \subset H' \cong H \subset E,$$

where H is a Hilbert space, densely contained in E, H' (isomorphic to H) is the dual of H (and E' is densely contained in H'), moreover the embedding of H in E (and of E' in H')is continuous. (We remark that this assumption is not strong, in fact it can be realized in great generality, see e.g. [301], [462], [428]).
For $u \in FC_b^\infty$, $k \in E'$ we have $\frac{\partial u}{\partial k}(z) = \underset{E'}{<} k, \nabla u(z) \underset{E}{>}$ (with $\underset{E'}{<}\ ,\ \underset{E}{>}$ the dualisation between E and E').

Definition 20. *For $u, v \in FC_b^\infty$ we define*

$$\overset{\circ}{\mathcal{E}}_\mu(u,v) \equiv \int_E < \nabla u, \nabla v >_H d\mu,$$

where $< , >_H$ is the scalar product in H.
Let K be an orthonormal basis in H consisting of elements in E'. We have:

$$\overset{\circ}{\mathcal{E}}_\mu(u,v) = \sum_{k\in K} \overset{\circ}{\mathcal{E}}_{\mu,k}(u,v).$$

We remark that $\overset{\circ}{\mathcal{E}}_\mu$ is a bilinear positive symmetric form. $\overset{\circ}{\mathcal{E}}_\mu$ is closable if all $\overset{\circ}{\mathcal{E}}_{\mu,k}$ are closable (we leave the verification of this as an exercise: one can use, e.g., Proposition 11 in Chapter 2). In this case the closure $\mathcal{E}_\mu$ of $\overset{\circ}{\mathcal{E}}_\mu$ exists and is a positive symmetric closed bilinear form. It is also easy to verify that $\mathcal{E}_\mu$ is a Dirichlet form. $\mathcal{E}_\mu$ is called the "classical Dirichlet form given by μ".

Remark 33. For $E = H = \mathbb{R}^d$ and FC_b^∞ replaced by $C_0^\infty(\mathbb{R}^d)$, $\mathcal{E}_\mu$ coincides with the classical Dirichlet form given by μ (over $\mathbb{R}^d$) as defined in Definition 15.

By Friedrichs' representation theorem to $\mathcal{E}_\mu$ there is a uniquely associated self-adjoint, negative operator L_μ s.t. $\mathcal{E}_\mu(u,v) = ((-L_\mu)^{\frac{1}{2}}u, (-L_\mu)^{\frac{1}{2}}v)$. L_μ is called the Dirichlet operator associated with μ.

Remark 34. $\mathcal{E}_\mu$ should not be confused with the "maximal Dirichlet form given by μ", $\mathcal{E}_\mu^m$, obtained from the closed extension of $\tilde{\mathcal{E}}_\mu$ with domain $D(\mathcal{E}_\mu^m) = \bigcap_{k\in K}\{u \in L^2(\mu) | \int |\frac{\partial u}{\partial k}|^2 d\mu < \infty\}$. In general $\mathcal{E}_\mu^m$ is a strict extension of $\mathcal{E}_\mu$, they coincide exactly when $\tilde{L}_\mu$is essentially self-adjoint on FC_b^∞, see, e.g., [119], [88], [359] for results of this type.

Exercise 6. Show that
$L_\mu = \overset{\circ}{L}_\mu$ on FC_b^∞, with $\overset{\circ}{L}_\mu = \sum_k \overset{\circ}{L}_{\mu,k}$ on FC_b^∞, where

$$\overset{\circ}{L}_{\mu,k} \equiv \frac{\partial^2}{\partial k^2} + \beta_{\mu,k}(\cdot)\frac{\partial}{\partial k},$$

$$\triangle_H \equiv \sum_{k\in K} \frac{\partial^2}{\partial k^2} \quad , < \beta_\mu(z), \nabla > u(z) = \sum_{k\in K} < \beta_\mu(z), k > \frac{\partial}{\partial k}u(z),$$

$u \in FC_b^\infty$, $z \in E$.
Moreover $\overset{\circ}{\mathcal{E}}_\mu(u,v) = \sum_{k\in K} (u, \overset{\circ}{L}_{\mu,k}v)$ $\forall u, v \in FC_b^\infty$.

Exercise 7. When $E = C_{(0)}([0,t];\mathbb{R})$, (Wiener space of continuous functions from $[0,t]$ to $\mathbb{R}$ vanishing at time zero) we have

$$H = H^{1,2}([0,t];\mathbb{R}) = \{w \in E | \int_0^t |\dot{w}(s)|^2 ds < \infty\},$$

where μ is the Wiener measure: show that β_μ is a linear function (cf. [172],[365], [396]).
We know by the general theory of Chapter 3.1 that L_μ generates a symmetric submarkovian C_0-contraction semigroup $L^2(\mu)$. We shall show that L_μ generates a diffusion process. For this we first describe shortly the general structure theory of Dirichlet operators and forms.

3.5 Beurling-Deny structure theorem for Dirichlet forms

We first consider the case of a locally compact separable metric space E (not necessarily a linear space). Let m be a Radon measure on E (see, e.g., [462] for Radon measures).

Definition 21. *A bilinear symmetric positive densely defined form $\mathcal{E}$ in $L^2_{\mathbb{R}}(E,m) \equiv L^2(m)$ is called regular if $D(\mathcal{E}) \cap C_0(E)$ is dense in $D(\mathcal{E})$with respect to the $\mathcal{E}_1$-norm and dense in $C_0(E)$ with respect to the supremum-norm.*

Definition 22. *A bilinear form $\mathcal{E}$ in $L^2(m)$ is called local if $\mathcal{E}(u,v) = 0$, for all $u, v \in D(\mathcal{E})$ s.t. supp$[u] \cap$ supp$[v] = 0$ (for some representatives $[u]$, $[v]$ of u resp. v as element of $L^2(m)$, with supp$[u]$, supp$[v]$ compact).*

Examples are:

a) $\mathcal{E}(u,v) = \int u\, v\, dm,\quad u, v \in C_0^\infty(\mathbb{R}^d)$,
b) $\mathcal{E}(u,v) = \int \nabla u \cdot \nabla v\, dm,\quad u, v \in C_0^\infty(\mathbb{R}^d)$.

Definition 23. *A bilinear form $\mathcal{E}$ in $L^2(m)$ is called strong local if $\mathcal{E}(u,v) = 0\ \forall u, v \in D(\mathcal{E})$ such that v is constant in a neigborhood of supp$[u]$.*

Remark 35. Strong local implies local, but not viceversa, e.g. in the examples above a) is local but not strong local.

Theorem 10. *(Beurling-Deny structure theorem)*
Let $\mathcal{E}$ be a regular Dirichlet form. Then $\mathcal{E}$ can be written as

$$\mathcal{E} = \mathcal{E}^c + \mathcal{E}^j + \mathcal{E}^k$$

on $D(\mathcal{E}) \cap C_0(E)$, with $\mathcal{E}^c$ strong local,

$$\mathcal{E}^j(u,v) = \int'_{E\times E} [u(x) - u(y)][v(x] - v(y)] J(dxdy),$$

where $(E \times E)' \equiv \{(x,y) \in E \times E \mid x \neq y\}$ and J is a symmetric Radon measure on $(E \times E)'$,

$$\mathcal{E}^k(u,v) = \int uv\, dk,$$

dk being a Radon measure on E. The parts $\mathcal{E}^c$ (diffusion part) , $\mathcal{E}^j$ (jump part) , $\mathcal{E}^k$ (killing part) are uniquely determined by $\mathcal{E}$.

Proof. For the proof see [244], [258]. □

If E is a manifold or $E = U$ with U an open subset of $\mathbb{R}^d$ then $\mathcal{E}^c$ can be further specified.

Theorem 11. *(Beurling-Deny structure theorem for an open subset of $\mathbb{R}^d$) Let U be an open subset of $\mathbb{R}^d$, m a Radon measure on U. Then any submarkovian symmetric positive bilinear form $\overset{\circ}{\mathcal{E}}$ on $L^2(U,m)$ with $D(\overset{\circ}{\mathcal{E}}) = C_0^\infty(U)$ s.t. $\overset{\circ}{\mathcal{E}}$ is closable and one has that $\mathcal{E} = \overline{\overset{\circ}{\mathcal{E}}}$ is a regular Dirichlet form and for $\mathcal{E}$ (and hence $\overset{\circ}{\mathcal{E}}$) on $C_0^\infty(\mathbb{R}^d)$ we have*

$$\overset{\circ}{\mathcal{E}} = \mathcal{E}^c + \mathcal{E}^j + \mathcal{E}^k,$$

with $\mathcal{E}^c(u,v) = \sum_{i,j=1}^d \int \frac{\partial u}{\partial x_i}(x) \frac{\partial v}{\partial x_j}(x) d\nu_{ij}(x)$, where $\nu_{ij}(\cdot)$ is a random measure (for all $i,j = 1,\cdots,d$), with $(\nu_{ij}(K))_{i,j=1}^d$ is a positive definite, symmetric matrix for all compact $K \subset U$.

Proof. Since $\overset{\circ}{\mathcal{E}}$ is submarkovian and closable, the closure is a Dirichlet form. That $\mathcal{E}$ is regular follows easily, see [244]. Then we can apply Theorem 1 in chapter 1 to get $\mathcal{E}^c, \mathcal{E}^j, \mathcal{E}^k$. For the formula for $\mathcal{E}^c$ see [244].

Remark 36. There is an extension of Theorem 10 to infinite dimensional spaces, with regularity replaced by "quasi-regularity" (a concept we shall discuss in Chapter 4) see [111], [112].

3.6 A remark on the theory of non symmetric Dirichlet forms

There exists an extension of the entire theory of Dirichlet forms to the case of bilinear forms which are closable and contract under suitable "contraction operations", but are not necessarily symmetric.

Definition 24. *A bilinear form $\mathcal{E}$ in a real Hilbert space H is said to satisfy the weak sector condition if $\exists k > 0$ s.t.*

$$|\mathcal{E}_1(u,v)| \leq k\mathcal{E}_1[u]^{\frac{1}{2}}\mathcal{E}_1[v]^{\frac{1}{2}}$$

$\mathcal{E}$ is a coercitive closed form if $D(\mathcal{E})$ is dense in H, $\mathcal{E}$ satisfies the weak sector condition and the symmetric part $\overset{\circ}{\mathcal{E}}$ of $\mathcal{E}$ (defined as $\overset{\circ}{\mathcal{E}}(u,v) \equiv \frac{1}{2}[\mathcal{E}(u,v) + \mathcal{E}(v,u)] \forall u,v \in D(\mathcal{E})$, with $D(\overset{\circ}{\mathcal{E}}) = D(\mathcal{E})$) is closed.

Analogs of the relation between symmetric forms, contraction semigroups and associated operators, resolvents, discussed in Chapter 2.3, exist also for coercive closed forms. The main difference consists in the fact that instead of a single semigroup $(T_t)_{t\geq 0}$ (resp. resolvent family $(G_\alpha)_{\alpha>0}$) there are two semigroups $(T_t, \hat{T}_t)_{t\geq 0}$ (resp. two resolvents $((G_\alpha, \hat{G}_\alpha)_{\alpha>0})$): in duality (i.e. adjoint to each other in the case of a space $L^2(m)$). See the table below. For a Beurling-Deny structure theorem in this case see [378].

Table 3

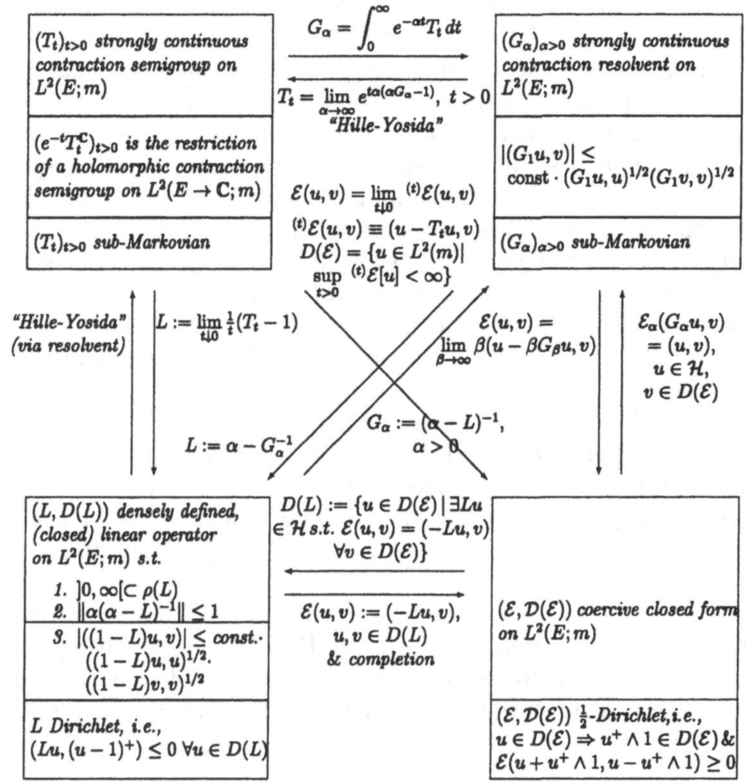

4 Potential Theory and Markov Processes associated with Dirichlet Forms

4.1 Motivations

Let E be a topological space, m be a σ-finite measure on E. Let $T \equiv (T_t)_{t\ge 0}$ be a sub-Markov semigroup in $L^2(m) \equiv L^2_{\mathbb{R}}(E,m)$. We shall discuss the following questions:
1) Is it possible to associate a "nice process" to T such that T is its transition semigroup (analogously as the Brownian motion process on $\mathbb{R}^d$ is associated with the heat semigroup as transition semigroup)?
2) Which kind of Dirichlet forms correspond to such "nice processes"?

As is well known, in the case where E is a locally compact separable metric space, an association of this type is surely possible if the transition

Originally published in: *Ecole d'Eté de Probabilités de Saint-Flour XXX – 2000*, Lecture Notes in Mathematics, Vol. **1816**, 43–50, DOI: 10.1007/3-540-44922-1_4, © Springer-Verlag Berlin Heidelberg 2003, Reprint by Springer-Verlag Berlin Heidelberg 2012

semigroup has the additional property of being a Feller semigroup, but what about the general case of the theory of Dirichlet forms?

Historically, one has analyzed first precisely the case of a locally compact separable metric space for the case where the Feller semigroup is replaced by the semigroup associated with a regular Dirichlet form, in which case the nice associated process is a Hunt process, see [243], [244], [258], [468], [469].

The next case that was historically treated was the one of non locally compact spaces, e.g. separable Banach spaces or rigged Hilbert spaces or conuclear spaces. These cases were discussed particularly in connection with the development of a mathematical theory of quantum fields, see [57], [58], [59], [61], [408] (for the case of rigged Hilbert spaces), [335] (for separable Banach spaces). A more general theory was then formulated in [118], [119], [120], some assumptions being further weakened by Schmuland [457] (see also [458], [459], [460], [461]). The setting of abstract Wiener spaces was particularly discussed in [172].
A non standard analytic setting was developed in [39], [36]. The central analytic concept which developed from all these approaches as being appropriate for the association of nice processes to general Dirichlet forms on general Hausdorff topological spaces E with a σ-finite measure m is that of quasi-regularity, first introduced in [105], [107], [108], [114], [109], [110], [113], see [367].

We shall here limit ourselves to a short sketch of the construction of nice processes starting from quasi regular Dirichlet forms, giving its main ideas. One basic idea is to replace continuous functions by quasi-continuous functions, as functions continuous modulo "small sets", and construct kernels acting on such functions.

4.2 Basic notions of potential theory for Dirichlet forms

$\mathcal{E}$-exceptional sets Let E be a Hausdorff topological space, $\mathcal{B}$ the σ-algebra of its Borel subsets, m a σ-finite measure.

Let $(\mathcal{E}, D(\mathcal{E}))$ be a Dirichlet form on $L^2(m) = L^2_{\mathbb{R}}(E, m)$.

Definition 25. *An increasing sequence $(F_k)_{k\in\mathbb{N}}$ of closed subsets of E is called an $\mathcal{E}$-nest if $\bigcup_k D(\mathcal{E})_{F_k}$ is $\tilde{\mathcal{E}}_1^{1/2}$-dense in $D(\mathcal{E})$ (where $D(\mathcal{E})_F \equiv \{u \in D(\mathcal{E}) \mid u = 0$ m-a.e. on the complement $F^c of F\}$; $\tilde{\mathcal{E}}_1^{1/2}$ is the norm given by the scalar product in $L^2(m)$ defined by $\tilde{\mathcal{E}}_1$).*

Definition 26. *$N \subset E$ is called "$\mathcal{E}$-exceptional" if $N \subset \bigcap_k F_k^c$ for some $\mathcal{E}$-nest $\{F_k\}_{k\in\mathbb{N}}$. We say that a certain property of points in E holds "$\mathcal{E}$-quasi everywhere" (q.e.) if it holds outside some $\mathcal{E}$-exceptional subset of E.*

There is an important relation between $\mathcal{E}$-exceptional sets and sets of small capacity; for this we first have to introduce the concept of capacity (an extension to our setting of the concept of capacity in classical potential theory, see, e.g. [162], [276], [160]).

Definition 27. *Let U be an open subset of E. Define the capacity cap U of U by*

$$cap\ U \equiv \inf_{u \in \mathcal{L}_U} \mathcal{E}_1[u]$$

where

$$\mathcal{L}_U \equiv \{u \in D(\mathcal{E}) \mid u \geq 1 \ m\text{-a.e. on } U\}$$

(with $\inf\{.\} = +\infty$ *if* $\mathcal{L}_U = \emptyset$*). For any $A \subset E$ define*

$$A \equiv \inf_{A \subset U, U \ open} cap\ U.$$

Remark 37. We have that cap $(U) = 0$ implies $m(U) = 0$.

Proof. We have

$$\mathcal{E}_1[u] \geq \|u\|^2 \geq m(U) \quad \forall u \in \mathcal{L}_U,$$

hence cap $U = \inf_{u \in \mathcal{L}_U} \mathcal{E}_1[u] \geq m(U)$. □

Proposition 14. *Let E, m, $\mathcal{E}$ as above. If $(A_n)_{n \in \mathbb{N}}$ is an increasing sequence of sets then 1) cap $(\bigcap_n A_n) = \sup_n$ cap A_n*

Moreover,

2) $(F_k)_{k \in \mathbb{N}}$ is an $\mathcal{E}$-next $\Longleftrightarrow \lim_{k \to \infty}$ cap $F_k^c = 0$

3) N is $\mathcal{E}$-exceptional $\Longleftrightarrow$ cap $N = 0$.

Proof. (Sketch):

1) is easy (see, e.g., [244]). 2) $\Rightarrow$ 3) is obvious. To prove 2) one uses the following

Lemma 2. *Let $U \subset E$ be open, s.t. $\mathcal{L}_U \neq \emptyset$, then*

a) there exists uniquely an element $e_U \in \mathcal{L}_U$ (called "equilibrium potential") s.t. $\mathcal{E}_1[e_U] = Cap\ U$.

b) $0 \leq e_U \leq 1$ m-a.e. on E and $U \mapsto e_U$ is monotone increasing.

Proof. For a) one uses that $\mathcal{L}_U$ is convex (which we leave as an exercise).

b) Take $u = (0 \vee e_U) \wedge 1$. $\mathcal{E}$ being a Dirichlet form, it contracts under the unit contraction, i.e. for $u \in D(\mathcal{E})$

$$\mathcal{E}_1[u] \leq \mathcal{E}_1[e_U] = \text{Cap } U$$

which implies, by the definition of capacity, that $u = e_U$ q.e., hence $0 \leq e_U \leq 1$ m-a.e.. That $u \mapsto e_U$ is monotone increasing is easily seen using $\mathcal{E}_\infty(\sqcap^+, \sqcap^-) \leq \prime$ for any $u \in D(\mathcal{E})$, see [244]. □

Definition 28. *Let $(F_k)_{k \in \mathbb{N}}$ be an $\mathcal{E}$-nest. Set*

$$C(\{F_k\}) \equiv \Big\{u : A \to \mathbb{R},\ for\ some\ A \subset E\ s.t.\ \bigcup_k F_k \subset A \subset E, u_{|F_k}$$
$$continuous\ \forall k \in \mathbb{N}\Big\}.$$

An $\mathcal{E}$-q.e. defined function u on E is called $\mathcal{E}$-quasi-continuous (q.c.) if there exists an $\mathcal{E}$-nest $\{F_k\}_{k \in \mathbb{N}}$ s.t. $u \in C(\{F_k\})$. $C(\{F_k\})$ is then called the set of quasi-continuous functions associated with the nest $(F_k)_{k \in \mathbb{N}}$.

Remark 38. One shows that u is $\mathcal{E}$-q.c. if for any $\epsilon > 0$ $\exists U \subset E$ open s.t. Cap $U < \epsilon$, $u_{|E-U}$ continuous, see, e.g. [367].

4.3 Quasi-regular Dirichlet forms

Definition 29. *Let E be a Hausdorff topological space, m a σ-finite measure on E, and let $\mathcal{B}$ the smallest σ-algebra of subsets of E with respect to which all continuous functions on E are measurable.*

A Dirichlet form $\mathcal{E}$ is called quasi-regular if

i) there exists an $\mathcal{E}$-nest $(F_k)_{k\in\mathbb{N}}$ of compact subsets of E;

ii) there exists an $\tilde{\mathcal{E}}_1^{1/2}$-dense subset of $D(\mathcal{E})$ whose elements have $\mathcal{E}$-q.c. m-versions.

iii) there exists $u_n \in D(\mathcal{E})$, $n \in \mathbb{N}$, with $\mathcal{E}$-q.c. m-versions $\tilde{u}_n$ and there exists an $\mathcal{E}$-exceptional subset N of E s.t. $\{\tilde{u}_n\}_{n\in\mathbb{N}}$ separates the points of $E - N$.

Remark 39. Thinking of Stone-Weierstrass type results, "point separation" means richness of elements...

Remark 40. We leave as an exercise to prove that if E is a locally compact separable metric space then $\mathcal{E}$ regular implies $\mathcal{E}$ quasi-regular but not viceversa (in general).

4.4 Association of "nice processes" with quasi-regular Dirichlet forms

$(E, \mathcal{B}, m)$ be as in the preceding section. Let $\mathcal{E}$ be a symmetric Dirichlet form in $L^2(m)$ and $(T_t = e^{tL})_{t\geq 0}$ the associated symmetric submarkovian C_0-contraction semigroup on $L^2(m)$.

Definition 30. *Let $(p_t)_{t\geq 0}$ be a submarkovian semigroup acting in $C_b(E)$ s.t. $(p_t u)(x) = \int p_t(x, dy)u(y), u \in C_b(E), p_t$ being a submarkovian semigroup of kernels i.e. $p_t(x, A) \in [0, 1], x \to p_t(x, A)$ is measurable $\forall x \in E, A \in \mathcal{B}$, $A \to p_t(x, A)$ is a measure on $(E, \mathcal{B})$ with $p_t(x, E) \leq 1$(see e.g., [142]).*
p_t is said to be associated with $\mathcal{E}$ (or $(T_t)_{t\geq 0}$ or with the infinitesimal generator L) if $p_t u$ is an m-version of $T_t u, \forall t > 0$, for all $u \in \mathcal{B}_b(E) \cap L^2(m)$.
Let $X \equiv (X_t)_{t\geq 0}$ be a (sub-) Markov process with state space E and transition semigroup $(p_t)_{t\geq 0}$, s.t.

$$(p_t u)(z) = E^z [u(X_t)]$$

$\forall z \in E, t \geq 0$ (with E^z the expectation for the process with start measure $\delta_z(.)$) is said to be associated with $\mathcal{E}$ (or $(T_t)_{t\geq 0}$ or the infinitesimal generator

L) if p_t is associated with $\mathcal{E}$.
X is said to be properly associated with $\mathcal{E}$ if in addition $p_t u$ is $\mathcal{E}$-quasi continuous (in the sense of the definition in IV.3) for all $t > 0$.

Remark 41. In a sense the $\mathcal{E}$-quasi-continuity of $p_t u$ is replaces the Feller-property (for the latter see, e.g., [162], [201], [220]).
The following theorem gives the basic relation between certain properties of symmetric Dirichlet forms and corresponding properties of associated (sub-) Markov processes; the meaning of the latter properties will be shortly discussed afterwards.

Theorem 12. *Let E be a topological Hausdorff space. Then:*

a) $\mathcal{E}$ is a quasi-regular Dirichlet form iff X is an m-tight special standard process.
b) $\mathcal{E}$ is a local quasi-regular Dirichlet form iff X is an m-tight special standard process and it is a diffusion i.e.
$P^z \{t \to X_t$ continuous on $[0,\zeta)\} = 1$ for all $z \in E$
(for some random variable ζ, with values $[0,+\infty]$, the life time of the process).

The basic concept on the r.h.s. is defined as follows:

Definition 31. *(right process)*
Let Δ be one pointset, disjoint from E, and define $E_\Delta \equiv E \cup \Delta$.

i) Let $X \equiv (X_t)_{t\geq 0}$ be a (family of) stochastic process(es) on a probability space $(\Omega, \mathcal{M}, P^z)_{z\in E_\Delta}$, with state space E, life time ζ, a measurable map $\Omega \to [0,+\infty]$ (if $\zeta = +\infty$ one can forget about Δ), s.t.

$$X_t(\omega) \in E \text{ for } t < \zeta(\omega),$$

$$X_t(\omega) \in \Delta \text{ for } t \geq \zeta(\omega).$$

$\forall\omega \in \Omega$. Assume $(t,\omega) \to X_t(\omega)$ is measurable.
Let $\mathcal{M}_t$ be a filtration in $\mathcal{M}$ s.t. X_t is $\mathcal{M}_t$- adapted.

ii) Assume that $z \to P^z(\mathcal{B})$ is measurable for all $\mathcal{B} \in \mathcal{M}$ and the Markov property holds:

$$P^z(X_{s+t} \in \mathcal{B}|\mathcal{M}_s) = P^{X_s}(X_t \in \mathcal{B}),$$

$\forall s,t \geq 0, P^z$-a.s. $, z \in E_\Delta, \mathcal{B} \in \mathcal{B}(E_\Delta)$.
X is then called a Markov process.

iii) X is called normal if $P^z(X_0 = z) = 1, \forall z \in E_\Delta$.
iv) X is said to be right continuous if $t \to X_t(\omega)$ is right continuous, $\forall t \geq 0, \forall\omega \in \Omega$.

v) X is said to be strong Markov if $\mathcal{M}_t$ is right continuous and

$$P^\mu(X_{\sigma+t} \in \mathcal{B} | \mathcal{M}_\sigma) = P^{X_\sigma}(X_t \in \mathcal{B}),$$

P^μ-a.s., $\forall \mathcal{M}_t$-stopping times σ, for any probability measure μ on E, and with $P^\mu(.) \equiv \int P^z(.)\mu(dz)$.

X is said to be a right process if i)-v) hold.
An "m-tight special standard process" is shortly described as a right process which is "concentrated on compacts" (m-tightness), has almost surely left limits and is almost-surely left quasi-continuous, see [367] for details.

Remark 42. 1) An analogon of the above theorem holds also for non symmetric Dirichlet forms $\mathcal{E}$ in the sense discussed in [367]. In the same way as to $\mathcal{E}$ there are associated two submarkov semigroups $(T_t, \hat{T}_t)_{t\geq 0}$ in duality, there are two corresponding (properly)- associated "nice processes" $X, \hat{X}$.
2) There also exists, in the same spirit as in the above theorem, an analytic characterization (in terms of Dirichlet forms) of Hunt processes (see [104], [367]).
3) The main consequence of the above theorem is that concepts of analytic potential theory, like capacity and equilibrium potential become related to concepts of probabilistic potential theory, like hitting distributions and entrance times, e.g. one has for an open set U:

$$\text{cap}(U) = \mathcal{E}_1[e_U]$$

with e_U the "1-equilibrium potential".
Moreover $\text{cap}(U) < \infty$ iff $e_U(z)$ is a quasi-continuous version of $E^z(e^{-\sigma_U})$, with σ_U the entrance time of the associated process in U.
In this way capacity gets related with hitting distribution and, e.g., $\text{cap}(N) = 0$ iff N is an $\mathcal{E}$-exceptional set and this is so iff $P^m(\sigma_{\tilde{N}} < \infty)$, for some Borel subset $\tilde{N} \supset N$ of E; see below and [244], [258], [367] for details.

A hint to the proof of Theorem 12:
From the experience with the construction of Markov processes in the simpler case of a locally compact separable space ("finite dimensional case"), we know it is easy to construct a (sub-) Markov process if one disregards detailed path properties. In contrast the proof of the existence of a version of it with "nice properties" is quite hard.
In fact the case of a general Hausdorff space E in the above theorem is indeed reduced to the case of E locally compact, separable, metric, which had been treated before by Fukushima [246] and Silverstein [468]. This reduction has to be done in such a way as to preserve, e.g., the quasi-regularity property (in fact it gets transformed into the regularity property under compactification). In the case where E is locally compact separable metric one realizes that the

Feller property of the transition semigroup leads to strong Markov processes and eventually to Hunt processes (cf. [162]).
On the other hand one shows:

Lemma 3. *Each $u \in D(\mathcal{E})$ (for $\mathcal{E}$ a regular Dirichlet form, in the sense of [246], [258]) has a quasi-continuous modification s.t. $u \restriction (E_\Delta - G)$ is quasi-continuous, for any open subset G of E (cf. Theorem 3.1.3 in [246])*

Proof. A) One has Cap $G^u_\lambda \leq \frac{1}{\lambda^2}\mathcal{E}_1[u], \forall\lambda > 0, u \in D(\mathcal{E}) \cap C(E))$ with $G^u_\lambda \equiv \{x \in E \| u(x) > \lambda|\}$
(seen by realizing that G^u_λ is open, $|u|/\lambda \in \mathcal{L}_{G^u_\lambda}$, for $u \in D(\mathcal{E}) \cap C(\mathcal{E})$, with $\mathcal{L}_{l\cdot\lambda}$ as in the definition of capacity, and using the "Dirichlet property" $\mathcal{E}_1[|u|] \leq \mathcal{E}_1[u]$.)

B) By the regularity of $\mathcal{E}$ and A) one can find $u_n \in D(\mathcal{E}) \cap C_0(E)$ s.t. $u_n \to u$. By passing if necessary to a subsequence, denoted again by u_n, one then has $\mathcal{E}_1[u_{n+1} - u_n] \leq \frac{1}{2^{3n}}$
Then by A): Cap$G^{u_{n+1}-u_n}_{2^n} \leq \frac{1}{2^n}$.
Hence $F_k \equiv \bigcap\limits_{l=k}^{\infty} \left(G^{u_{n+1}-u_n}_{2^n}\right)^c$ is an $\mathcal{E}$-nest.
But $|u_n(x) - u_m(x)| \leq \sum\limits_{\nu=N+1}^{\infty} |u_{\nu-1}(x) - u_\nu(x)|$, for any $x \in F_k$, $n, m > N \geq k$.
Setting $\tilde{u}(x) \equiv \lim\limits_{n\to\infty} u_n(x), x \in \bigcup\limits_k F_k$ we have that $\tilde{u} \restriction F_k$ is continuous and $\tilde{u}$ is quasi-continuous. □

The next observation consists in showing that one can construct a countable set $B_0 \subset D(\mathcal{E}) \cap C_0(E)$ dense in sup-norm in $C_0(E)$ s.t. B_0 is linear and closed under $|.|$ (as seen by approximation, using the regularity of $\mathcal{E}$, see [Fu], proof of Lemma 6.1.2).
Set $H_0 \equiv \bigcup\limits_{t\in\mathbb{Q}_+} T_t(B_0) \cup G_1(B_0)$ (where $(T_t)_{t\geq 0}$ is the Markov semigroup associated with $\mathcal{E}$ and G_1 is the corresponding resolvent G_α evaluated at $\alpha = 1$).
Let $u \in H_0$ and let $\tilde{u}$ be its quasi-continuous modification given by the above Lemma.
Set $\tilde{H}_0 = \{\tilde{u} | u \in H_0\}$. One shows (using that $\tilde{H}_0$ is countable!) that there exists a regular nest $\{F^0_k\}$ on E s.t. $\tilde{H}_0 \subset C(\{F^0_k\})$ (where $C(\{F^0_k\})$) denotes the functions which are continuous in $(F^0_k)^c$). Let $\bigcup F^0_k \equiv Y_0$. One sets for $u \in L^2(m), x \in Y_0, t \in \mathbb{Q}_+ : \tilde{p}_t u(x) \equiv T_t u(x)$.
$\tilde{p}_t$ is not yet a semigroup, but has a submarkovian kernel $\tilde{p}_t(x, B), x \in Y_0$, $B \in \mathcal{B}$ (cfg. [246], proof of Lemma 6.1.2).
One extends $\tilde{p}_t$ to E by setting $\tilde{p}_t(x, B) = 0, \forall x \in E - Y_0$. One shows $\tilde{p}_t C_\infty(E) \subset C(\{F^0_k\})$, and that $\tilde{p}_t$ is a quasi-continuous version of $T_t u$, for any $u \geq 0$, Borel, $u \in L^2(m), t \in \mathbb{Q}_+$ (cf. Lemma 6.1.2 in [246]), and that $\tilde{p}_t$ is a semigroup of Markovian kernels on $(E, \mathcal{B})$.
One then uses $\tilde{p}_t$ to get by Ionescu-Tulcea-Kolmogorov's construction a

Markov process on X_Δ. The crucial step then consists in showing that this Markov process has a càdlàg version on some Borel subset $\tilde{Y}, \tilde{Y} \subset Y_0$, s.t. $\text{Cap}(E - \tilde{Y}) = 0$. This relies on an interplay of analytic and probabilistic methods, where regularity again plays an important role (see [246], Lemma 6.2.3). The same ingredients then enable us to show that the process is a Hunt process.
By the reduction of the infinite dimensional case to the locally compact case, one then completes the construction in the general case, see [114], [367] for details.

Remark 43. 1) The m-tight special standard process X properly associated with a quasi-regular symmetric Dirichlet form is m-symmetric (in the sense that its transition semigroup is symmetric in $L^2(m)$), and has m as an invariant measure (by construction).
In general it has a finite life time ζ, but if $T_t 1 = 1$ (T_t being the Markov semigroup associated with $\mathcal{E}$) and $1 \in L^2(m)$, then the life time is infinite. The process X can always be taken as a canonical process (cfg. [367]).
2) Let X be a right process properly associated with a quasi-regular Dirichlet form $\mathcal{E}$ (in the general case where E is a topological Hausdorff space and m is a σ-finite measure on E).
One shows that there exists an $\mathcal{E}$-nest (F_k), with F_k compact measurable subsets in E and a locally compact separable metric space $E^\#$ containing densely $Y \equiv \bigcup_k F_k$ with $\mathcal{B}(Y) \equiv \{A \in \mathcal{B}(E^\#) | A \subset Y\}$ and, moreover, there exists a Hunt process $\overline{X}$ on $E^\#$, the "natural extension" of $X \upharpoonright (E - N), N \subset E$,
N invariant, $\mathcal{E}$-exceptional, s.t. $\overline{X}$ is properly associated with the regular Dirichlet form $\mathcal{E}^\#$, the image of $\mathcal{E}$ in $L^2(E^\#, m^\#)$. This observation is exploited in the "regularization method", see [244], [114], [465].

4.5 Stochastic analysis related to Dirichlet forms

Let $(E, \mathcal{B}, m)$ be as in chapter 4.1

Definition 32. *(Additive functional)*
Let $X = (X_t)_{t \geq 0}$ be a right process associated with a quasi regular Dirichlet form $\mathcal{E}$ (according to Theorem 12 in chapter 4.4).
$A \equiv (A_t)_{t \geq 0}$ is called an additive functional associated with X if A_t is $\mathcal{M}_t$-measurable, càdlàg and such that $A_{t+s}(\omega) = A_t(\omega) + A_s(A_t(\omega))$ for all $s, t \geq 0, \omega$ in the underlying probability space Ω.
A is called a continuous additive functional if $t \to A_t(\omega)$ is continuous $\forall \omega \in \Omega$.

Definition 33. *Let A be an additive functional. The energy of A is by definition*

$$e(A) \equiv \lim_{t \downarrow 0} \frac{1}{2t} E^m(A_t^2),$$

if the limit exists, m-a-s.

Definition 34. *Let* $\mathfrak{M} = \{M | M$ *additive functional,* $E^z(M_t^2) < \infty$, $E^z(M_t) = 0, \mathcal{E} - q.e., z \in E, \forall t \geq 0\}$.
One shows that any $M \in \mathfrak{M}$ *is a martingale under* P^x, *for any* $x \in E - N$, N *being a properly exceptional set (depending on* M, *in general). Thus the elements of* $\mathfrak{M}$ *are called martingale additive functionals.*
One shows ([246] p.135, by a method of P.A. Meyer) that for any martingale additive functional M, *there exists uniquely a positive continuous additive functional* $\langle M \rangle$ *s.t. for any* $t > 0$:
$E^z(\langle M \rangle_t) = E^z(M_t^2)$, *q.e.* $z \in E, \langle M \rangle_t$ *is then by definition the quadratic variation of* M.
Let $\mathfrak{N}^c = \{N$ *is a continuous additive functional of zero energy i.e.* $e(N) = 0$ *and s.t.* $E^z(|N_t|) < \infty$ *for q.e.* $z \in E\}$.

Theorem 13. *(Fukushima's decomposition)*
Let X *be the right process associated with a quasi regular symmetric Dirichlet form on a Hausdorff topological space. If* $u \in D(\mathcal{E})$ *then there exists uniquely a martingale additive functional of finite energy* $M^{[u]}$ *and an element of zero energy* $N^{[u]} \in \mathfrak{N}^c$ *s.t. for any quasi-continuous version* $\tilde{u}$ *of* u:

$$\tilde{u}(X_t) = \tilde{u}(X_0) + M_t^{[u]} + N_t^{[u]}.$$

Remark 44. $N^{[u]}$ is not necessarily of bounded variation (but it has zero energy).
For the proof of this theorem see [244], [258], [367], [114].

5 Diffusions and stochastic differential equations associated with classical Dirichlet forms

5.1 Diffusions associated with classical Dirichlet forms

We consider the example discussed in chapter 4, 4.2 of the classical Dirichlet form $\mathcal{E}_\mu$, associated with a probability measure μ on a separable Banach space E s.t.

$$E' \subset \mathcal{H}' \cong \mathcal{H} \subset E$$

We suppose as in chapter 4,4.2 that there exists $\beta_\mu \in L^2(\mu)$ s.t. the integration by parts formula holds, i.e.

$$\int \frac{\partial u}{\partial k} d\mu = - \int u \beta_{\mu,k} d\mu \qquad \forall k \in K \subset E', u \in FC_b^\infty,$$

Originally published in: *Ecole d'Eté de Probabilités de Saint-Flour XXX – 2000*, Lecture Notes in Mathematics, Vol. **1816**, 51–63, DOI: 10.1007/3-540-44922-1_5, © Springer-Verlag Berlin Heidelberg 2003, Reprint by Springer-Verlag Berlin Heidelberg 2012

K consisting of elements which form an orthonormal basis in $\mathcal{H}$.
Let $\beta_\mu : E \to E$ s.t., for all $z \in E$:

$$\underset{E'}{\langle} k, \beta_\mu(z) \underset{E}{\rangle} = \beta_{\mu,k}(z),$$

where $\underset{E}{\langle}, \underset{E'}{\rangle}$ is the dualization between E and E'.
Let L_μ be the self-adjoint operator with $-L_\mu \geq 0$ associated with $\mathcal{E}_\mu$ (Dirichlet operator associated with μ). Then, on FC_b^∞:

$$L_\mu = \Delta_\mathcal{H} + \beta_\mu(z) \cdot \nabla_\mathcal{H},$$

where $\Delta_\mathcal{H} = \sum_{k \in K} \partial_k^2$ is the Gross-Laplacian associated with $\mathcal{H}$, $\nabla_\mathcal{H}$ the natural gradient associated with $\mathcal{H}$, so that

$$\beta_\mu(z) \cdot \nabla_\mathcal{H} = \sum_{k \in K} \beta_{\mu,k}(z) \partial_k.$$

Proposition 15. *The classical Dirichlet form $\mathcal{E}_\mu$ given by μ is quasi-regular.*

Proof. One has to verify the properties i),ii),iii) in Definition 29 of quasi-regularity for Dirichlet forms. For i) it is enough to show that there exist compacts $F_k \uparrow E$ with Cap $(E - F_k) \downarrow 0$ ("tightness of the capacity"), we leave this as an exercise (cf. [367]).

ii) The subset FC_b^∞ is $\widetilde{\mathcal{E}}_1$-dense in $D(\mathcal{E}_\mu)$ by the construction of $\mathcal{E}_\mu$ as the closure of $\overset{\circ}{\mathcal{E}}_\mu$. Its elements are continuous.
iii) FC_b^∞ separates the points of E (and hence also of $E - N$) since E is a separable Banach space (use, e.g., the theorem of Hahn-Banach, cf. [367], p.119)

□

Proposition 16. *$\mathcal{E}_\mu$ is local.*

Proof. We have to show that $\mathcal{E}_\mu(u, v) = 0$ if $\operatorname{supp} u, \operatorname{supp} v$ are compact, $\operatorname{supp} u \cap \operatorname{supp} v = \emptyset$.
This is obvious for $u, v \in FC_b^\infty$. Now for arbitrary $u, v \in D(\mathcal{E}_\mu)$ we can find (by the $\widetilde{\mathcal{E}}_1$-density of FC_b^∞) $u_n, v_n \in FC_b^\infty$ s.t. $\mathcal{E}_1([u_n - u]) \to 0$, $\mathcal{E}_1([v_n - v]) \to 0$, as $n \to \infty$, and the Proposition is proven. □

By the general theory of association, cf. Chapter 4, to $\mathcal{E}_\mu$ there is properly associated an m-tight special standard process $X_t, 0 \leq t \leq \zeta$, which by the locality of $\mathcal{E}_\mu$ is a diffusion process.
Since the C_0-contraction Markov semigroup $T_t = e^{tL}, t \geq 0$, associated with $\mathcal{E}_\mu$ moreover satisfies $T_t 1 = 1$, it follows that $X = (X_t)_{t \geq 0}$ is a μ-symmetric conservative process s.t. $\zeta = +\infty$.
An application of Fukushima's decomposition theorem (Chapter 4, Theorem 12) to the present case yields:

Theorem 14. *For any* $u \in D(\mathcal{E}_\mu), t \geq 0$*:*

a) if $u \in D(L_\mu)$ *then*

$$N_t^{[u]} = \int_0^t (L_\mu u)(X_s) ds.$$

b) $\langle M_t^{[u]} \rangle_t = 2 \int_0^t \langle \nabla u(X_s), \nabla u(X_s) \rangle_{\mathcal{H}} ds$, *where* $\langle \cdot \rangle_t$ *denotes the quadratic variation process (so that* $E^z\left(\langle M \rangle_t\right) = E^z(M_t^2)$ *,q.e.* $z \in E, t > 0, M_t$ *a martingale additive functional.)*

Proof. a) The proof is based on an extension to quasi-regular Dirichlet forms on general Hausdorff topological spaces (cf. [367]) of the following Lemma:

Lemma 4. *Let* $\mathcal{E}$ *be a regular Dirichlet form on* $L^2(m)$ *on a locally compact separable space* E*, and let* X *be a properly associated right process. Then for any* $g \in L^2(m)$ *:*
$\int_0^t g(X_s)ds$ *is a continuous additive funtional of zero energy.*

The proof of the Lemma is left as an exercise (hint: use the Markov property of X).
Now take $u \in D(L_\mu)$, so that $u = G_1 f, f \in L^2(m)$. Set $g \equiv u - f$
Then:

$$\int_0^t g(X_s)ds = \int_0^t (u(X_s) - f(X_s))\, ds$$
$$= \int_0^t L_\mu u(X_s) ds,$$

where for the latter equality we have used

$$L_\mu u = [L_\mu - 1 + 1]\, u = -f + G_1 f.$$

Applying the above Lemma we get that
$\int_0^t L_\mu u(X_s)ds$ is a continuous additive functional of 0 energy and by the uniqueness in Fukushima's decomposition theorem this is then $N_t^{[u]}$.

b) To give an idea of the proof of this point, let us look at the finite dimensional case $E = U, U$ an open subset of $\mathbb{R}^d$, with a classical pre-Dirichlet form $\overset{\circ}{\mathcal{E}}_\mu$.

From the finite dimensional theory we know that if X is the Markov process properly associated with the closure $\mathcal{E}_\mu$ of $\overset{\circ}{\mathcal{E}}_\mu$, then by the Beurling-Deny decomposition

$$\langle M^{[u]}\rangle_t = 2\int_0^t \langle \nabla u(X_s), \nabla u(X_s)\rangle ds$$

For the infinite dimensional case see [367].

□

Remark 45. a) holds for any quasi-regular Dirichlet form.
b) can be generalized to

$$M_t^{[u]} = \int_0^t \rho(X_s)\, ds, \text{ with}$$

$\rho(x) = L_\mu u^2(x) - 2u(x)L_\mu u(x)$.
In this form the proof of b) can found in [208] for the case of locally compact spaces and [126], [451], [452], [453] for quasi regular Dirichlet forms.

Remark 46. For $E = \mathbb{R}^d, \mu$ the Lebesgue measure on U, so that $\mathcal{E}_\mu$ is the classical Dirichlet form uniquely associated with $-\Delta$ we have, taking $u(x) \equiv u_i(x) \equiv x_i (\in L^2(\mu_K))$ (with μ_K the restriction of μ to the interior $\overset{\circ}{K}$ of a compact subset K in $\mathbb{R}^d$). But for any $u \in D(\mathcal{E}_{\mu_K}) \exists w \in D(\mathcal{E}_\mu)$ s.t. $w = u$ m-a.e. on $\overset{\circ}{K}$ and $M_t^{[w_i]} = M_t^{[u_i]}$ for $t < \sigma_{\mathbb{R}^d - \overset{\circ}{K}}$ (the hitting time of $\mathbb{R}^d - \overset{\circ}{K}$). On the other hand

$$\langle M^{[u_i]}\rangle_t = 2\int_0^t ds = 2t.$$

By a local version of Levy's characterization theorem we then have $M^{[u_i]} = \mathcal{W}^i$, with $\mathcal{W}^i$ the i-th component of a Brownian motion in $\mathbb{R}^d$.

Remark 47. In particular we see from the preceding remark that the finite energy additive functional $M^{[u_i]}$ is just the i-th component of Brownian motion.
In general, finite energy additive functionals of a quasi-regular Dirichlet form can be represented by stochastic integrals, see [367].

5.2 Stochastic differential equations satisfied by diffusions associated with classical Dirichlet forms

Proposition 17. *Let μ be a probability measure on $\mathcal{S}'(\mathbb{R}^d)$, as in Chapter 5.5.1, s.t.*

$$u_k(\cdot) \equiv \underset{\mathcal{S}}{\langle} k, \cdot \underset{\mathcal{S}'}{\rangle} \in L^2(\mu).$$

Then $u_k \in D(L_\mu) \subset D(\mathcal{E}_\mu)$ (where $\mathcal{E}_\mu$ is the classical Dirichlet form given by μ). Moreover $L_\mu u_k = \beta_{\mu,k}(\mu\text{-a.s.})$

Proof. We have, for any $v \in FC_b^\infty$:

$$\mathcal{E}_\mu(u_k, v) = ((-L_\mu)u_k, v) = \int \langle k, \nabla v \rangle_{\mathcal{H}} d\mu,$$

where we used the relation between $\mathcal{E}_\mu$ and L_μ, the definition of $\beta_{\mu,k}$, and the integration by parts formula. □

Theorem 15. *Let $X \equiv (X_t)_{t\geq 0}$ be the diffusion process associated with the classical Dirichlet form given by μ as in Proposition 17. Then X satisfies "componentwise", in the weak probabilistic sense, the stochastic differential equation:*

$$\langle k, X_t \rangle = \langle k, X_0 \rangle + \int_0^t \beta_{\mu,k}(X_s) ds + w_t^k, t \geq 0, P^z\text{-a.s., q.e } z \in E.$$

Hereby $(w_t^k, \mathcal{F}_t, P^z)_{t\geq 0}$ is a 1-dimensional Brownian motion starting at 0 (for $\|k\|_{\mathcal{H}}$ suitably normalized).

Proof. By the above Fukushima decomposition formula we have

$$\langle M^{[u_k]} \rangle_t = 2 \int_0^t \langle \nabla_{u_k}(X_s), \nabla_{u_k}(X_s) \rangle_{\mathcal{H}} ds = 2t\|k\|_{\mathcal{H}}^2$$

(because of $\nabla_{u_k} = \langle k, . \rangle$).
Hence by Levy's characterization of Brownian motion:

$$\left(M^{[u_k]}\right)_t = w_t^k.$$

Moreover:

$$N_t^{[u_k]} = \int_0^t L_\mu u_k(X_s) ds$$
$$= \beta_{\mu,k}$$

because $L_\mu u_k(X_s) = (\Delta_{\mathcal{H}} + \beta_\mu \cdot \nabla_{\mathcal{H}})u_k = \beta_{\mu,k}$,
where in the latter step we have used that $u_k(.) = \langle k, . \rangle$ is linear. □

Let us now vary k along an orthonormal basis K in $E' \subset \mathcal{H}$. Then $w_t^k, w_t^{k'}$ are independent for $k \neq k'$.
<u>Assume</u> that there exists a probability measure μ_t on E s.t., for all $k \in \mathcal{H}$:

$$\hat{\mu_t}(k) = \exp\left(-\frac{1}{2}t\|k\|^2_{\mathcal{H}}\right),$$

(in which case there exists a Brownian motion on E with unit covariance given by the scalar product in $\mathcal{H}$: see, e.g., [334], [428], [118]).
Then we have the following

Theorem 16. *Under above assumption about the existence of μ_t there exist maps $W, N : \Omega \to C([0,\infty], E)$ s.t. $t \to W_t(\omega), t \to N_t(\omega)$ are $\mathcal{F}_t$-B-measurable, for all non negative t.*
W_t is such that for q.e. $z \in E$, under P^z, it is an $\mathcal{F}_t$- Brownian motion starting at 0, with covariance given by the scalar product in $\mathcal{H}$.
Moreover,

$$\langle k, N_t\rangle = N_t^{[u_k]} = \int_0^t \beta_{\mu,k}(X_s)ds.$$

One has:
$X_t = z + W_t + N_t, t \geq 0, P^z\text{-a.s.}, q.e.\ z \in E.$

Remark 48. X also solves the martingale problem for $D \subset D(L_\mu) \subset L^2(E_\mu)$ in the sense that X is a μ-symmetric, right process s.t.

$$\tilde{u}(X_t) - \tilde{u}(X_0) - \int_0^t L_\mu u(X_s)ds, t \geq 0$$

is $\mathcal{F}_t$-measurable under P^z, for some quasi-continuous, right continuous modification $\tilde{u}$, independent of N, and independent of the μ-version of the class $L_\mu u \in L^2(\mu)$.

Remark 49. A particular case of the above results concerns

$$E = C_{(0)}([0,t];\mathbb{R})$$

(Wiener space),

$$\mathcal{H} = H^{1,2}([0,t];\mathbb{R})$$

(Cameron-Martin-space),
μ is the standard Wiener measure on E.
In this case

$$\overset{\circ}{\mathcal{E}}_\mu(u,v) = \int \langle\nabla u, \nabla v\rangle_{\mathcal{H}} d\mu,$$

$$\mathcal{E}_\mu(u,v) = \int \langle \overline{\nabla} u, \overline{\nabla} v \rangle_{\mathcal{H}} d\mu,$$

with $\overline{\nabla}$ Malliavin's closed gradient (the verification of the latter is left as an exercise; see also, e.g. [396],[365], [329],[57], and references therein).
In this case we have:

$$\beta_{\mu,k}(.) = \langle k, . \rangle$$

(since $\int \frac{\partial u}{\partial k} d\mu = -\int u\beta_{\mu,k} d\mu, u \in FC_b^\infty$, as seen from the following computation

$$\begin{aligned}\int \frac{\partial}{\partial s} u(. + sk)|_{s=0} d\mu &= \int u(.) \frac{d\mu(. - sk)}{d\mu(.)} d\mu(.) \\ &= \int v(.) e^{-\langle k,. \rangle} e^{-\frac{1}{2}\|k\|^2_{\mathcal{H}}} d\mu(.))\end{aligned}$$

Incidentally: the computation in parenthesis of the Cameron-Martin density under translations of Wiener measure is the basis of a corresponding computation for the quasi-invariance of a natural measure on loop-groups, see [57], [373], [68] and is used in an essential way in the representation of related infinite dimensional Lie groups (cf. [68]).

Remark 50. A similar computation can be done for other Gaussian measures, of the form $\mu = N(0; A^{-1}), A \geq c1, c > 0, A$ a Hilbert Schmidt operator in the Hilbert space $\mathcal{H}$.
In this case we have

$$\beta_{\mu,k} = \langle Ak, . \rangle,$$

see [119], [269]. See also [165], [134], [318] for other results on infinite dimensional Ornstein-Uhlenbeck processes.

5.3 The general problem of stochastic dynamics

Given a (probability) measure on some space one can ask the ("inverse problem") question whether there exists a Markov process X with corresponding transition semigroups P_t, μ-symmetric (in the sense that $P_t^* = P_t$ in $L^2(\mu)$), having μ as P_t-invariant measure, in the sense that

$$\int T_t u d\mu = \int u d\mu$$

for all $u \in L^2(\mu)$.
One then says that X is the "stochastic dynamics" (or "Glauber dynamics") associated with μ.

Remark 51. 1) If μ is a probability measure then we have:
μ is P_t-invariant iff $P_t 1 = 1$
(with 1 the function identically 1 in $L^2(\mu)$)
(we leave the proof as an exercise).

2) There is a notion of measure μ infinitesimal invariant with respect to $(T_t)_{t\geq 0}$, e.g. this has been discussed in connection with hydrodynamics (cf. [35]):
μ is namely called infinitesimal invariant under a C_0-semigroup $T_t = e^{tL}, t \geq 0$ if $\int Lu\, d\mu = 0, \forall u \in D_0 \subset D(L), D_0$ dense in $L^2(\mu)$.
In general μ infinitesimal invariant is strictly weaker than μ T_t- invariant , unless T_t is μ-symmetric and $1 \in D(L), L1 = 0$ (because then from $\int Lu\, d\mu = 0$ one deduces $\int L^n u\, d\mu = 0$, for all n, hence $\int T_t u\, d\mu = 0$)
For recent work on invariant and infinitesimally invariant measures see, e.g., [21], [37], [38], [95], [128], [166], [167], [168], [201].

The converse problem to the above "inverse problem" is the following "direct problem": given a Markov process X find a probability measure μ s.t. μ is an invariant measure for X.
In this case one says that μ is the invariant measure to the stochastic dynamics described by X.
Connected with this direct problem are the following ones:

1) Existence of the classical Dirichlet form $\mathcal{E}_\mu$ associated with μ (closability problem for the pre-Dirichlet form $\overset{\circ}{\mathcal{E}}_\mu$ in $L^2(\mu)$). If this is solved then one can construct a diffusion Y having μ as invariant measure (Y in general can be different from X).
2) Does the logarithmic derivative $\beta_\mu = (\beta_{\mu,k})_{k\in K}$ of μ exist e.g. as an element in $L^2(\mu)$?
3) Does X satisfy a stochastic differential equation?

Further associated questions are, e.g.:

4) What is the asymptotic behavior for $t \to \infty$ of X_t, and of the semigroup $T_t = e^{tL_\mu}, t \geq 0$, associated with $\mathcal{E}_\mu$?
5) Is a solution of the martingale problem for L_μ on a closed domain D strictly contained in $D(L_\mu)$ already uniquely determined by the knowledge of L_μ on D? This is the "Markov uniqueness problem", cf. [147], [223], [88], [119] and, for the related "strong uniqueness problem", i.e. the essential-self-adjointness resp. maximal dissipativity of (L_μ, D) see these references and, e.g., [85], [86], [359].

Other problems are, e.g.:

6) When does T_t have the Feller property and thus permit a more direct construction of an associated "nice process"? (see, e.g., [82], [214], [262], [444])
7) Is the invariant measure μ to X unique?

Problems of this type are often encountered, e.g., in the study of processes associated with "Gibbs measures", e.g., in quantum field theory, statistical mechanics (on lattices and in the continuum, in problems connected with the geometry and analysis of configuration spaces), quantum statistical mechanics (and connected problems of geometry and analysis on loop spaces), in the study of self-intersection functionals of diffusion processes and polymer models, in models of population dynamics, see below, chapter 6, for further discussion and for the construction of stochastic dynamics in some examples. First we shall briefly discuss in general the large time asymptotics of processes associated with Dirichlet forms.

5.4 Large time asymptotics of processes associated with Dirichlet forms

Let E be a topological Hausdorff space, m a σ-finite measure on E, as in chapter 3-4 on Dirichlet forms.

Definition 35. *A general Dirichlet form $\mathcal{E}$ in $L^2(m)$ is said to be irreducible if $u \in D(\mathcal{E})$ and $\mathcal{E}[u] = 0$ imply that m is constant m-a.e.*

Definition 36. *Let $(T_t), t \geq 0$ be a submarkovian C_0-contraction semigroup in $L^2(m)$. T_t is called irreducible if $T_t(uf) = uT_tf, \forall t > 0, \forall f \in L^\infty(m)$ implies $u =$ const m-a.e.*

Definition 37. *Let $(T_t)_{t \geq 0}$ be a (submarkovian) C_0-contraction semigroup in $L^2(m)$. $(T_t)_{t \geq 0}$ is said to be $L^2(m)$-ergodic if $T_t u \to \int u dm$ as $t \to \infty$ in $L^2(m), \forall u \in L^2(m)$.*

In [90] (see also [91], [64]) the following Theorem is proven:

Theorem 17. *For symmetric Dirichlet forms $\mathcal{E}$ and associated symmetric submarkovian C_0-contraction semigroups $T_t = e^{tL}, t \geq 0$, in $L^2(m)$, the following are equivalent:*

a) $\mathcal{E}$ is irreducible
b) $(T_t)_{t \geq 0}$ is irreducible
c) $T_t u = u \forall t > 0, u \in L^2(m)$ implies $u =$ const m-a.e.
d) $(T_t)_{t \geq 0}$ is $L^2(m)$-ergodic
e) $u \in D(L)$ and $Lu = 0$ imply $u =$ const m-a.e.

Proof. *b)* $\to$ *a)* is immediate, using the contraction property of $\mathcal{E}$. The rest is left as an exercise (cf. [90]). □

It is also interesting to connect asymptotic properties of semigroups with corresponding properties of associated processes.

Definition 38. *Let X be a right process on a topological Hausdorff space, properly associated with a quasi-regular Dirichlet form $\mathcal{E}$. Let for any $\mu \in \mathcal{P}(E)$ (the linear space of probability measures on E):*

$$P^\mu \equiv \int_E P^z \mu(dz),$$

where P^z is the probability measure on the paths of X corresponding to a starting point $z \in E$.
(X, P^μ) is said to be time-ergodic if for any $G : \Omega \to \mathbb{R}$, s.t. G is Θ_t invariant $\forall t \geq 0$ one has $G = const$, P^μ-a.s. (that G is Θ_t-invariant means that $G(\Theta_t\omega) = G(\omega) \forall t \geq 0, \Theta_t$ being the natural shift in path space given by $\Theta_t\omega(s) = \omega(t+s), \forall 0 \leq s, t \leq \zeta(\omega)$).

One has then the following

Theorem 18. *("Fukushima's ergodic theorem")*
The classical Dirichlet form $\mathcal{E}_\mu$ on a topological Hausdorff space (X, μ) (with μ σ-finite) is irreducible in $L^2(\mu)$ iff (X, P^μ) is time-ergodic.
Moreover, the transition semigroup P_t to X (so that X_t is properly associated to P_t, $t \geq 0$) is such that $P_t u \to \int u d\mu$ as $t \to +\infty, \mathcal{E}_\mu$-q.e., $\forall u \in B_b(E)$.

Proof. The proof is given in [87] (for previous work see, e.g., [57], [390], [244]). □

Corollary 3. *Let $\mathcal{E}_\mu$ be as in above theorem. Then $\mathcal{E}_\mu$ is irreducible if μ is the only P_t-invariant probability measure on $B(E)$ which does not charge $\mathcal{E}_\mu$-exceptional subsets of E.*

Proof. See [244]. □

For measures μ which are quasi-invariant with respect to suitable subspaces K of E, which we shall call "space quasi-invariant", one has an interesting relation between above time ergodicity of E and "space ergodicity", i.e. ergodicity with respect to K. This is the context of next section.

5.5 Relations of large time asymptotics with space quasi-invariance and ergodicity of measures

Let E be a Hausdorff topological space, which is a locally convex topological vector space with the topology of a Souslin-space. (cf. [462] for this concept, E can be, e.g., a Banach space, or a space like $\mathcal{S}'(\mathbb{R}^d)$).
Assume there is a Hilbert space $\mathcal{H}$ s.t. $E' \subset \mathcal{H} \subset E$ where the embeddings are dense and continuous.

Definition 39. *A probability measure μ on E is said to be K-quasi invariant if $\mu(.) \ll \mu(. + tk), \forall k \in K, \forall t \in \mathbb{R}$ (where $\ll$ means absolutely continuous).*

Remark 52. An example of a quasi invariant measure is given by $E = C_0, ([0,t]; \mathbb{R}^d), \mu$ the Wiener measure on E, with $K = H^{1,2}([0,t]; \mathbb{R}^d)$ the Cameron-Martin space, cf. Chapter 5.5.2.
As clearly mentioned in Chapter 5.5.2, the non commutative analogon of this setting, with $\mathbb{R}^d$ replaced by a compact Lie group, is the basis for the representation theory of loop groups and algebras, see, e.g., [57], [68].

Definition 40. *A probability measure μ on E is said to be K-ergodic if μ is K-quasi invariant and for any $u \in L^2(\mu)$ one has that $u(z+tk) = u(z)$ μ-a.e. ($\forall z \in E, t \in \mathbb{R}$) implies $u = const$, μ- a.e.*

Given a probability measure μ on E, one might ask whether there is a relation between the irreducibility of the corresponding classical Dirichlet form $\mathcal{E}_\mu$ and the K-ergodicity of μ. This has already been discussed in [57]. The surprise is that a close relation of this type involves another Dirichlet form $\mathcal{E}_\mu^{\max}$ (with larger domain than $\mathcal{E}_\mu$), rather than $\mathcal{E}_\mu$.
In order to define $\mathcal{E}_\mu^{\max}$ let us first define its domain:

$$D(\mathcal{E}_\mu^{\max}) \equiv \left\{ u \in \bigcap_{k\in K} D(\mathcal{E}_{\mu,k}), \sum_k \mathcal{E}_{\mu,k}[u] < \infty \right\}$$

One sees then easily that

$$D(\mathcal{E}_\mu^{\max}) \supset D(\mathcal{E}_\mu).$$

In general however one can have $D(\mathcal{E}_\mu^{\max}) \neq D(\mathcal{E}_\mu)$ (see e.g. [246] for finite dimensional examples, with E replaced by a bounded subset of $\mathbb{R}^d$).

Remark 53. $D(\mathcal{E}_\mu^{\max})$ is an infinite dimensional weighted space analogue of the Sobolev space $H^{1,2}(\mu)$ whereas $D(\mathcal{E}_\mu)$ is an infinite dimensional weighted space analogue of $H_0^{1,2}(\mu)$.
One defines $\mathcal{E}_\mu^{\max}$ to be equal to $\mathcal{E}_\mu$ on $D(\mathcal{E}_\mu)$. One can then show that $\mathcal{E}_\mu^{\max}$ so defined has a unique closed extension to a Dirichlet form with domain exactly equal to $D(\mathcal{E}_\mu^{\max})$ as defined above, see [99], [101], [223].

Remark 54. It is an important open problem to establish whether $\mathcal{E}_\mu^{\max}$ is quasi-regular in general, hence whether to it there can be properly associated a right process.
The advantage of $\mathcal{E}_\mu^{\max}$ over $\mathcal{E}_\mu$ is that irreducibility for it implies K-ergodicily of μ i.e. the following theorem holds

Theorem 19. *If $\mathcal{E}_\mu^{max}$ is irreducible then μ is K-ergodic.*

Proof. Let $u : E \to \mathbb{R}$ be $B(E)$-measurable and in $L^2(\mu)$. Suppose u is k-invariant, $k \in K$. Then one can show that $\frac{\partial}{\partial k} u = 0$, see [119], hence $u \in D(\mathcal{E}_{\mu,k})$.
By the definition of $D(\mathcal{E}_\mu^{\max})$ this implies $u \in D(\mathcal{E}_\mu^{\max})$ and $\mathcal{E}_\mu^{\max}[u] = 0$.
By the irreducibility of $\mathcal{E}_\mu^{\max}$ this implies $u =$ const, μ-a.e., which by the definition of K-ergodicity of μ yields that μ is K-ergodic. □

Remark 55. One has $\mathcal{E}_\mu^{\max}$ irreducible $\Rightarrow \mathcal{E}_\mu$ irreducible (but the converse is not true in general, see, e.g. [119]).

Remark 56. The question whether $\mathcal{E}_\mu^{\max} = \mathcal{E}_\mu$ for a given setting $(E, \mu, \mathcal{H}, K)$ is called the "Markov uniqueness question". One can namely show in general:

$\mathcal{E}_\mu^{\max} = \mathcal{E}_\mu \Leftrightarrow$ the only Dirichlet form extending $(\mathcal{E}_\mu, D(\mathcal{E}_\mu))$ is $\mathcal{E}_\mu$
$\Leftrightarrow$ if $\mathcal{E}$ is a Dirichlet form and $\mathcal{E} = \mathcal{E}_\mu$ on FC_b^∞ then $\mathcal{E} = \mathcal{E}_\mu$
$\Leftrightarrow$ Let $(T_t)_{t \geq 0}$ be the submarkov semigroup with generator coinciding on FC_b^∞ with the classical Dirichlet operator L_μ given by μ (i.e. L_μ is the operator associated to $\mathcal{E}_\mu$ in the sense of the representation theorem) Then $T_t = e^{tL_\mu}$.

In general it is known that Markov uniqueness is weaker than "strong uniqueness" or "L^2-uniqueness", which is the property that L_μ is essentially self-adjoint on FC_b^∞ on $L^2(\mu)$.
The Markov and strong uniqueness problems are thoroughly discussed in [223]. We mention here some further basic work by [85], [86], [89], [499], [471], [185], [37], [38], [440], [441] (connected with applications in various domains).
To give an idea of these connections let us mention shortly what happens in the finite dimensional case $E = \mathbb{R}^d$: for $\mu(dx) = \rho(x)dx$, $\sqrt{\rho} \in H^{1,2}_{\text{loc}}$ one has Markov uniqueness in general (see [359]).
For U a bounded region and $\rho = 1, \mathcal{E}_\mu^{\max}$ is the Dirichlet form describing reflected Brownian motion, $\mathcal{E}_\mu$ describes absorbing Brownian motion and there are infinitely many other forms between $\mathcal{E}_\mu, \mathcal{E}_\mu^{\max}$ describing Brownian motion with other types of boundary behaviour.
It is also known that there are other closed symmetric positive bilinear forms with associated generators of symmetric C_0-contraction semigroups in $L^2(\mu)$ with generators coinciding with L_μ (here Δ) on $C_0^\infty(U)$ but which are <u>not</u> submarkovian, e.g. the Krein extension of $\Delta \restriction C_0^\infty(U)$, see [244].
A "concrete" (probabilistic and analytic) classification of all extensions in the infinite dimensional case is a very interesting open problem.

Remark 57. There exists a partial converse to the previous theorem.

Theorem 20. *If μ is K-quasi invariant and a "strictly positive" measure on E (in the sense that its Radon-Nikodym derivatives in the directions of K are strictly positive) and moreover μ is K-ergodic, then $\mathcal{E}_\mu^{max}$ is irreducible*

Proof. See [87]. □

We shall now see that for special μ called "Gibbs measures", one has a close relation between irreducibility and K-ergodicity.
Let

$$\mathcal{G}^b \equiv \{\mu \in P(E) | \mu \text{ satisfies } (IP)^b\},$$

where $(IP)^b$ is the following "integration by parts formula with resp. to μ and the direction b":

$$\int \frac{\partial u}{\partial k} d\mu = - \int u b_k d\mu$$

$\forall u \in FC_b^\infty, \forall k \in K$.
Any element μ of $\mathcal{G}^b$ is called a "b-Gibbs state" " and $b \equiv (b_k)_k \in K$ is the logarithmic derivative of μ.

Remark 58. We shall see below how to relate this definition of b-Gibbs state with other definitions of Gibbs states.

Remark 59. Let us look at a probability measure on $E = \mathbb{R}^n$ of the form

$$d\mu(z) = Z^{-1}e^{-S(z)}dz,$$

where dz is Lebesgue measure on E, S is a lower bounded measurable function on $\mathbb{R}^n$, Z a normalization constant s.t. μ is a probability measure on $\mathbb{R}^n$. The $(IP)^b$-formula holds with $K = \mathbb{R}^n$ and $b_k = -d_k S(z)$, d_k being the derivative in the direction k, i.e. b_k is the logarithmic derivative of the measure μ.
In this sense it is often inspiring to think of μ also in the case of an infinite dimensional E as a measure of the above form (of course there is no good analogue of Lebesgue measure on infinite dimensional spaces, so this way of thinking has to be understood "cum grano salis", e.g. as limit of finite dimensional measures, see, e.g., [16]).

Remark 60. a) $\mathcal{G}^b$ is a convex set, in the sense that any $\mu \in \mathcal{G}^b$ can be written as an integral with respect to $\nu \in (\mathcal{G}^b)_{\text{ex}}$, with $(\mathcal{G}^b)_{\text{ex}}$ the set of extreme elements in $\mathcal{G}^b$, see [90].

We have the following

Theorem 21. *Let μ be in $\mathcal{G}^b$.*
Consider the following statements:

i) $\mu \in \mathcal{G}^b_{ex}$
ii) $\mathcal{E}_\mu^{max}$ *irreducible*
iii) $\mathcal{E}_\mu$ *irreducible*
iv) (X, P^μ)*-time ergodic (with X a right process properly associated with $\mathcal{E}_\mu$)*

Then: i) $\leftrightarrow$ ii) $\rightarrow$ iii) $\leftrightarrow$ iv)
If $\mathcal{E}_\mu^{max} = \mathcal{E}_\mu$ (i.e. one has Markov uniqueness) then i),ii),iii),iv) are all equivalent.

Proof. *ii)* $\rightarrow$ *iii)* is clear
iii) $\leftrightarrow$ *iv* was discussed above. For the rest of the proof see [90], [91]. □

Remark 61. $\mathcal{E}_\mu$ acts as a rate function for the large deviation of occupation densities of X from the ergodic behaviour, as shown in [390].

6 Applications

The applications of the theory of Dirichlet forms are so numerous and belong to so many different areas that it would be impossible to give here even a sketchy but balanced overview.

We shall restrict ourselves to some examples, mainly taken from physics, which illustrate some of the basic advantages of the approach and where the analysis has been pushed forward most intensively in recent years, in particular concerning the stochastic processes involved, which are difficult to obtain (if at all) by other methods, and where in any case the theory of Dirichlet forms has played a pioneering role.

In most of the cases we shall consider the processes which have invariant measures of the form of "Gibbs measures", i.e. measures heuristically given by the formula

$$d\mu(z) = Z^{-1} e^{-S(z)} dz, \quad z \in E \tag{1}$$

(E being the state space, cf. Remark 59 in Chapter 5). For somewhat complementary references where problems connected with the ones discussed here see also, e.g., [2], [3], [9], [116].

6.1 The stochastic quantization equation and the quantum fields

Let us consider a classical relativistic scalar field (as a simpler analogue of the classical electromagnetic vector field) over the d-dimensional Minkowski space-time ($d = s + 1, s =$ space dimension, the physical case being for $s = 3$). φ is the (real-valued) solution of the non-linear Klein-Gordon (or massive wave) equation:

$$\Box\varphi + m^2\varphi + v'(\varphi) = 0 \tag{2}$$

(with $\Box = \frac{\partial 2}{\partial t^2} - \Delta_{\vec{x}}$ the d'Alembert wave operator, $m \underset{(-)}{>} 0$ being the mass parameter, v a real valued differentionable function on $\mathbb{R}$ called "(self-)interaction", $t \geq 0$ is the time variable, $\vec{x} \in \mathbb{R}^s$ is the space variable).

Inspired by Feynman's heuristic "path integrals" quantization procedures (we refer to [60], [23], [24], [25], [26], [73], [16], [53], [278], [279], [280], [49] for work implementing this in related situations), Symanzik formulated a program of constructing a quantization of the solution of (2), in terms of a measure of the heuristic form (1) with $S(z)$ an "action functional" of the form

$$S(z) = S_0(z) + \int_{\mathbb{R}^d} v(z(x))dx,$$

with

Originally published in: *Ecole d'Eté de Probabilités de Saint-Flour XXX – 2000*, Lecture Notes in Mathematics, Vol. **1816**, 64–73, DOI: 10.1007/3-540-44922-1_6, © Springer-Verlag Berlin Heidelberg 2003, Reprint by Springer-Verlag Berlin Heidelberg 2012

$$S_0(z) = \frac{1}{2} \int_{\mathbb{R}^d} (-z(x)) \Delta z(x) dx + m^2 \int_{\mathbb{R}^d} |z(x)|^2 dx,$$

with $z = z(x), x = (t, \vec{x}) \in \mathbb{R} \times \mathbb{R}^s = \mathbb{R}^d$. In this case then E should be a space of maps from $\mathbb{R}^d$ into $\mathbb{R}$. The reason for this is that the moment functions $\int z(x_1)..z(x_n)\mu(dz)$ of μ heuristically defined by (1) with such an S give, after analytic continuation $x = (t, \vec{x}) \to (it, \vec{x})$, the "correlation functions in the vacuum"

$$\langle \varphi_Q(t_1, \vec{x_1}) ... \varphi_Q(t_n, \vec{x_n}) \rangle$$

of the relativistic quantum field $\varphi_Q(t, \vec{x})$ corresponding to the classical Klein-Gordon field $\varphi(t, \vec{x})$.
From the perspective of Chapter 5.3, the construction of μ is related to the construction of a process $(X_\tau)_{\tau \geq 0}$ on E s.t.

$$dX_\tau = \beta_\mu(X_\tau) d\tau + dw_\tau$$

with w_τ a Brownian motion on E with covariance given by a suitable Hilbert space $\mathcal{H}$, with $\beta_\mu(z) = -\nabla_{\mathcal{H}} S(z)$. For $\mathcal{H} = L^2(\mathbb{R}^d)$ we get heuristically

$$-\nabla_{\mathcal{H}} S(z(x)) = -(-\Delta_x + m^2) z(x) - v'(z(x)),$$

so X_τ satisfies heuristically

$$dX_\tau(x) = (\Delta_x - m^2) X_\tau(x) d\tau - v'(X_\tau(x)) d\tau + \eta(\tau, x) d\tau \qquad (3)$$

with $\eta(\tau, x)$ a Gaussian white noise in all variables $\tau \in \mathbb{R}, x \in \mathbb{R}^d$, s.t. heuristically,

$$\frac{d}{d\tau} w_\tau(x) = \eta(\tau, x)$$

with $(w_\tau(\cdot))$ a (cylindrical) Brownian motion on $\mathcal{S}'(\mathbb{R}^d)$ with covariance given by the scalar product in $L^2(\mathbb{R}^d)$.

(3) is called the "stochastic quantization equation". It has been discussed by Parisi-Wu as a computational, Monte-Carlo type method for the construction of μ (τ being a "computer time"). This equation has since received a lot of attention, both in physics and mathematics, after the pioneering work of Jona-Lasinio and coworkers [302].
As for the definition of the measure $\mu \equiv \mu^v$, heuristically given by (1) with S as above, one starts from the case $v = 0$. In this case, as realized by E. Nelson [393] (see also, e.g. [10], [15], [428],[429] for other connections) $\mu^0 \equiv \mu^{v=0}$ is realized rigorously as the normal distribution with mean zero and covariance $(-\Delta_x + m^2)^{-1}$ (which, by Minlos theorem, is a well defined measure, e.g., in $\mathcal{S}'(\mathbb{R}^d)$, with support e.g. in $\mathcal{H}^{-1,2}(\mathbb{R}^d)$) (this is called Nelson's free field measure).
For $d = 1, \mu^v$ has been constructed for large classes of v as weak limit as

$\Lambda \uparrow \mathbb{R}$ of measures of the form $\mu_\Lambda^v(dz) = Z_\Lambda^{-1} e^{-\int_\Lambda v(z(x))dx} \mu^0(dz)$ (see [293], [450], [197]).
A direct analogous procedure for $d = 2$ fails, since μ_Λ^v is ill defined, since for $z \in \operatorname{supp}\mu^0$, $\int_{\mathbb{R}^d} v(z(x))dx$ is infinite μ^0-a.s. (this is due to the singularity of the covariance of μ^0 on the diagonal $(x = y)$ of the type $|x-y|^{-(d-2)}$ for $d \geq 3$ and $-2\pi \ln|x-y|$ for $d = 2$), with :: being the Wick ordering.
But, for $d = 2$, replacement of v by $:v:$ (so that e.g. for $v(y) = y^n$, $\langle k, :v:(z)\rangle, k \in \mathcal{S}(\mathbb{R}^2)$ is an element in the n-th chaos subspace in $L^2(\mu_0)$) yields (by a fundamental estimate of Nelson, see, e.g. [466]) a well-defined probability measure μ_Λ^v (heuristically given by $Z^{-1} e^{-\int_\Lambda :v:(z(x))dx} \mu^0(dz)$), and one shows then that μ_Λ^v converges weakly, under some assumptions on v, for $\Lambda \uparrow \mathbb{R}^2$, to a well defined probability measure μ^v on $\mathcal{S}'(\mathbb{R}^2)$ (see [466], [265], [39]). μ^v is then by definition the "$v(\varphi)_2$-model" of (Euclidean) quantum field theory (for v a polynomial P one has the "$P(\varphi)_2$-model").

Remark 62. The problem whether the coordinate process X with distribution μ^v is a global Markov field was open for a long time and was solved in works by Albeverio, Gielerak, Høegh-Krohn, Zegarlinski, see, e.g., [132] and references therein.

Looked upon as an $\mathcal{S}'(\mathbb{R}^{d-1})$-valued symmetric Markov process $t \to X_t(f), f \in \mathcal{S}(\mathbb{R}^{d-1}), t \geq 0$, it has a generator which coincides on a dense set, e.g. FC_b^∞, with the $\mathcal{S}'(\mathbb{R}^{d-1})$-valued diffusion process $X_t(f), t \geq 0$ associated with the classical Dirichlet form given by μ_0^v, where μ_0^v is the restriction of μ^v to the σ-algebra $\sigma(X_0(f), f \in \mathcal{S}(\mathbb{R}^{d-1})$. Wether $X_t(f)$ and $\widetilde{X_t}(f)$ have generators coinciding on their full domain is an open question for $v \neq 0$ ("Markov uniqueness" for the process associated with μ_0^v) (for $v = 0, X_t(f) = \widetilde{X_t}(f) =$ Nelson's free field at time t and with test function f. Its generator is the Hamiltonian of the relativistic free field). The corresponding problem for the diffusion generated by the analogue $\mu_{(0),\Lambda_0}^v$, of $\mu_{(0)}^v$ in a bounded region Λ_0 of $\mathbb{R}^d$ has been solved in [359].

Let us now come back to the stochastic quantization equation (SQE): it has been verified in [119], [255], [421] that μ^v is (for $d = 2$ and a large class of v's) such that the classical Dirichlet form μ^v given by it exists and that the properly associated diffusion $X = (X_\tau)_{\tau \geq 0}$ indeed solves the SQE (3), componentwise (in fact $\beta_{\mu,k} \in L^2(\mu)$) and on E itself.
Recent work on pathwise solutions of the SQE is in [52] and [199], see also [384] for a discussion of the impossibility to use a Girsanov transformation to produce solutions, even on a bounded domain Λ in $\mathbb{R}^2$.

Remark 63. a) The problem of the necessity of the renormalization $v \to :v:$ in order to avoid "triviality" is discussed in [42], [52], [51], [47].

b) For a discussion of the Markov resp. strong uniqueness problem for μ_Λ^v see [359], [499] (the case where $\Lambda = \mathbb{R}^2$, i.e for μ^v is still open , see [121] for a partial result and [37], [38] for a related problem in hydrodynamics).

c) Despite the unproven Markov uniqueness, the space of Gibbs states for μ^v (in the sense discussed, e.g., in [72]) can be identified with $\mathcal{G}^b$, and b given on FC_b^∞ by the expression β_{μ^v}, i.e.

$$b_k(z) = {}_{\mathcal{S}}\langle(\Delta_x - m^2)k, z\rangle_{\mathcal{S}'} - {}_{\mathcal{S}}\langle k, : v'(z) : \rangle_{\mathcal{S}'}$$

(with ${}_{\mathcal{S}}\langle\ ,\ \rangle_{\mathcal{S}'}$, the dualization between $\mathcal{S}(\mathbb{R}^2)$ and $\mathcal{S}'(\mathbb{R}^2)$). It is important to realize that b is independent of the Gibbs state, it only depends on μ^0 and v in the support of k.

Using the ergodic theory briefly exposed in Chapter 5.4, it has been shown in [90], [91] that for $\mu \in \mathcal{G}_{\text{ex}}^b$ one has that $\mathcal{E}_\mu$ is irreducible and the solution X of the SQE (3) is time-ergodic. This is a result which has been hard to obtain, and holds, e.g., for $v(y)$ an even degree polynomial, with leading term of the form $\lambda^2 y^n, \lambda > 0$ sufficiently small $(y \in \mathbb{R})$.

d) More work has been done on a stochastic quantization equation with regularization $\varepsilon > 0$, denoted by $(\text{SQE})_\varepsilon$, obtained from (3) by replacing on the r.h.s. X_τ by $A^{1-\varepsilon}X_\tau, w_\tau$ by $A^{-\frac{\varepsilon}{2}}w_\tau$ and $: v : (X_\tau) \text{by} A^{-\varepsilon} : v' : (X_\tau)$, with $A \equiv -\Delta + m^2$. μ^v is heuristically still invariant for $(\text{SQE})_\varepsilon$, for any $\varepsilon > 0$, this has been shown rigorously in [302], [169], [120], [199] (see also [384] and references therein). Markov uniqueness for $\mu_\Lambda^{v_\varepsilon}, \Lambda \subset \mathbb{R}^d$ has been also shown in [121], [123]; L^p-uniqueness of $\mu_\Lambda^v, \Lambda \Subset \mathbb{R}^d$ in the sense of [223] (the case $p = 2$ being strong uniqueness) has been shown in [359], [200], see also [223]. The problem of corresponding uniqueness results for $\mathbb{R}^d$ instead of Λ is still open.

e) Log-Sobolev inequalities for $\mu^{v_\varepsilon}, (\varepsilon \geq 0)$ would yield exponential ergodicity of X, but this is still an open problem (even for $d = 1$) (the analogous problem in lattice statistical mechanics is solved in [89], see Section 2 below).

Remark 64. For $d = 3$ only a construction of the analogue of μ^v works for $v \neq 0$ in the special case $v(y) = y^4$. It is not known whether one can associate a Markov process to any of the $\mu^v, \mu_\Lambda^v, \mu_0^v, \mu_{0,\Lambda_0}^v, \Lambda \subset \mathbb{R}^3, \Lambda_0 \subset \mathbb{R}^2$, for a negative result see [133].

For further discussions of these topics and related ones see also [11], [12], [13], [15], [54], [55], [56], [196], [198], [201], [219], [221], [228], [260], [241], [272], [291], [303], [310], [318], [328], [496], [497], [498], [499], [500], [360], [394], [395], [430], [431], [432], [433], [435], [437], [438], [439], [440], [441], [442], [478], [479], [494].

6.2 Diffusions on configuration spaces and classical statistical mechanics

We shall present here shortly a probabilistic construction of diffusion processes on configuration spaces, see [92], [93], [94] for details (and, e.g., [325], [202], [436], [369], [401], [434] for continuation of the latter work; for previous related work on stochastic analysis related to Poisson processes see, e.g., [179], [305], [420]).

Let M be a connected oriented C^∞ Riemanian manifold such that $\mathrm{Vol}(M) = +\infty$ (where Vol is the volume measure). Let

$$\begin{aligned} \Gamma &\equiv \text{configuration space (of locally finite configurations) over } M \\ &= \{\gamma \subset M \| \gamma \cap K| < \infty \text{ for each compact } K \subset M\}. \end{aligned}$$

$\gamma \in \Gamma$ can be identified with the $\mathbb{Z}_+$-valued Radon measure $\sum\limits_{x\in\gamma} \varepsilon_x$, we shall not distinguish in the following γ and the corresponding Radon measure $\sum\limits_{x\in\gamma} \varepsilon_x$.

Any $f \in C_0^\infty(M)$ can be lifted to the map from Γ to $\mathbb{R}$ given by

$$\langle f, \gamma \rangle = \sum_{x\in\gamma} f(x) = \int_M f d\gamma.$$

One can "lift the geometry from M to Γ", e.g., given $v \in V_0(M) \equiv \{\text{smooth vector fields on } M\}$, one gets a flow ϕ_t^v on M, and this flow is lifted to the flow $\tilde{\phi}_t^v$ on Γ, defined by

$$\tilde{\phi}_t^v(\gamma) = \{\phi_t^v(x) | x \in \gamma\}.$$

Let $T_x M$ be the tangent space to M at $x \in M$ and let TM be the tangent bundle $(T_x M)_{x\in M}$. Let ∇_v^M the derivation on M given by

$$\nabla_v^M f(x) = \langle \nabla^M f(x), v(x) \rangle_{T_x M},$$

∇^M being the gradient operator associated with M.

One can also lift the operations $\nabla_v^M, \nabla^M, \tilde{\nabla}_v^M, \tilde{\nabla}^M$ from quantities associated with M to quantities associated with Γ by defining first the space FC_b^∞ of smooth bounded cylinder functions on Γ by

$$FC_b^\infty \equiv \{u : \Gamma \to \mathbb{R} | u(\gamma) = g(\langle f_1, \gamma \rangle, \ldots, \langle f_n, \gamma \rangle) | \exists n \in \mathbb{N}, f_i \in C_0^\infty(M), g \in C_b^\infty(\mathbb{R}^n)\}$$

and setting for u as in the definition of FC_b^∞, $\tilde{\nabla}_u^M(\gamma) = \int \sum\limits_i \partial_i g(x) \nabla^M f_i(x) \gamma(dx)$

Moreover:

$$\tilde{\nabla}_v^\Gamma u(\gamma) = \langle \tilde{\nabla}^M u(\gamma, x), v(x) \rangle_{L^2(M \to TM; \gamma)}.$$

In this way we let also correspond to the tangent bundle TM the tangent bundle $T\Gamma \equiv (T_\gamma \Gamma)_{\gamma \in \Gamma}$ with metric given by the inner product in $L^2(M \to$

$TM;\gamma)$. This gives then a lift of ∇^M as acting on smooth functions on M to $\tilde{\nabla}^\Gamma$ as acting on smooth cylinder functions on Γ.
Similarly one can define a lift div $^\Gamma$ of div^M to an operation on vector fields on Γ. Defining then $\Delta^\Gamma \equiv \mathrm{div}^\Gamma \nabla^\Gamma$ on FC_b^∞, we have a lift of the Laplace-Beltrami operator Δ^M on functions on M to a Laplace-Beltrami operator on functions over Γ.
A natural question here is for which measures μ on Γ does one have $\mathrm{div}^\Gamma = (-\nabla^\Gamma)^*$, the adjoint being taken in $L^2(\mu)$. The following theorem was proven in [94].

Theorem 22. *For μ a probability measure μ on Γ with finite first absolute moments, i.e., s.t.*

$$\int |\langle f, \cdot \rangle| d\mu(\cdot) < \infty \qquad \forall f \in C_b^\infty(M)$$

the following are equivalent:

i) $\mathit{div}^\Gamma = (-\nabla^\Gamma)^*$,
ii) *μ is a mixed Poisson measure, i.e. there exists a σ-finite measure λ on $\mathbb{R}_+$ s.t. $\mu = \int_0^\infty \pi_{z\sigma(\cdot)} \lambda(z)$, where $\sigma \equiv Vol(\cdot)$, and π_σ is the Poisson measure on Γ with intensity measure σ so that*

$$\hat{\pi}_\sigma(f) = \int_\Gamma e^{i\langle f, \gamma \rangle} \pi_\sigma(d\gamma) = e^{\int_M (e^{if(\cdot)} - 1) d\sigma(\cdot)}. \qquad \forall f \in C_0^\infty(M)$$

Remark 65. It follows for μ as in the above theorem:

1) μ is the volume measure on Γ (in the natural sense of being a "flat measure" on Γ)
2) μ is quasi-invariant with respect to $\gamma \to \phi(\gamma), \phi \in \mathrm{Diff}_0(M)$ (the diffeomorphisms which are identically the unit outside some compact subset of M).

A stochastic dynamics can be associated with the classical (quasi regular local) Dirichlet form $\mathcal{E}_\mu$, in the form of a diffusion process, satisfying a differential equation of the form given in Theorem 14, in Chapter 5, with a drift coefficient in $L^2(\mu)$. This process is generated by Δ^Γ, moreover, ergodicity and strong uniqueness hold, see [87].
There is an extension of this work to the case where μ is replaced by a "Gibbs measure" (see, e.g., [263]), called again μ, describing a system of particles, in the sense of classical statistical mechanics, for a general class of interactions including "physically realistic" ones, see [93], [333], [325], [403], [505], [48].
Correspondingly in this case one has $\mathrm{div}^\Gamma = (-\nabla^\Gamma)^*$ but with $*$ taken with respect to the "non flat" measure μ, i.e. $(\nabla^\Gamma)^* = \nabla^\Gamma - \beta_\mu$, with β_μ a drift term (the logarithmic derivative of μ).

Also the case where $\mathbb{R}^d$ is replaced by a (non compact) manifold M has been handled and a corresponding Hodge type L^2-cohomology theory has been developed, see [29], [30], [34], [32], [33]. For relations with representation theory of infinite dimensional groups see, e.g., [292], [349].

6.3 Other applications

In this section we briefly mention some other applications of the method of classical Dirichlet forms and associated diffusions for defining and studying stochastic processes.

Classical spin systems In this case one studies random variables associated with points on a lattice $\mathbb{Z}^d$ (or other discrete structures), with values in $\mathbb{R}$ (or, e.g., a compact Lie group M), with distributions of the "Gibbs type", i.e. of the form

$$\mu(dz) = \text{“}Z^{-1}e^{-S(z)}dz\text{”}$$

with $z = (z_k)_{k\in\mathbb{Z}^d}, z_k \in \mathbb{R}$ (or M).

$S(z)$ describes the interaction between the spins in the "spin configuration" z. Also in this case the diffusion properly associated with the classical Dirichlet form $\mathcal{E}_\mu$ satisfies a stochastic differential (SDE) equation with drift in $L^2(\mu)$, and it is ergodic, if μ is an extreme state. As opposite to the cases discussed before, in Sect. 6.1, 6.2, log-Sobolev inequalities for classical spin systems have been proven, so that exponential ergodicity holds.
The solution process to the corresponding SDE has a drift in $L^2(\mu)$. A dynamical theory of phase transitions can be developed. See [98], [82], [83], [84], [89], [189], [213], [271], [356], [446], [447] for references and also for current work.

Natural measures and diffusion processes associated with individual and lattice loop spaces Let (M, g) be a (compact) Riemannian manifold. Let $E = LM = C(S^1, M)$ be the corresponding free loop space, and let $L_xM = \{\gamma \in LM | \gamma(0) = x \in M\}$ be the corresponding x-based loop space. Let μ resp. μ_x be the pinned Wiener measure on LM resp. L_xM, associated with a Brownian loop in M, with initial distribution the Høegh-Krohn-Bismut measure $\text{Vol}(\cdot)P_t(x, x)$ resp. the Dirac measure in $x \in M$. On $L^2(\nu)$ ($\nu = \mu$ resp. μ_x) we consider the classical Dirichlet form given by ν:

$$\mathcal{E}_\nu = \overline{\mathring{\mathcal{E}}_\nu}, \text{ with } \mathring{\mathcal{E}}_\nu(u, v) = \int \langle \nabla u, \nabla v \rangle_{\mathcal{H}} \, d\nu,$$

$u, v \in FC_b^\infty$, (FC_b^∞ being defined as an analogon of the smooth bounded cylinder functions on M) and $\mathcal{H}$ the Cameron-Martin space associated with E, consisting of loops with finite kinetic energy. This diffusion on E has been constructed and discussed in [103] (full loops) and [218] (based loops).

For a long time the problem of log-Sobolev inequalities has been open, see, e.g. [270], it was discussed recently in the negative by Eberle [222], [224] (for positive results for the case where a "potential" is added see [267], [268], [270] and references therein).
For the problem of uniqueness see [4], [8], [5], [6], [7], [412].
For other problems related to loop spaces and strings see [350], [352], [351], [353], [473].
For applications to quantum statistical mechanics see [78], [79], [80], [81], [97], [321], [322], [326], [357].

6.4 Other problems, applications and topics connected with Dirichlet forms

We mention here some topics that - although of great interest - have unfortunately not been covered in these lectures. They illustrate other aspects of the usefulness and power of the method of Dirichlet forms.

Polymers The construction and study of diffusions with polymer measures as invariant measures has been made possible using methods of Dirichlet forms in [124], resp. [125] (for the case of polymer measures of the Edwards-Westwater-type in 2 resp. 3 dimensions). An open problem here is the ergodicity of the process constructed in the 3-dimensional case. One notes that in two dimensions the drift is in $L^2(\mu)$ but this is not so in three dimensions. The stochastic differential equation satisfied by the diffusion is studied by other methods in [27].

Non-symmetric Dirichlet forms and generalized Dirichlet forms Although the theory of non symmetric resp. generalized Dirichlet forms could only be mentioned shortly in this course, it has lead to important new developments in the theory of singular (finite and infinite dimensional) processes. The main attention has been given to the local forms associated with diffusion processes, see, e.g. [110], [112], [314], [346], [348], [347], [383], [367].
Whereas non symmetric Dirichlet forms have first order terms essentially dominated by the symmetric part, generalized Dirichlet forms allow the inclusion of general first order terms in the generators [474], [475], [476], [477].
The latter lead to non proper associated processes which have found striking applications in the study of stochastic PDE's (see [?] for the Gaussian noise and [355] for Lévy noise. See also [492] for further developments). It should also be mentioned that the theory of generalized Dirichlet forms include also time dependent Dirichlet forms, cf. 6.4, below.

Complex-valued Dirichlet forms A theory of such Dirichlet forms has been developed in [117], [131], with applications to quantum theory. It has also lead to a new approach to some aspects of non symmetric Dirichlet forms [386]. See also [317] for further developments in connections with "open system".

Invariant measures for singular processes A theory of such invariant measures has been developed especially in the case of diffusions, e.g. [21],[330]. For jump processes see [130],[128], [129].

Subordination of diffusions given by Dirichlet forms A theory of subordination of diffusions given by Dirichlet forms has been developed in [127] (see also [129], based on Lévy processes on Banach spaces [126]). For previous work on subordination and Dirichlet forms see [294], [295], [296], [170], [171], [230], [231], [232], [299], [229], [298], [284], [285], [287], [288], [289], [382], [400]; for relations with relativistic Schrödinger operators see [181], [464]. For other relation to jump processes see [311], [376], [397].

Time dependent Dirichlet forms A theory of time-dependent Dirichlet forms leads in particular to processes which satisfy S(P)DE's with time dependent coefficients see [406], [404], [405], [474], [475]. The case of Nelson's diffusions is covered in [474], [475], [173]. For related work see also [300].

Differential operators and processes with boundary conditions Examples of processes described by Dirichlet forms in finite dimensions, including complicated boundary behavior, are given in [258], see also, e.g., [159], [177], [204] (and [188a] for systems of elliptic equations). In infinite dimensions not so many examples have been developed until now, see however [507].

Convergence of Dirichlet forms The problem of when a sequence of Dirichlet forms converges in such a way that the limit is again a Dirichlet form has been discussed originally, in the symmetric case, in [70], [64], [102], [143], [343], [417], [409], [443], [442], [486], [493], [398], [397]. The study of such questions in the non symmetric case has been initiated in [377], [378], [379], [380].

Dirichlet forms and geometry In the sections 6.2 and 6.3 in Chapter 6 we already mentioned some work involving Dirichlet forms and geometry (loop spaces, configuration spaces). For work in other directions, in particular in connection with differential geometry in finite dimensions resp. on special infinite dimensional manifolds see [225], [139], [154], [156], [187], [180], [29], [144], [146], [247], [250], [251], [254], [161], [148], [281], [282], [286], [319], [226], [283], [186], [261], [374].

Further problems involving classical Dirichlet forms For relations with hyperbolic problems see [320] and for scattering problems [324], [323], [175], [182], [184], [327]. For control problems see Nagai [392], [391]. For problems of filter theory see [385]. For Dirichlet forms associated with Lévy Laplacian see [1], [17]. For problems of homogenization theory see [18], [257]. For an inverse problem in stochastic differential equations see [19]. For a small time asymptotics for Dirichlet forms see [422]. For a Girsanov transformation for Dirichlet forms on infinite dimensional spaces see [122]. Structural questions about Dirichlet forms and associated spaces are discussed in [472], [149], [163], [178], [215], [216], [240], [252], [253], [248], [274], [339], [340], [341]. For local Dirichlet forms in relation to problems of classical continuum mechanics see [153], [155], [387], [388], [345], [399]. For questions of infinite dimensional diffusion processes and Dirichlet forms see [354], [361], [362], [365], [366], [370], [371], [418].

Dirichlet forms and processes on fractals, discrete structures and metric measure spaces Important work has been done for constructing and studying processes on fractals in [338], [387], [389], [336], [332], [331], [141], [249], [275], [304], [315], [414]. For the study of Dirichlet forms and processes on p-adic spaces with relation to certain trees, see [74],[75]. A theory of hyperfinite Dirichlet forms (in the sense of non standard analysis) has been developed in [39],[36]. The construction of local Dirichlet forms and diffusion processes on metric measure spaces was carried out in [482]. The important particular case of Alexandrov spaces was studied in great detail in [342]. See also [445], [454], [487], [488], [489], [490], [491], [504], [508], [509].

Harmonic mappings, non linear Dirichlet forms Dirichlet form techniques turned out to be a powerful tool in the study of generalized harmonic mappings with values in metric spaces. Jost [307] pointed out how to define the energy of mappings from the state space of a Dirichlet form into a metric space. This leads to the concept of nonlinear or generalized Dirichlet forms, [308], [481]. The stochastic counterparts are nonlinear Markov operators and martingales in metric spaces [485], [484], [480]. For work on nonlinear Dirichlet forms see [264], [297], [381], [151].

Non commutative and supersymmetric Dirichlet forms and processes The study of non commutative Dirichlet forms has been initiated in [63] (see also [64], [66]) in the symmetric case. This was extended to the nonsymmetric case in [206], [358], [273]. Associated processes have also been studied in [69], [67], [45]. For recent further work, also connected to non commutative geometry, see [365], [195], [190], [191], [192], [194], [306], [363], [273] (see also [50], [402], [410], [411]). Supersymmetric Dirichlet forms have been considered in [138], [77].

References

[1] Accardi L., Bogachev V.I. (1997), "The Ornstein- Uhlenbeck process associated with the Lévy Laplacian and its Dirichlet forms", Probab. Math. Statist. **17**, no. 1, 95-114

[2] Accardi L., Heyde C.C. (1998), "Probability Towards 2000", Springer Verlag

[3] Accardi L., Lu Y.G., Volovich I.V. (2000), "A white-noise approach to stochastic calculus", Acta Appl. Math. **63**, no. 1-3, 3-25

[4] Acosta E. (1994), "On the essential self-adjointness of Dirichlet operators on group-valued path space", Proc. AMS **122**, 581-590

[5] Aida S. (1998), "Differential Calculus on Path and Loop Spaces II. Irreducibility of Dirichlet Forms on Loop Spaces", Bull. Sci. math. **122**, 635-666

[6] Aida S. (2000), "Logarithmic Derivatives of Heat Kernels and Logarithmic Sobolev Inequalities with Unbounded Diffusion Coefficients on Loop Spaces", Journal of Functional Analysis **174**, 430-477

[7] Aida S. (2000), "On the irreducibility of Dirichlet forms on domains in infinite-dimensional spaces", Osaka J. Math. **37**, no. 4, 953-966

[8] Aida S., Shigekawa I. (1994), "Logarithmic Sobolev inequalities and spectral gaps: perturbation theory", Journal of Functional Analysis **126**, no. 2, 448-475

[9] Albeverio S. (1985), "Some points of interaction between stochastic analysis and quantum theory", Springer Verlag, Lect. Notes Control Inform. Sciences **78**, 1-26

[10] Albeverio S. (1989), "Some new developments concerning Dirichlet forms, Markov fields and quantum fields", B. Simon et al, eds., IX. Intern. congr. Math. Phys., Adam Hilger, Bristol

[11] Albeverio S. (1993), "Mathematical physics and stochastical analysis - a round table report", Proc. Round Table, St. Chéron, Bull Sci. Math. **117**, 125-152

[12] Albeverio S. (1996), "Loop groups, random gauge fields, Chern-Simons models, strings: some recent mathematical developments", Espace des lacets Proc. Conf. Loop spaces '94, Eds. R.Léandre, S.Paycha, T.Wurzbacher, Strasbourg, 5-34

[13] Albeverio S. (1997), "A survey of some developments in loop spaces: associated stochastic processes, statistical mechanics, infinite dimensional Lie groups, topological quantum fields", Proc. Steklov Inst. Math. **217**, no. 2, 203-229

[14] Albeverio S. (1997), "A survey of some developments in loop spaces: associated stochastic processes, statistical mechanics, infinite dimensional Lie groups, topological quantum fields", Proc. Ste. Inst. Maths. **217**, 203-229

[15] Albeverio S. (1997), "Some applications of infinite dimensional analysis in mathematical physics", Helv. Phys. Acta **70**, 479-506

[16] Albeverio S. (1997), "Wiener and Feynman-Path integrals and their Applications", AMS, Proceedings of Symposia in Applied Mathematics **52**, 163-194

Originally published in: *Ecole d'Eté de Probabilités de Saint-Flour XXX – 2000*, Lecture Notes in Mathematics, Vol. **1816**, 75–106, DOI: 10.1007/3-540-44922-1, © Springer-Verlag Berlin Heidelberg 2003, Reprint by Springer-Verlag Berlin Heidelberg 2012

[17] Albeverio S., Belopolskaya Ya., Feller M. (2002), "Lévy Dirichlet forms", in preparation

[18] Albeverio S., Bernabei M. S. (2002), "Homogenization in random Dirichlet forms", Preprint SFB 611 no. 7, submitted to Forum Mathematicum

[19] Albeverio S., Blanchard Ph., Kusuoka S., Streit L. (1989), "An inverse problem for stochastic differential equations", J. Stat. Phys **57**, 347-356

[20] Albeverio S., Blanchard Ph., Ma Z.M. (1991), "Feynman-Kac semigroups in terms of signed smooth measures", Birkhäuser, Random Partial Differential Equations, 1-31

[21] Albeverio S., Bogachev V., Röckner M. (1999), "On uniqueness of invariant measures for finite- and infinite-dimensional diffusions", Communications on Pure and Applied Mathematics **52**,325-362

[22] Albeverio S., Brasche J., Röckner M. (1989), "Dirichlet forms and generalized Schrödinger operators", Springer Verlag, Lect. Notes in Phys. **345**, 1-42

[23] Albeverio S., Brzeźniak Z. (1993), "Finite dimensional approximations approach to oscillatory integrals in infinite dimensions", J. Funct. Anal. **113**, 177-244

[24] Albeverio S., Brzeźniak Z. (1995), "Oscillatory integrals on Hilbert spaces and Schrödinger equation with magnetic fields", J. Math. Phys. **36**, 2135-2156

[25] Albeverio S., Brzeźniak Z., Boutet de Monvel-Berthier A.M. (1995), "Stationary phase in infinite dimensions by finite dimensional approximations: applications to the Schrödinger equation", Pot. Anal. **4**, 469-502

[26] Albeverio S., Brzeźniak Z., Boutet de Monvel-Berthier A.M. (1996), "The trace formula for Schrödinger operators from infinite dimensional oscillatory integrals", Math. Nachr. **182**, 21-65

[27] Albeverio S., Brzeźniak Z., Daletskii A. (2002), "Stochastic differential equations on product loop manifolds", Bonn preprint

[28] Albeverio S., Cruzeiro A.B. (1990), "Global flows with invariant (Gibbs) measures for Euler and Navier-Stokes two dimensional fluids", Comm. Math. Phys. **129**, 431-444

[29] Albeverio S., Daletskii A. (1998), "Stochastic equations and quasi-invariance on infinite product groups", Infinite Dimensional Analysis, Quantum Probability and Related Topics, Vol. **2**, No. 2, 283-288

[30] Albeverio S., Daletskii A., Kondratiev Y. (1998), "Some examples of Dirichlet operators associated with the actions of infinite dimensional Lie groups", Meth. Funct. Anal. and Top. **2**, 1-15

[31] Albeverio S., Daletskii A., Kondratiev Y. (2000), " DeRham complex over product manifolds: Dirichlet forms and stochastic dynamics", in Mathematical physics and stochastic analysis (Lisbon, 1998), 37-53

[32] Albeverio S., Daletskii A., Kondratiev Y., Lytvynov E. (2001), "Laplace operators in deRham complexes associated with measures on configuration spaces", Bonn preprint

[33] Albeverio S., Daletskii A., Lytvynov E. (2000), "Laplace operator and diffusions in tangent bundles over Poisson spaces", in [CHNP], 1-25

[34] Albeverio S., Daletskii A., Lytvynov E. (2001), "DeRham cohomology of configuration spaces with Poisson measure", Journal of Functional Analysis **185**, 240-273

[35] Albeverio S., De Faria M., Høegh-Krohn R. (1979), "Stationary measures for the periodic Euler flow in two dimensions", J. Stat. Phys. **20**, 585-595

[36] Albeverio S., Fan R.Z. (2002), "Hyperfinite Dirichlet forms and stochastic processes", preprint Bonn in preparation

[37] Albeverio S., Ferrario B. (2002), "Uniqueness results for the generators of the two-dimensional Euler and Navier-Stokes flows. The case of Gaussian invariant measures", J. Funct. Anal. **193**, No.1, 77-93

[38] Albeverio S., Ferrario B. (2002), "2D vortex motion of an incompressible ideal fluid: the Koopman-von Neumann approach", Infinite Dimens. Anal. Quantum Prob., in press

[39] Albeverio S., Fenstad J.E., Høegh-Krohn R., Lindstrøm T. (1986), "Nonstandard methods in stochastic analysis and mathematical physics", Academic Press, New York

[40] Albeverio S., Fukushima M., Karwowski W., Streit L. (1981), "Capacity and quantum mechanical tunneling", Comm. Math. Phys. **80**, 301-342

[41] Albeverio S., Fukushima M., Hansen W., Ma Z.M., Röckner M. (1992), "An invariance result for capacities on Wiener space", J. Funct. Anal. **106**, 35-49

[42] Albeverio S., Gallavotti G., Høegh-Krohn R. (1979), "Some results for the exponential interaction in 2 or more dimensions", Comm. Math. Phys. **70**, 187-192

[43] Albeverio S., Gesztesy F., Karwowski W., Streit L. (1985), "On the connection between Schrödinger and Dirichlet forms", J. Math. Phys. **26**(10), 2546-2553

[44] Albeverio S., Gesztesy F., Høegh-Krohn R., Holden H. (1998), "Solvable Models in Quantum Mechanics", Springer Verlag, Berlin

[45] Albeverio S., Goswami D. (2002), "A Remark on the structure of symmetric quantum dynamical semigroups on von Neumann algebras", to appear in IDAQP (2002)

[46] Albeverio S., Gottschalk H., Wu J.L. (1997), "Models of local relativistic quantum fields with indefinite metric (in all dimensions)", Commun. Math. Phys. **184**, 509-531

[47] Albeverio S., Gottschalk H., Yosida M. (2001), "Representing Euclidean quantum fields as scaling limit of particle systems", Bonn preprint, to appear in J. Stat. Phys.

[48] Albeverio S., Grothaus M., Kondratiev Y., Röckner M. (2001), "Stochastic dynamics of fluctuations in classical continuous systems", Journal of Functional Analysis **185**, 129-154

[49] Albeverio S., Guatteri G., Mazzucchi S. (2002), "Phase Space Feynman Path Integrals", J. Math. Phys. **43**, no. 6, 2847-2857

[50] Albeverio S., Guido D., Ponosov A., Scarlatti S. (1996), "Singular traces and compact operators", J. Funct. Anal. **137**, no. 2, 281-302

[51] Albeverio S., Haba Z., Russo F. (1996), "On non-linear two-space-dimensional wave equation perturbed by space-time white noise", in Stochastic analysis: random fields and measure-valued processes (Ramat Gan, 93/95), Israel Math. Conf. Proc. **10**, Bar-Ilan Univ., Ramat Gar, 1-25

[52] Albeverio S., Haba Z., Russo F. (2001), "A two-space dimensional semilinear heat equation perturbed by (Gaussian) white noise", Probab. Theory Relat. Fields **121**, 319-366

[53] Albeverio S., Hahn A., Sengupta A. (2002), "Chern-Simons theory, Hida distributions, and state models", Bonn preprint, to appear in Special Issue on "Geometry and Stochastic Analysis", IDAQP

[54] Albeverio S., Hida T., Potthof J., Streit L. (1989), "The vacuum of the Høegh-Krohn model as a generalized white noise functional", Phys. Lett. B **217**, no. 4, 511-514

[55] Albeverio S., Hida T., Potthof J., Röckner M., Streit L. (1990), "Dirichlet forms in terms of white noise analysis I - Construction and QFT examples", Rev. Math. Phys **1**, 291-312

[56] Albeverio S., Hida T., Potthof J., Röckner M., Streit L. (1990), "Dirichlet forms in terms of white noise analysis II - Closability and Diffusion Processes", Rev. Math. Phys **1**, 313-323

[57] Albeverio S., Høegh-Krohn R. (1974), "A remark on the connection between stochastic mechanics and the heat equation", Journal of Mathematical Physics **15**, 1745-1747

[58] Albeverio S., Høegh-Krohn R. (1975), "Homogeneous random fields and statistical mechanics", Journal of Functional Analysis **19**, 242-272

[59] Albeverio S., Høegh-Krohn R. (1976), "Quasi invariant measures, symmetric diffusion processes and quantum fields", Editions du CNRS, Proceedings of the International Colloquium on Mathematical Methods of Quantum Field Theory, Colloques Internationaux du Centre National de la Recherche Scientifique, No. **248**, 11-59

[60] Albeverio S., Høegh-Krohn R. (1976), "Mathematical Theory of Feynman Path Integrals", Lect. Notes Maths. **523**, Springer Berlin

[61] Albeverio S., Høegh-Krohn R. (1977), "Dirichlet forms and diffusion processes on rigged Hilbert spaces", Zeitschrift für Wahrscheinlichkeitstheorie und verwandte Gebiete **40**, 1-57

[62] Albeverio S., Høegh-Krohn R. (1977), "Hunt processes and analytic potential theory on rigged Hilbert spaces", Ann. Inst. H. Poincaré (Probability Theory) **13**, 269-291

[63] Albeverio S., Høegh-Krohn R. (1977), "Dirichlet forms and Markov semigroups on C^*-algebras", Commun. Math. Phys. **56**, 173-187

[64] Albeverio S., Høegh-Krohn R. (1979), "The method of Dirichlet forms", Springer Verlag, Lecture Notes in Physics **93**, 250-258

[65] Albeverio S., Høegh-Krohn R. (1981), "Stochastic Methods in Quantum Field Theory and Hydrodynamics", Physic Reports (Review Section of Physics Letters) **77**, no. 3, 193-214

[66] Albeverio S., Høegh-Krohn R. (1982), "Some remarks on Dirichlet forms and their applications to quantum mechanics", Springer Verlag, Lect. Notes in Math. **923**, 120-132

[67] Albeverio S., Høegh-Krohn R. (1985), "A remark on dynamical semigroups in terms of diffusion processes", Springer Verlag, Lect. Notes in Math. **1136**, 40-45

[68] Albeverio S., Høegh-Krohn R., Marion J., Testard D., Torresani B. (1993), "Non commutative distributions - Unitary representation of gauge groups and algebras", M. Dekker, New York

[69] Albeverio S., Høegh-Krohn R., Olsen G. (1980), "Dynamical semigroups and Markov processes on C^*-algebras", Crelle's J. Reine und Ang. Math.**319**, 25-37

[70] Albeverio S., Høegh-Krohn R., Streit L. (1977), "Energy forms, Hamiltonians, and distorted Brownian paths", J. Math. Phys. **18**, 907–917

[71] Albeverio S., Høegh-Krohn R., Streit L. (1979), "Regularization of Hamiltonians and processes", J. Math. Phys. **21**(7), 1636-1642

[72] Albeverio S., Høegh-Krohn R., Zegarlinski B. (1989), "Uniqueness of Gibbs states for general $P(\varphi)_2$−weak coupling models by cluster expansion", Commun. Math. Phys. **121**, 683-697

[73] Albeverio S., Johnson G.W., Ma Z.M. (1996), "The Analytic Operator-Valued Feynman Integral via Additive Functionals of Brownian Motion", Acta Applicandae Mathematicae **42**, 267-295

[74] Albeverio S., Karwowski W. (1991), "Diffusion on p-adic numbers", Proc. Third Nagoya Lévy Seminar, Gaussian Random Fields, Ed. Ito, Hida

[75] Albeverio S., Karwowski W., Yasuda K. (2002), "Trace formula for p-adics", Acta Appl. Math. **71**, no. 1, 31-48

[76] Albeverio S., Karwowski W., Zhao X.L. (1999), "Asymptotics and spectral results for random walks on p-adics", Stoch. Proc. Appl. **83**, 39-59

[77] Albeverio S., Kondratiev Y. (1995), "Supersymmetric Dirichlet operators", Ukrainian Math. J. **47**, 583-592

[78] Albeverio S., Kondratiev Y., Kozitsky Y., Röckner M. (2001), "Euclidean Gibbs States of Quantum Lattice Systems", BiBoS preprint, Universität Bielefeld, No. 01-03-03, to appear in Rev. in Math. Phys

[79] Albeverio S., Kondratiev Y., Minlos R., Shchepan'uk G. (2000), "Uniqueness Problem for Quantum Lattice Systems with Compact Spins", Letters in Mathematical Physics **52**, 185-195

[80] Albeverio S., Kondratiev Y., Minlos R., Shchepan'uk G. (2000), "Ground State Euclidean Measures for Quantum Lattice Systems on Compact Manifolds", Rep. Math. Phys. **45**, 419-429

[81] Albeverio S., Kondratiev Y., Pasurek T., Röckner M. (2001), "Euclidean Gibbs States of Quantum Crystals", Moscow Mathematical Journal **1**, No. 3, 307-313

[82] Albeverio S., Kondratiev Y., Pasurek T., Röckner M. (2002), "Euclidean Gibbs measures on loop lattices: Existence and a priori estimates", BiBoS Preprint, Universität Bielefeld, No. 02-05-086

[83] Albeverio S., Kondratiev Y., Pasurek T., Röckner M. (2002), "A priori estimates and existence for Euclidean Gibbs measures", BiBos-Preprint N02-06-089, (submitted to Trans. Moscow Math. Soc.)

[84] Albeverio S., Kondratiev Y., Pasurek T., Röckner M. (2002), "A priori estimates and existence for quantum Gibbs states: approach via Lyapunov functions"

[85] Albeverio S., Kondratiev Y. Röckner M. (1992), "An approximative criterium of essential selfadjointness of Dirichlet operators", Potential Anal. **1**, no. 3, 307-317

[86] Albeverio S., Kondratiev Y. Röckner M. (1993), "Addendum to the paper 'An approximative criterium of essential self-adjointness of Dirichlet operators'", Potential Anal. **2**, 195-198

[87] Albeverio S., Kondratiev Y., Röckner M. (1994), "Infinite dimensional diffusions, Markov fields, quantum fields and stochastic quantization", Kluwer, Dordrecht, Stochastic Analysis and Applications, 1-94

[88] Albeverio S., Kondratiev Y., Röckner M. (1995), "Dirichlet Operators via Stochastic Analysis", Journal of Functional Analysis **128**, 102-138

[89] Albeverio S., Kondratiev Y., Röckner M. (1995), "Uniqueness of the stochastic dynamics for continuous spin systems on a lattice", J. Funct. Anal. **133**, 10-20

[90] Albeverio S., Kondratiev Y., Röckner M. (1996), "Ergodicity of L^2-Semigroups and Extremality of Gibbs States", Journal of Functional Analysis **144**, No. 2, 394-423

[91] Albeverio S., Kondratiev Y., Röckner M. (1997), "Ergodicity for the Stochastic Dynamics of Quasi-invariant Measures with Applications to Gibbs States", Journal of Functional Analysis **149**,No. 2, 415-469

[92] Albeverio S., Kondratiev Y., Röckner M. (1998), "Analysis and geometry on configuration spaces", J. Funct. Anal. **154**, 444-500

[93] Albeverio S., Kondratiev Y., Röckner M. (1998), "Analysis and geometry on configuration spaces: The Gibbsian case", J. Funct. Anal. **157**, 242-291

[94] Albeverio S., Kondratiev Y., Röckner M. (1999), "Diffeomorphism groups and current algebras: Configuration spaces analysis in quantum theory", Rev. Math. Phys. **11**, 1-23

[95] Albeverio S., Kondratiev Y., Röckner M. (2002), "Symmetryzing Measures for Infinite Dimensional Diffusions: An Analytic Approach", BiBoS Preprint, Universität Bielefeld, No. 02-05-087, to appear in Lect. Notes Math., "Geometric Analysis and unlinear partial differential equations", S. Hildebrandt et. al., eds, Springer, Berlin

[96] Albeverio S., Kondratiev Y., Röckner M. (2002), "Strong Feller Properties for Distorted Brownian Motion and Applications to Finite Particle Systems with Singular Interactions", BiBoS preprint, Universität Bielefeld, No. 02-05-084, to appear in AMS-Volume dedicated to L. Gross

[97] Albeverio S., Kondratiev Y., Röckner M., Tsikalenko T. (1997), "Dobrushin's Uniqueness for Quantum Lattice Systems with Nonlocal Interaction", Commun. Math. Phys. **189**, 621-630

[98] Albeverio S., Kondratiev Y., Röckner M., Tsikalenko T. (2000), "A Priori Estimates for Symmetrizing Measures and Their Applications to Gibbs States", Journal of Functional Analysis **171**, 366-400

[99] Albeverio S., Kusuoka S. (1992), "Maximality of infinite-dimensional Dirichlet forms and Høegh-Krohn's model of quantum fields", Cambridge Univ. Press, in "Ideas and methods in quantum statistical physics",301-330

[100] Albeverio S., Kusuoka S. (1995), "A basic estimate for two-dimensional stochastic holonomy along Brownian bridges", J. Funct. Anal. **127**, no. 1, 132-154

[101] Albeverio S., Kusuoka S., Röckner M. (1990), "On partial integration in infinite dimensional space and applications to Dirichlet forms", J. London Math. Soc. **42**, no. 1, 122-136

[102] Albeverio S., Kusuoka S., Streit L. (1986), "Convergence of Dirichlet forms and associated Schrödinger operators", J. Funct. Anal. **68**, 130-148

[103] Albeverio S., Léandre R., Röckner M. (1993), "Construction of a rotational invariant diffusion on the free loop spaces", C.R. Acad. Sci **316**, Ser. I, 1-6

[104] Albeverio S., Ma Z.M. (1991), "Perturbation of Dirichlet forms - lower semiboundedness, closability and form cones", J. Funct. Anal. **99**, 332-356

[105] Albeverio S., Ma Z.M. (1991), "Necessary and sufficient conditions for the existence of m-perfect processes associated with Dirichlet forms", Springer Verlag, Lect. Notes Math. **1485**, 374-406

[106] Albeverio S., Ma Z.M. (1992), "Additive Functionals, nowhere Radon and Kato Class Smooth Measures Associated with Dirichlet Forms", Osaka J. Math. **29**, 247-265

[107] Albeverio S., Ma Z.M. (1992), "Characterization of Dirichlet forms associated with Hunt processes", World Scientific Singapore, Proc. Swansea Conf. Stochastic Analysis, 1-25

[108] Albeverio S., Ma Z.M. (1992), "A general correspondence between Dirichlet forms and right processes", Bull. Am. Math. Soc. **26**, 245-252

[109] Albeverio S., Ma Z.M., Röckner M. (1992), "Non-symmetric Dirichlet forms and Markov processes on general state space", CRA Sci., Paris **314**, 77-82

[110] Albeverio S., Ma Z.M., Röckner M. (1992), "Regularization of Dirichlet spaces and applications", CRAS, Ser. I, Paris **314**, 859-864

[111] Albeverio S., Ma Z.M., Röckner M. (1992), "A Beurling-Deny structure theorem for Dirichlet forms on general state space", Cambridge University Press, R. Høegh-Krohn's Memorial Volume, Vol **1**, 115-123

[112] Albeverio S., Ma Z.M., Röckner M. (1993), "Local property of Dirichlet forms and diffusions on general state spaces", Math. Annalen **296**, 677-686

[113] Albeverio S., Ma Z.M., Röckner M. (1993), "Quasi-regular Dirichlet forms and Markov processes", J. Funct. Anal. **111**, 118-154

[114] Albeverio S., Ma Z.M., Röckner M. (1995), "Potential theory of positivity preserving forms without capacity", Walter de Gruyter & Co., Dirichlet Forms and Stochastic Processes, 47-53

[115] Albeverio S., Ma Z.M., Röckner M. (1995), "Characterization of (non-symmetric) Dirichlet forms associated with Hunt processes", Random Operators and Stochastic Equations **3**, 161-179

[116] Albeverio S., Ma Z.M., Röckner M. (1997), "Partition of unity for Sobolev spaces in infinite dimensions", J. Functional Anal. **143**, 247-268

[117] Albeverio S., Morato L.M., Ugolini S. (1998), "Non-Symmetric Diffusions and Related Hamiltonians", Potential Analysis **8**, 195-204

[118] Albeverio S., Röckner M. (1989), "Classical Dirichlet forms on topological vector spaces - the construction of the associated diffusion process", Prob. Theory and Rel. Fields **83**, 405-434

[119] Albeverio S., Röckner M. (1990), "Classical Dirichlet forms on topological vector spaces – closability and a Cameron-Martin formula.", J. Funct. Anal. **88**, 395–436

[120] Albeverio S., Röckner M. (1991), "Stochastic differential equations in infinite dimension: solution via Dirichlet forms", Prob. Th. Rel. Fields **89**, 347-386

[121] Albeverio S., Röckner M. (1995), "Dirichlet form methods for uniqueness of martingale problems and applications",in "Stochastic Analysis",Edts. M.C. Cranston et al., AMS, Providence,513-528
[122] Albeverio S., Röckner M., Zhang T.S. (1993), "Girsanov transform of symmetric diffusion processes in infinite dimensional space", Ann. of Prob. **21**, 961-978
[123] Albeverio S., Röckner M., Zhang T.S. (1993), "Markov uniqueness and its applications to martingale problems, stochastic differential equation and stochastic quantization", C.R. Math. Rep. Acad. Sci. Canada XV, no. 1, 1-6
[124] Albeverio S., Röckner M., Zhou X.Y. (1999), "Stochastic quantization of the two dimensional polymer measure", Appl. Math. Optimiz. **40**, 341-354
[125] Albeverio S., Röckner M., Zhou X.Y. (2002), "Stochastic quantization of the three-dimensional polymer measure", in preparation
[126] Albeverio S., Rüdiger B. (2002), "The Lévy-Ito decomposition theorem and stochastic integrals, on separable Banach spaces", BiBoS preprint, Universität Bielefeld, No. 02-01-071
[127] Albeverio S., Rüdiger B. (2002), "Infinite dimensional Stochastic Differential Equations obtained by subordination and related Dirichlet forms", BiBoS preprint, Universität Bielefeld, No. 02-03-077
[128] Albeverio S., Rüdiger B., Wu J.L. (2000), "Invariant Measures and Symmetry Property of Lévy Type Operators", Potential Analysis **13**, 147-168
[129] Albeverio S., Rüdiger B., Wu J.L. (2001), "Analytic and Probabilistic Aspects of Lévy Processes and Fields in Quantum Theory", in "Lévy Processes: Theory and Applications", Eds. O.E. Barndorff-Nielsen et al., Birkhäuser, Basel, 187-224
[130] Albeverio S., Song S. (1993), "Closability and resolvent of Dirichlet forms perturbed by jumps", Pot. Anal. **2**, 115-130
[131] Albeverio S., Ugolini S. (2000), "Complex Dirichlet Forms: Non Symmetric Diffusion Processes and Schrödinger Operators", Potential Analysis **12**, 403-417
[132] Albeverio S., Zegarlinski B. (1992), "Global Markov property in quantum field theory and statistical mechanics: a review on results and problems", in R.Høegh-Krohn's Memorial Volume **2**, Edts. S.Albeverio, J.E.Fenstad, H.Holden, T.Lindstrøm, Cambridge University Press, 331-369
[133] Albeverio S., Zegarlinski B. (2002), in preparation
[134] Albeverio S., Zhang T.S. (1998), "Approximations of Ornstein-Uhlenbeck processes with unbounded linear drifts", Stoch. and Stoch. Repts. **63**, no. 3-4, 303-312
[135] Albeverio S., Zhao X.L. (2002), "Stochastic processes on p-adics", book in preparation
[136] Albeverio S., Ru-Zong F., Röckner M., Stannat W. (1995), "A remark on coercive forms and associated semigroups", Operator Theory: Adv. and Appl. **78**, 1-8
[137] Aldous D., Evans S.N. (1999), "Dirichlet forms on totally disconnected spaces and bipartite Markov chains", J. Theoret. Probab. **12**, no. 3, 839-857

[138] Arai A., Mitoma I. (1991), "De Rham-Hodge-Kodaira decomposition in ∞-dimensions", Math. Ann. **291**, 51-73

[139] Bakry D., Qian Z. (2000), "Some new results on eigenvectors via dimension, diameter, and Ricci curvature", Adv. Math. **155**, no. 1, 98-153

[140] Bally V., Pardoux E., Stoica L. (2002), "Backward stochastic differential equations associated to a symmetric Markov process", preprint

[141] Barlow M.T., Kumagai T. (2001), "Transition density asymptotics for some diffusion processes with multifractal structures", Electron. J. Probab. **6**, no. 9, 23pp.

[142] Bauer H. (2002), "Wahrscheinlichkeitstheorie", Walter de Gruyter & Co., Berlin

[143] Baxter J., Dal Maso G., Mosco U. (1987), "Stopping times and Γ-convergence", Trans. Amer. Math. Soc. **303**, no. 1, 1-38

[144] Bendikov A. (1995), "Potential Theory on Infinite- Dimensional Abelian Groups", de Gruyter, Studies in Mathematics **21**

[145] Bendikov A., Léandre R. (1999), "Regularized Euler-Poincaré number of the infinite dimensional torus", Infinite Dimensional Analysis, Quantum Probability and Related Topics **2**, 617-626

[146] Bendikov A., Saloff-Coste L. (2000), "On- and Off-Diagonal Heat Kernel Behaviours on Certain Infinite Dimensional Local Dirichlet Spaces", American Journal of Mathematics **122**, 1205-1263

[147] Berezansky Y.M., Kondratiev Y.G. (1995), "Spectral Methods in Infinite-Dimensional Analysis", Kluwer, Vol. **1,2**, first issued in URSS, 1988

[148] Berg C., Forst G. (1973), "Non-symmetric translation invariant Dirichlet forms", Invent. Math. **21**, 199-212

[149] Bertoin J. (1986), "Les processus de Dirichlet en tant qu'espace de Banach", Stochastics **18**, 155-168

[150] Beurling A., Deny J. (1958), "Espaces de Dirichlet", Acta Math. **99**, 203-224

[151] van Beusekom P. (1994), "On nonlinear Dirichlet forms", Ph. D. Thesis, Utrecht

[152] Biroli M. (2000), "Weak Kato measures and Schrödinger problems for a Dirichlet form", Rend. Accad. Naz. Sci. XL Mem. Mat. Appl. (5) **24**, 197-217

[153] Biroli M., Mosco U. (1992), "Discontinuous media and Dirichlet forms of diffusion type", Developments in partial differential equations and applications to mathmatical physics (Ferrara, 1991), 15-25, Plenum, New York

[154] Biroli M., Mosco U. (1993), "Sobolev inequalities for Dirichlet forms on homogeneous spaces", RMA Res. Notes Appl. Math. **29**, 305-311

[155] Biroli M., Mosco U. (1995), "A Saint-Venant principle for Dirichlet forms on discontinuous media", Ann. Mat. Pura Appl. **169**, 125-181

[156] Biroli M., Mosco U. (1995), "Sobolev and isoperimetric inequalities for Dirichlet forms on homogeneous spaces", Atti. Accad. Naz. Lincei Cl. Sci. Fis. Mat. Natur. Rend. Lincei(9) Mat. Appl. **6**, no. 1, 37-44

[157] Biroli M., Mosco U. (1999), "Kato space for Dirichlet forms", Potential Anal. **10**, no. 4, 327-345

[158] Biroli M., Picard C., Tchou N. (2001), " Error estimates for relaxed Dirichlet problems involving a Dirichlet form", Adv. Math. Sci. Appl. **11**, no. 2, 673-684

[159] Biroli M., Tersian S. (1997), "On the existence of nontrivial solutions to a semilinear equation relative to a Dirichlet form", Istit. Lombardo Accad. Sci. Lett. Rend. A **131**, no. 1-2, 151-168

[160] Bliedtner J., Hansen W. (1986), "Potential Theory", Springer Verlag

[161] Bloom W.R. (1987), "Non-symmetric translation invariant Dirichlet forms on hypergroups", Bull. Austral. Math. Soc. **36**, no.1, 61-72

[162] Blumenthal R.M., Getoor R.K. (1968), "Markov processes and Potential theory", Academic Press, New York

[163] Boboc N., Bucur G. (1998), "Characterization of Dirichlet forms by their excessive and co-excessive elements", Potential Anal. **8**, no. 4, 345-357

[164] Bogachev V., Krylov N., Röckner M. (1997), "Elliptic regularity and essential self-adjointness of Dirichlet Operators on $\mathbb{R}^n$", Annali Scuola Norm. Super. di Pisa Cl. Sci. (4) **24**, no. 3, 451-461

[165] Bogachev V., Röckner M. (1995), "Mehler Formula and Capacities for Infinite Dimensional Ornstein-Uhlenbeck Processes with General Linear Drift", Osaka J. Math. **32**, 237-274

[166] Bogachev V., Röckner M., Stannat W. (1999), "Uniqueness of invariant measures and maximal dissipativity of diffusion operators on L^1", Preprint SFB 343, Bielefeld, 99-119, 19pp.

[167] Bogachev V., Röckner M., Wang F.Y. (2001), "Elliptic Equations for Invariant Measures on Finite and Infinite Dimensional Manifolds", J. Math. Pures Appl. **80** No. 2, 177-221

[168] Bogachev V., Röckner M., Zhang T.S. (2000), "Existence and uniqueness of invariant measures: an approach via sectorial forms", Appl. Math. Optim. **41**, no. 1, 87-109

[169] Borkar V.S., Chari R.T., Mitter S.K. (1988), "Stochastic quantization of field theory in finite and infinite volume", J. Funct. Anal. **81**, no. 1, 184-206

[170] Bouleau N. (1984), "Quelques resultats probabilistes sur la subordination au sens de Bochner", Lect. Notes in Math. **1061**, 54-81

[171] Bouleau N. (2001), "Calcul d'erreur complet lipschitzien et formes de Dirichlet", J. Math. Pures Appl. (9) **80**, no. 9, 961-976

[172] Bouleau N., Hirsch F. (1991), "Dirichlet Forms and Analysis on Wiener Space", deGruyter, Studies in Mathematics **14**

[173] Bouslimi M. (2000), "Subordination operators of Dirichlet forms", BiBoS preprint, Universität Bielefeld, No. 00-07-21

[174] Brasche J.F. (1989), "Dirichlet forms and nonstandard Schrödinger operators, standard and nonstandard", World Sci. Publishing, Teaneck, NJ, 42-57

[175] Brasche J.F. (2002), "On singular perturbations of order $s, s \leq 2$, of the free dynamics: Existence and completeness of the wave operators", Göteborg Preprint

[176] Brasche J.F. (2002), "On eigenvalues and eigensolutions of the Schödinger equation on the complement of a set with classical capacity zero", in preparation

[177] Brasche J.F., Karwowski W. (1990), "On boundary theory for Schrödinger operators and stochastic processes. Order, disorder and chaos in quantum systems", Birkhäuser, Basel, Oper. Theory Adv. Appl. 46

[178] Briane M., Tchou N. (2001), "Fibered microstructures for some nonlocal Dirichlet forms", Ann. Scuola Norm. Sup. Pisa Cl. Sci. (4) **30**, no. 3-4, 681-711

[179] Carlen E.A., Pardoux E. (1990), "Differential calculus and integration by parts on Poisson space", Kluwer Academic Publishers, Stochastics,Algebra and Analysis in Classical and Quantum Dynamics, 63-73

[180] Carlen E.A., Kusuoka S., Stroock D.W. (1987), "Upper bounds for symmetric Markov transition functions", Ann. Inst. H. Poincaré Probab. Statist. **23**, no. 2, 245-287

[181] Carmona R., Masters W.C. (1990), "Relativistic Schrödinger Operators: Asymptotic Behaviour of the Eigenfunctions", Journal of Functional Analysis **91**, 117-142

[182] van Casteren J.A. (1999), "Feynman-Kac Semigroups, Martingales and Wave Operators", J. Korean Math. Soc. **38**, no. 2, 227-274

[183] van Casteren J.A. (2000), "Some problems in stochastic analysis and semigroup theory", Progress in Nonlinear Differential Equations and their Applications **42**, 43-60

[184] van Casteren J.A., Demuth M. (2000), "Stochastic spectral theory for selfadjoint Feller operators. A functional integration approach", Probability and its Applications, Birkhäuser, Basel

[185] Cattiaux P., Fradon M. (1997), "Entropy, reversible diffusion processes, and Markov uniqueness", J. Funct. Anal. **134**, 243-272

[186] Changsoo B., Chul K.K., Yong M.P. (2002), "Dirichlet forms and symmetric Markovian semigroups on CCR algebras with respect to quasifree states", to appear in J. Math. Phys.

[187] Chen, M.F. (1989), "Probability metrics and coupling methods", Pitman Res. Notes Math. Ser. **200**

[188] Chen Z.Q., Fitzsimmons P.J., Takeda M., Ying J., Zhang T.S. (2002), "Absolute continuity of symmetric Markov processes", University of Washington, Seattle - preprint

[188a] Chen Z.Q., Zhao Z. (1994), "Switched diffusion processes and systems of elliptic equations: a Dirichlet space approach", Proc. Roy. Soc. Edinburgh Sect. A **124**, no. 4, 673-701

[188b] Chen Z.Q., Zhang T.S. (2002), "Girsanov and Feynman-Kac type transformations for symmetric Markov processes", Ann. I.H. Poincaré PR **38**, 475-505

[189] Choi V., Park Y.M., Yoo H.J. (1998), " Dirichlet forms and Dirichlet operators for infinite particle systems: essential self-adjointness", J. Math. Phys. **39**, no. 12, 6509-6536

[190] Cipriani F. (1997), "Dirichlet forms and Markovian semigroups on standard forms of von Neumann algebras", J. Funct. Anal. **147**, no. 2, 259-300

[191] Cipriani F. (1998), "The Variational Approach to the Dirichlet Problem in $\mathbb{C}^*$-Algebras", Quantum Probability Banach Centre Publications, Vol. **43**, 135-146

[192] Cipriani F. (1999), "Perron Theory for Positive Maps and Semigroups on von Neumann Algebras", Stochastic processes, physics and geometry: new interplays, II, CMS COnf. Proc., 29, 115-123

[193] Cipriani F. (2000), "Estimates for capacities of nodal sets and polarity criteria in recurrent Dirichlet spaces", Forum Math. **12**, 1-21

[194] Cipriani F. (2002), "Noncommutative potential theory and the sign of the curvature operator in Riemannian geometry", preprint 494/P, Politecnico di Milano

[195] Cipriani F., Sauvageot J.L. (2000), "Derivations as Square Roots of Dirichlet Forms", Preprint, Politecnico di Milano

[196] Clément Ph., den Hollander F., van Neerven J., de Pagter B. (2000), "Infinite Dimensional Stochastic Analysis", Roy. Neth. Ac. of Arts & Sci. Amsterdam, ISBN: 90-6984-296-3 60-06

[197] Courrège P., Renouard P. (1975), "Oscillateur anharmonique, mesures quasi-invariantes sur $C(\mathbb{R}, \mathbb{R})$ et theéorie quantique des champs en dimension $d = 1$", Astérique, no. 22-23, 3-245

[198] Da Prato G., Debussche A. (2000), "Maximal dissipativity of the Dirichlet operators corresponding to the Burgers equation", CMS Conf. Proc **28**, 85-98

[199] Da Prato G., Debussche A. (2002), "Strong solutions to the stochastic quantization equations", Prepublication de l'IRMAR 02-18

[200] Da Prato G., Tubaro L. (2000), "Self -adjointness of some infinite dimensional elliptic operators and applications to stochastic quantization", Prob. Theory Related Fields **118**, No. 1, 131-145

[201] Da Prato G., Zabczyk J. (1992), "Stochastic equations in infinite dimensions", Cambridge University Press, Encyclopedia of Mathematics and Applications **44**

[202] Da Silva J., Kondratiev Y. Röckner M. (2001), " On a relation between intrinsic ans extrinsic Dirichlet forms for interacting particle systems", Math. Nachr. **222**, 141-157

[203] Dai Pra P., Roelly S., Zessin H. (2002), " A Gibbs variational principle in space-time for infinite-dimensional diffusions", Prob. Theory Related Fields **122**, no. 2, 289-315

[204] Dal Maso G., De Cicco V., Notarantonio L, Tchou N. (1998), "Limits of variational problems for Dirichlet forms in varying domains", J. Math. Pures Appl. (9) **77**, no. 1, 89-116

[205] Daletsky Yu., Fomin S.V. (1991), "Measures and Differential Equations in Infinite Dimensional Space", Kluwer, Dordrecht

[206] Davies B.E., Lindsay M.J. (1992), "Noncommutative symmetric Markov semigroups", Math. Z. **210**, no. 3, 379-411

[207] Dellacherie C., Meyer P.A. (1975), "Probabilités et Potentiel", Hermann Paris, Chapitres I à IV

[208] Dellacherie C., Meyer P.A. (1975), "Probabilités et Potentiel", Hermann Paris, Chapitres V à VIII

[209] Dellacherie C., Meyer P.A. (1983), "Probabilités et Potentiel", Hermann Paris, Théorie discrète du potentiel, Chapitres IX à XI

[210] Dellacherie C., Meyer P.A. (1975), "Probabilités et Potentiel", Hermann Paris, Théorie du potentiel associée à une résolvante, Théorie des processus de Markov, Ch. XII-XVI

[211] Deny J. (1970), "Méthodes Hilbertiennes et théorie du potentiel", Potential Theory, Centro Internazionale Matematico Estivo, Edizioni Cremonesi, Roma

[212] Deuschel J.D., Stroock D.W. (1989), "A function space large deviation principle for certain stochatic integrals", Probab. Theory Related Fields **83**, no. 1-2, 279-307

[213] Deuschel J.D., Stroock D.W. (1990), "Hypercontractivity and spectral gap of symmetric diffusions with applications to the stochastic Ising models", J. Funct. Anal. **92**, no. 1, 30-48

[214] Dohmann J.M.N. (2001), "Feller-type Properties and Path Regularities of Markov Processes", BiBos Preprint E02-04-081

[215] Dong Z. (1998), "Quasi-regular topologies for Dirichlet forms on arbitrary Hausdorff spaces", Acta Math. Sinica **14**, 683-690

[216] Dong Z., Gong F.Z. (1997), "A necessary and sufficient condition for quasiregularity of Dirichlet forms corresponding to resolvent kernels", Acta Math. Appl. Sinica **20**, no. 3, 378-385

[217] Doob J.L. (1984), "Classical Potential Theory and its Probabilistic Counterpart", Springer-Verlag, Grundl. d. math. Wiss. **262**

[218] Driver B., Röckner M. (1992), "Construction of diffusions on path and loop spaces of compact Riemmanian manifolds", C.R. Acad. Sci. Paris **315**, 603-608

[219] Dynkin E.B. (1980), "Markov processes and random fields", Bull. Amer. Math. Soc. **3**, no. 3, 975-999

[220] Dynkin E.B. (1982), "Green's and Dirichlet spaces associated with fine Markov processes", J. Funct. Anal. **47**, 381-418

[221] Dynkin E.B. (1984), "Gaussian and non-Gaussian random fields associated with Markov processes", J. Funct. Anal. **55**, no. 3, 344-376

[222] Eberle A. (1997), "Diffusions on path and loop spaces: existence, finite-dimensional approximation and Hölder continuity", Probab. Theory Related Fields **109**, no. 1, 77-99

[223] Eberle A. (1999), "Uniqueness and Non-Uniqueness of Semigroups generated by Singular Diffusion Operators", Springer Verlag, Lect. Notes Maths. **1718**

[224] Eberle A. (2001), "Poincaré inequalities on loop spaces", BiBos Preprint 01-07-049

[225] Elworthy K.D., Le Jan Y., Li X.M. (1999), "On the geometry of diffusion operators and stochastic flows", Lect. Notes in Math. **1720**

[226] Elworty K.D., Ma Z.M. (1997), "Vector fields on mapping spaces and related Dirichlet forms and diffusions", Osaka J. Math. **34**, no. 3, 629-651

[227] Ethier S., Kurtz Th.G., "Markov processes – characterization and convergence", J. Wiley, New York (1985)

[228] Fabes E., Fukushima M., Gross L., Kenig C., Röckner M., Stroock D.W. (1993), " Dirichlet forms", Eds. C. Dell'Antonio, U. Mosco Springer Verlag, Lecture Notes in Math. **1563**

[229] Farkas W., Jacob N. (2001), "Sobolev spaces on non-smooth domains and Dirichlet forms related to subordinate reflecting diffusions", Math. Nachr. **224**, 75-104

[230] Feyel D. (1979), "Propriétés de permanence du domaine d'un générateur infinitésimal", Springer Verlag, Lecture N. in Math. **713**, No. 4, 38-50

[231] Feyel D., de La Pradelle A. (1978), "Processus associés a une classe d'espaces de Banach fonctionnels", Z. Wahrsch. verw. Geb. **43**, 339-351

[232] Feyel D., de La Pradelle A. (1979), "Dualité des quasi-résolvantes de Ray", Springer Berlin, Lect. Notes in Math. **713**, 67-88

[233] Finkelstein D.L., Kondratiev Y.G., Röckner M., Konstantinov A.Y. (2001), "Gauss formula and symmetric extensions of the Laplacian on configuration spaces", Inf. Dim. Anal. Quant. Prob. **4**, 489-510

[234] Fitzsimmons P.J. (1989), "Markov processes and nonsymmetric Dirichlet forms without regularity", J. Funct. Anal. **85**, 287-306

[235] Fitzsimmons P.J. (1997), "Absolute continuity of symmetric diffusions", Ann. Prob. **25**, 230-258

[236] Fitzsimmons P.J. (2000), "Hardy's inequality for Dirichlet forms", J. Math. Anal. Appl. **250**, no. 2, 548-560

[237] Fitzsimmons P.J. (2001), "On the quasi-regularity of semi-Dirichlet forms", Potential Anal. **15**, no. 3, 151-185

[238] Fitzsimmons P.J., Getoor R.K. (1988), "On the potential theory of symmetric Markov processes", Math. Ann. **281**, 495-512

[239] Fitzsimmons P.J., Getoor R.K. (2002), "Additive Functionals and Characteristic Measures", Stochastic Analysis and Related Topics, RIMS International Project 2002, 23-24

[240] Föllmer H. (1981), "Dirichlet processes", Lect. Notes in Math. **851**, 476-478

[241] Föllmer H. (1984), "Von der Brownschen Bewegung zum Brownschen Blatt: Einige neuere Richtungen in der Theorie der stochastischen Prozesse", Birkhäuser Verlag, Perspectives in Mathematics Anniversary of Oberwolfenbach, 159-189

[242] Fritz J. (1987), "Gradient dynamics of infinite point systems", Ann. Probab. **15**, no. 2, 478-514

[243] Fukushima M. (1971), "Dirichlet spaces and strong Markov processes", Trans. Amer. Math. Soc. **162**, 185-224

[244] Fukushima M. (1980), "Dirichlet Forms and Markov Processes", North Holland, Amsterdam

[245] Fukushima M. (1981), "On a stochastic calculus related to Dirichlet forms and distorted Brownian motion", Phys. Rep. **77**, 255–262

[246] Fukushima M. (1982), "On absolute continuity of multidimensional symmetrizable diffusions", Lect. Notes in Math. **923**, 146-176

[247] Fukushima M. (1984), "A Dirichlet form on the Wiener space and properties of Brownian motion", Springer Verlag, Lecture N. in Math. **1096**, 290-300

[248] Fukushima M. (1987), "Energy forms and diffusion processes", World Scientific Publishing, Mathematics and Physics **1**, 65-97

[249] Fukushima M. (1992), "Dirichlet forms, diffusion processes and spectral dimensions for nested fractals", Eds. S. Albeverio et al., Ideas and Methods in Quantum and Statistical Physics, Cambridge University Press

[250] Fukushima M. (1997), "Dirichlet forms, Caccioppoli sets and Skorohod equation", Progr. Systems Control Theory **23**, 59-66

[251] Fukushima M. (2000), "BV Functions and Distorted Ornstein Uhlenbeck Processes over the Abstract Wiener Space", Journal of Functional Analysis Vol. **174**, No. 1, 227-249

[252] Fukushima M. (2001), "Decompositions of Symmetric Diffusion Processes and Related Topics in Analysis", Sugaku Expositions Vol. **14**,No. 1, 1-13

[253] Fukushima M. (2002), "Function spaces and symmetric Markov processes", Stochastic Analysis and Related Topics, RIMS International Project 2002, 21-22

[254] Fukushima M., Hino M. (2001), "On the space of BV functions and a related stochastic calculus in infinite dimensions", J. Funct. Anal. **183**, 245-268

[255] Fukushima M., LeJan Y. (1991), "On quasi-support of smooth measures and closability of pre-Dirichlet forms", Osaka J. Math. **28**, 837-845

[256] Fukushima M., Ying J. (2002), "A note on regular Dirichlet spaces", Kansai Preprint

[257] Fukushima M., Nakao S., Takeda M. (1987), "On Dirichlet forms with random data-recurrence and homogenization", Springer Verlag, Lecture Notes in Math. **1250**

[258] Fukushima M., Oshima Y., Takeda M. (1994), "Dirichlet Forms and Symmetric Markov Processes", Walter de Gruyter

[259] Fukushima M., Uemura (2002), "On Sobolev and capacitary inequalities for Besov spaces over d-sets", preprint

[260] Funaki T. (1992), "On the stochastic partial differential equations of Ginzburg-Landau type", Lecture Notes in Control and Inform. Sci. **176**

[261] Funaki T., Nagai H. (1993), "Degenerative convergence of diffusion process toward a submanifold by strong drift", Stoch. Stoch. Rep. **44**, no. 1-2, 1-25

[262] Galakhov E.I. (1996), "Sufficient Conditions for the Existence of Feller Semigroups", Mathematical Notes **60**(3), 328-330

[263] Georgii H.O. (1979), "Canonical Gibbs Measures", Springer Verlag, Lecture Notes in Math. **760**

[264] Gianazza U. (1997), "Existence for a nonlinear problem relative to Dirichlet forms", Rend. Accad. Naz. Sci. XL Mem. Mat. Appl. (5) **21**, 209-234

[265] Glimm J., Jaffe A. (1981), "Quantum Physics", Springer, New-York

[266] Glover J., Rao M., Šikić H., Song R. (1994), "Quadratic Forms Corresponding to the Generalized Schrödinger Semigroups", Journal of Functional Analysis **125**, 358-378

[267] Gong F.Z., Ma Z.M. (1998), "The Log-Sobolev inequality on loop space over a compact Riemannian manifold", J. Funct. Anal. **157**, 599-623

[268] Gong F., Röckner M., Liming W. (2001), "Poincaré Inequality for Weighted First Order Sobolev Spaces on Loop Spaces", Journal of Functional Analysis **185**, 527-563

[269] Gross L. (1976), "Logarithmic Sobolev inequalities", Amer. J. Math. **97**, 1061-1083

[270] Gross L. (1992), "Logarithmic Sobolev Inequalities and Contractivity Properties of Semigroups", Springer Verlag, Lecture Notes in Math. **1563**

[271] Grothaus M. (1998), "New Results in Gaussian Analysis and their Applications in Mathematical Physics", Ph. D. thesis, University Bonn

[272] Guerra F., Rosen L., Simon B. (1975), "The $P(\phi)_2$ euclidean field theory as classical statistical mechanics II", Ann. of Math. (2) **110**, 111-189

[273] Guido D., Isola T., Scarlatti S. (1996), "Non-symmetric Dirichlet forms on semifinite von Neumann algebras", J. Funct. Anal. **135**, no. 1, 50-75

[274] Hagemann B.U. (1997), "Eine Klasse von Pseudodifferentialoperatoren die als Erzeuger regulärer Dirichlet-Formen auftreten", Diplomarbeit Bochum

[275] Hambly B.M., Kigami J., Kumagai T. (2002), "Multifractal formalisms for the local spectral and walk dimensions", Math. proc. Cambridge Philos. Soc. **132**, no. 3, 555-571

[276] Heyer H. (1979), "Einführung in die Theorie Markoffscher Prozesse", BI Mannheim

[277] Hida T., Hitsuda M. (1993), "Gaussian processes", Translations of Mathematical Monographs **120**, xvi+183pp

[278] Hida T., Kuo H.H., Potthoff J., Streit L. (1993), "White Noise, An Infinite Dimensional Calculus", Kluwer Academic Publishers

[279] Hida T., Potthoff J., Streit L. (1988), "Dirichlet forms and white noise analysis", Comm. Math. Phys. **116**, 235-245

[280] Hida T., Potthoff J., Streit L. (1989), "Energy forms and white noise analysis. New Methods and results in nonlinear field equations", Lect. Notes in Phys. **347**, 115-125

[281] Hino M. (2002), "On Dirichlet spaces over convex sets in infinite dimensions", Contemp. Math.

[282] Hino M. (2002), "A weak version of the extension theorem for infinite dimensional Dirichlet spaces", Stochastic Analysis and Related Topics, RIMS International Project 2002, 27-29

[283] Hino M., Ramirez J.A. (2002), "Smalltime Gaussian behaviour of symmetric diffusion semigroups", Kyoto Preprint

[284] Hirsch F. (1984), "Générateurs étendus et subordination au sens de Bochner", Seminar on potential theory, paris, no. 7, Lect. Notes in Math. **1061**, 134-156

[285] Hirsch F. (2002), "Intrinsic metrics and Lipschitz functions", preprint

[286] Hirsch F., Song S. (2001), "Markovian uniqueness on Bessel space", Forum Math. **13**, no. 2, 287-322

[287] Hoh W. (1993), "Feller semigroups generated by pseudo differential operators", de Gruyter, Dirichlet Forms and Stochastic Processes, 199-206

[288] Hoh W. (2002), "Perturbation of pseudodifferential operators with negative definite symbol", Appl. Math. Optim. **45**, no. 3, 269-281

[289] Hoh W., Jacob N. (1992), "Pseudo Differential Operators, Feller Semigroups and the Martingale Problem", Stochastic Monographs **7**,95-103

[290] Huang Z.Y., Yan J.A (2000), "Introduction to infinite dimensional stochastic analysis", Science Press, Kluwer, Beijing, Dordrecht

[291] Ikeda N. (1990), "Probabilistic methods in the study of asymptotics", Lecture Notes in Math. **1427**, 195-325

[292] Ismagilov R.S. (1996), "Representations of Infinite-Dimensional Groups", Amer. Math. Soc., Providence

[293] Iwata K. (1985), "Reversible measures of a $P(\phi)_1$-time evolution", Probabilistic methods in mathematical physics, 195-209

[294] Jacob N. (1996), "Pseudo-Differential Operators and Markov Processes", Akademie Verlag, Mathematical Research **94**

[295] Jacob N. (2001), "Pseudo-Differential Operators and Markov Processes. Vol. I. Fourier analysis and semigroups", Imperial College Press, London

[296] Jacob N. (2002), "Pseudo-Differential Operators and Markov Processes. Vol. II. Generators and their potential theory", Imperial College Press, London

[297] Jacob N., Moroz V. (2000), "On the semilinear Dirichlet problem for a class of nonlocal operators generating Dirichlet forms", Progr. Nonlinear Differential Equations Appl. **40**, 191-204

[298] Jacob N., Moroz V. (2002), "On the log-Laplace equation for nonlocal operators generating sub-Markovian semigroups", Appl. Math. Optim. **45**, no. 2, 237-250

[299] Jacob N., Schilling R.L. (1999), "Some Dirichlet spaces obtained by subordinate reflected diffusions", Revista Matemática Iberoamericana **15**,No. 1, 59-91

[300] Jacob N., Schilling R.L. (2000), "Fractional derivatives, non-symmetric and time-dependent Dirichlet forms and the drift form", Z. Anal. Anwendungen **19**, no. 3, 801-830

[301] Janson S. (1997), "Gaussian Hilbert spaces", Cambridge University Press

[302] Jona-Lasinio G., Mitter P.K. (1985), "On the stochastic quantization of field theory", Comm. Math. Phys. **101**, 409-436

[303] Jona-Lasinio G., Sénéor R. (1996), "Study of stochastic differential equations by constructive methods I.", J. of Statistical Physics **83**, 1109-1148

[304] Jonsson A. (2000), "Dirichlet forms and Brownian motion penetrating fractals", Pot. Anal. **13**, 69-80

[305] Jørgensen P.T. (1985), "Monotone convergence of operator semigroups and the dynamics of infinite particle systems", J. Approx. Theory **43**, no. 3, 205-230

[306] Jørgensen P.T. (1993), "Spectral theory for self-adjoint operator extensions associated with Clifford algebras", Contemp. Math. **143**, 131-150

[307] Jost J. (1997), "Generalized Dirichlet forms and harmonic maps", Calc. Var. Partial Differ. Equ. **5**, no. 1, 1-19

[308] Jost J. (1998), "Nonlinear Dirichlet forms", Stud. Adv. Math. **8**, 1-47

[309] Jost J., Kendall W., Mosco U., Röckner M., Sturm K.T. (1998), "New Directions in Dirichlet Forms", American Mathematical Society

[310] Kallianpur G., Perez-Abreu V. (1988), "Stochastic evolution equations driven by nuclear-space-valued martingales", Appl. Math. Optim. **17**, 237-272

[311] Kassmann M. (2001), "On Regularity for Beurling-Deny Type Dirichlet Forms", preprint SFB 256, University Bonn, No. 729

[312] Kato T. (1980), "Perturbation Theory for Linear Operators", Grundlehren der mathematischen Wissenschaften **132**, Springer-Verlag, Berlin, Heidelberg, New York,

[313] Kendall W.S. (1994), "Probability, Convexity, and Harmonic Maps II. Smoothness via Probabilistic Gradient Inequalities", Journal of Functional Analysis **126**, 228-257

[314] Kiefer S. (1995), "Beispiele für Nicht-Symmetrische Dirichlet Formen", Diplomarbeit, Ruhr Universität Bochum

[315] Kigami J. (2000), "Markov property of Kusuoka-Zhou's Dirichlet forms on self-similar sets", J. Math. Sci. Univ. Tokyo **7**, no. 1, 27-33

[316] Kim D.H., Kim J.H, Yun Y.S. (1998), "Non- symmetric Dirichlet forms for oblique reflecting diffusions", Math. Japon. **47**, no. 2, 323-331

[317] Kolokoltsov V.N. (2000), "Semiclassical Analysis for Diffusions and Stochastic Processes", LN Math. **1724**, Springer, Berlin

[318] Kolsrud T. (1988), "Gaussian random fields, infinite dimensional Ornstein-Uhlenbeck processes, and symmetric Markov processes", Acta Appl. Math. **12**, 237-263

[319] Kondratiev Y. (1985), "Dirichlet operators and the smoothness of solutions of infinite -dimensional elliptic equations", Soviet Math. Dokl. **31**, no. 3, 269-273

[320] Kondratiev Y. (1986), "Infinite-dimensional hyperbolic equations corresponding to Dirichlet operators", Soviet Math. Dokl. **33**, no. 3, 82-85

[321] Kondratiev Y., Barbuljak V. (1991), "Functional integrals and quantum lattice systems: III. Phase transitions", ibid., no. 10, 19-22

[322] Kondratiev Y., Globa S.A. (1990), "The construction of Gibbs states of quantum lattice systems", Selecta Math. Sovietica **9**, no. 3, 297-307

[323] Kondratiev Y., Konstantinov A.Y. (1988), "The scattering problem for special perturbations of harmonic systems", in "Limit problems for differential equations", Kiev, Institute of Mathematics. English translation (1994): Selecta Math. Sov. **13**, 217-224

[324] Kondratiev Y., Koshmanenko V.D. (1982), "Scattering problem for Dirichlet operators", Soviet Math. Dokl. **267**, no. 2, 285-288

[325] Kondratiev Y., Kuna T. (1999), "Harmonic analysis on configuration spaces I. General theory", Preprint SFB 256, University Bonn, submitted to Infinite Dimensional Analysis, Quantum Theory and Related Topics

[326] Kondratiev Y., Roelly S., Zessin H. (1996), "Stochastic dynamics for an infinite system of random closed strings: a Gibbsian point of view", Stoch. Processes Appl. **61**, no. 2, 223-248

[327] Kondratiev Y., Tsycalenko T.V. (1986), "Infinite dimensional hyperbolic equations for Dirichlet operators", Soviet Math. Dokl. **228**, no. 4, 814-817

[328] Kondratiev Y., Tsycalenko T.V. (1991), "Dirichlet operators and associated differential equations", Selecta Math. Sovietica **10**, no. 4, 345-397

[329] Krée M., Krée P. (1983), "Continuité de la divergence dans les espaces de Sobolev relatifs á l'espace de Wiener ", CRAS, Paris, **296**, no. 20, 833-836

[330] Krylov N., Röckner M., Zabczyk J. (1999),"Stochastic PDE's and Kolmogorov equations in infinite dimensions", Springer Verlag, Lecture Notes in Mathematics **1715**

[331] Kumagai T. (1993), "Regularity, closedness and spectral dimensions of the Dirichlet forms on P.C.F. self-similar sets", J. Math. Kyoto Univ. **33**, no. 3, 765-786

[332] Kumagai T. (2002), "Sub-Gaussian estimates of heat kernels on a class of fractal-like graphs and the stability under rough isometries", RIMS, Kyoto University

[333] Kuna T. (1999), "Studies in Configuration Space Analysis and Applications", Ph.D. thesis, University Bonn

[334] Kuo H. (1975), "Gaussian measures on Banach space", Springer Verlag, Lecture Notes in Math. **463**

[335] Kusuoka S. (1982), "Dirichlet forms and diffusion processes on Banach spaces", J. Fac. Sc. Univ. Tokyo **29**, 79-95

[336] Kusuoka S. (1989), "Dirichlet forms on fractals and products of random matrices", Publ. RIMS, Kyoto Univ. **25**, 659-680

[337] Kusuoka S. (2000), "Term structure and stochastic partial differential equations. Advances in mathematical economics", Adv. Math. Eco. **2**,67-85

[338] Kusuoka S., Zhou X.Y. (1992), "Dirichlet form on fractals: Poincaré constant and resistance", Prob. Theo. Rel. Fields **93**, 169-196

[339] Kuwae K. (1994), "Permanent sets of measures charging no exceptional sets and the Feynman-Kac formula", preprint 1994, to appear in Forum Math.

[340] Kuwae K. (2000), "On a strong maximum principle for Dirichlet forms", CMS Conf. Proc. **29**, 423-429

[341] Kuwae K. (2002), "Reflected Dirichlet forms and the uniqueness of Silverstein's extension", Potential Anal. **16**, no. 3, 221-247

[342] Kuwae K., Machigashira Y., Shioya T. (2001), "Sobolev spaces, Laplacian, and heat kernel on Alexandrov spaces", Math. Z. **238**, no. 2, 269-316

[343] Kuwae K., Shioya T. (2002), "Convergence of spectral structures: a functional analytic theory and its applications to spectral geometry", submitted to Communications in Analysis & Geometry

[344] Lang R. (1977), "Unendlich-dimensionale Wienerprozesse mit Wechselwirkung II. Die reversiblen Masse sind kanonische Gibbs-Masse", Z. Wahrscheinlichkeitstheorie und Verw. Gebiete **39**, no. 4, 277-299

[345] Laptev A., Weidl T. (1999), "Hardy inequalities for magnetic Dirichlet forms", Oper. Theory Adv. Appl. **108**, 299-305

[346] Le Jan Y. (1978), "Mesures associées à une forme de Dirichlet", Bull. Soc. Math. France **106**, 61-112

[347] Le Jan Y. (1983), "Quasicontinuous functions and Hunt processes", J. Math. Soc. Japan **35**, no. 1, 37-42

[348] Le Jan Y. (1995), "New examples of Dirichlet spaces", Dirichlet forms and stochastic processes, de Gruyter, Berlin, 253-256

[349] Léandre R. (2001), "Stochastic diffeology and homotopy", Progr. Probab. **50**, 51-57

[350] Léandre R. (2002), "An example of a brownian non linear string theory", preprint

[351] Léandre R. (2002), "Full stochastic Dirac-Ramond operator over the free loop space", preprint Institut Elie Cartan

[352] Léandre R. (2002), "Super Brownian motion on a loop group", preprint Institut Elie Cartan

[353] Léandre R., Roan S.S. (1995), "A stochastic approach to the Euler-Poincaré number of the loop space of developable orbifold", J. Geom. Phys. **16**, 71-98

[354] Leha G., Ritter G. (1993), "Lyapunov-type conditions for stationary distributions of diffusion processes on Hilbert spaces", Stoch. and Stoch. Rep. **48**, 195-225

[355] Lescot, Röckner M. (2002), "Pertubations of generalized Mehler semigroups and applications to stochastic heat equations with Lévy noise and singular drift", BiBos Preprint 02-03-076

[356] Lim H.Y., Park Y.M., Yoo H.J. (1997), " Dirichlet forms, Dirichlet operators and log-Sobolev inequalities for Gibbs measures of classical unbounded spin systems", J. Korean Math. Soc. **34**, no. 3, 731-770

[357] Lim H.Y., Park Y.M., Yoo H.J. (1998), " Dirichlet forms and Dirichlet operators for Gibbs measures of quantum unbouded spin systems: essential self-adjointness and log-Sobolev inequality", J. Statist. Phys. **90**, no. 3-4, 949-1002

[358] Lindsay J.M. (1993), "Non-commutative Dirichlet forms", de Gruyter, Dirichlet Forms and Stochastic Processes, 257-270

[359] Liskevich V., Röckner M. (1998), "Strong Uniqueness for Certain Infinite Dimensional Dirichlet Operators and Applications to Stochastic Quantization", Ann. Scuola Norm. Pisa **27**

[360] Löbus J.U. (1993), "Generalized diffusion operators", Akademie Verlag, Berlin

[361] Lyons T.J., Röckner M., Zhang T.S. (1996),"Martingale decomposition of Dirichlet processes on the Banach space $C_0[0,1]$", Stochastic Process. Appl. **64**, no. 1, 31-38

[362] Lyons T.J., Zhang T.S. (1994), "Decomposition of Dirichlet processes and its application", Ann. Prob. **22**, no. 1, 494-524

[363] Lyons T.J., Zheng W.A. (1988), "A crossing estimate for the canonical process on a Dirichlet space and a tightness result", Société Math. de France, Astérisque No. 157-158, 249-271

[364] Ma Z.M. (1990), "Some new results concerning Dirichlet forms, Feynman-Kac semigroups and Schrödinger equations", in Probability Theory in China (Contemp. Math.), Amer. Math. Soc., Providence, RI

[365] Ma Z.M. (1995), "Quasi-regular Dirichlet forms and Applications", Birkhäuser Basel, Proceedings of the International Congress of Mathematicans Vol **1,2**, 1006-1016

[366] Ma Z.M., Overbeck L., Röckner M. (1995), "Markov processes associated with Semi-Dirichlet forms", Osaka J. of Math. **32**, 97-119

[367] Ma Z.M., Röckner M. (1992), "An Introduction on the Theory of (Non-symmetric) Dirichlet Forms", Springer-Verlag, Berlin

[368] Ma Z.M., Röckner M. (1995), "Markov Processes Associated with Positivity Preserving Coercive Forms", Canadian J. Math. **47**, 817-840

[369] Ma Z.M., Röckner M. (2000), "Construction of diffusions on configuration spaces", Osaka J. Math. **37**, No. 2, 273-314

[370] Ma Z.M., Röckner M, Sun W. (2000), "Approximation of Hunt processes by multivariate Poisson processes", Acta Appl. Math. **63**, 233-243

[371] Ma Z.M., Röckner M., Zhang T.S. (1998), "Approximations of arbitrary Dirichlet processes by Markov chains", Ann. Inst. Henri Poincaré **34**, No. 1, 1-22

[372] Malliavin P. (1997), "Stochastic Analysis", Springer Verlag

[373] Malliavin M.P., Malliavin P. (1990), "Integration on loop groups. I. Quasi invariant measures", J. Funct. Anal. **93**, 207-237

[374] Malý J., Mosco U. (1999), "Remarks on measure-valued Lagrangians on homogeneous spaces", Ricerche Mat. **48**, 217-231

[375] Manavi A., Voigt J. (2002), "Maximal operators associated with Dirichlet forms perturbed by measures", Potential Anal. **16**, no. 4, 341-346

[376] Marchi S. (1997), "Influence of the nonlocal terms on the regularity of equations involving Dirichlet forms", Istit. Lombardo Accad. Sci. Lett. Rend. A **131**, no. 1-2, 189-199

[377] Mataloni S. (1999), "Representation formulas for non-symmetric Dirichlet forms", Z. Anal. Anwendungen **18**, no. 4, 1039-1064

[378] Mataloni S. (1999), "On a type of convergence for non-symmetric Dirichlet forms", Adv. Math. Sci. Appl. **9**, no.2, 749-773

[379] Mataloni S. (2001), "Quasi-linear relaxed Dirichlet problems involving a Dirichlet form", Rend. Accad. Naz. Sci. XL Mem. Mat. Appl.(5) **25**, 67-96

[380] Mataloni S., Tchou N. (2001), "Limits of relaxed Dirichlet problems involving a nonsymmetric Dirichlet form", Ann. Mat. Pura Appl. (4) **179**, 65-93

[381] Matzeu M. (2000), "Mountain pass and linking type solutions for semilinear Dirichlet forms", Progr. Nonlinear Differential Equations Appl. **40**, 217-231

[382] McGilivray I. (1997), "A recurrence condition for some subordinated strongly local Dirichlet forms", Forum Math. **9**, no. 2, 229-246

[383] Menendez S.C. (1975), "Processus de Markov associé à une forme de Dirichlet non symétrique", Z. Wahrscheinlichkeitstheorie verw. Gebiete **33**, 139-154

[384] Mikulievikius R., Rozovskii B.L. (1999), "Martingale problems for stochastic PDEs", Stochastic Partial Differential Equations. Six Perspectives, AMS Rhode Island, 251-333

[385] Mitter S. (1989), "Markov random fields, stochastic quantization and image analysis", Math. Appl. **56**, 101-109

[386] Morato L. (2002), in preparation

[387] Mosco U. (1994), "Composite Media and Asymptotic Dirichlet Forms", Journal of Functional Analysis **123**, no. 2, 368-421

[388] Mosco U. (1998), "Dirichlet forms and self-similarity. New directions in Dirichlet forms", AMS/IP Stud. Adv. Math. **8**, 117-155

[389] Mosco U. (2000), "Self-similar measures in quasi-metric spaces", Birkhäuser, Basel, Progr. Nonlinear Differential Equations Appl. **40**, 233-248

[390] Mück S. (1993), "Large deviations w.r.t. quasi-every starting point for symmetric right processes on general state spaces", Prob. Theor. Rela. Fields **99**, 527-548

[391] Nagai H. (1986), "Stochatic control of symmetric Markov processes and nonlinear variational inequalities", Stochastics **19**, no. 1-2, 83-110

[392] Nagai H. (1992), "Ergodic control problems on the whole Euclidean space and convergence of symmetric diffusions", Forum Math. **4**, no. 2, 159-173

[393] Nelson E., "Analytic vectors", Ann. Math. **70**, 572-615 (1959)

[394] Nelson E., "Dynamical theories of Brownian motion", Princeton Univ. Press

[395] Nelson E. (1973), "The free Markov field", J. Funct. Anal. **12**, 217-227

[396] Nualart D. (1996), "The Malliavin Calculus and Related Topics", Springer Berlin, Probability and its Applications

[397] Ogura Y., Tomisaki M., Tsuchiya M. (2001), "Esitence of a strong solution for an integro-differential equation and superposition of diffusion processes", Birkhäuser, Boston, Stochastics in finite and infinite dimensions, 341-359

[398] Ogura Y., Tomisaki M., Tsuchiya M. (2002), "Convergence of local Dirichlet forms to a non-local type one", Ann. I. H. Poincaré **30**, no. 4, 507-556

[399] Oksendal B. (1988), "Dirichlet forms, quasiregular functions and Brownian motion", Invent. math. **91**, 273-297

[400] Ôkura H. (2002), "Recurrence and transience criteria for subordinated symmetric Markov processes", Forum Math. **14**, 121-146

[401] Oliveira M.J. (2002), "Configuration Space Analysis and Poissonian White Noise Analysis", Ph. D. thesis, Universidade de Lisboa

[402] Olkiewicz R., Zegarlinski B. (1999), "Hypercontractivity in Noncommutative $\mathbb{L}_p$ Spaces", Journal of Functional Analysis **161**, 246-285

[403] Osada H. (1996), "Dirichlet form approach to infinite-dimensional Wiener process with singular interactions", Comm. Math. Phys. **176**, 117-131

[404] Oshima Y. (1992), "On a construction of Markov Processes associated with time dependent Dirichlet spaces", Forum Mathematicum **4**, 395-415

[405] Oshima Y. (1992), "Some properties of Markov processes associated with time dependent Dirichlet forms", Osaka J. Math. **29**, 103-127

[406] Oshima Y. (1993), "Time dependent Dirichlet forms and its application to a transformation of space-time Markov processes", de Gruyter, Dirichlet Forms and Stochastic Processes, 305-320

[407] Ouhabaz E.M. (1992), "Propriétés d'ordre et de contractivité des semigroupes avec applications aux opérateures elliptiques", Ph.D. thesis, Besançon

[408] Paclet Ph. (1978), "Espaces de Dirichlet et capacités fonctionelles sur les triplets de Hilbert-Schmidt", Séminare Equations aux Dérivées Partielles en Dimension Infinite, No. 5

[409] Pardoux E., Veretennikov A.Y. (2001), "On the Poisson equation and diffusion approximation I.", Ann. Probab. **29**, no. 3, 1061-1085

[410] Park Y.M. (2000), "Construction of Dirichlet forms on standard forms of von Neumann algebras", Infin. Dim. Anal. Quantum Probab. Theory Related Top. **3**, no. 1, 1-14

[411] Park Y.M. (2002), "Dirichlet forms and symmetric Markovian semigroups on von Neumann algebras", Proceedings of the Third Asian Mathematical Conference, 2000 (Diliman), 427-443

[412] Park Y.M., Yoo H.J. (1997), "Dirichlet operators on loop spaces: Essential self-adjointness and log-Sobolev inequality", J. Math. Phys. **38**, 3321-3346

[413] Pazy A. (1983), "Semigroups of linear operators and applications to partial diffential equations", Applied Mathematical Sciences, 44. New York, Springer

[414] Peirone R. (2000), "Convergence and uniqueness problems for Dirichlet forms on fractals", Boll. Unione Mat. Ital. Sez. B Artic. Ric. Mat. (8) **3**, no. 2, 431-460

[415] Pontier M. (2000), "A Dirichlet form on a Poisson space", Potential Anal. **13**, no. 4, 329-344

[416] Popovici A. (2001), "Forward interest rates: a Hilbert space SDE approach", preprint BiBoS,

[417] Posilicano A. (1996), "Convergence of Distorted Brownian Motions and Singular Hamiltonians", Potential Anal. **5**,no. 3, 241-271

[418] Potthoff J., Röckner M. (1990), "On the contraction property of energy forms for infinite-dimensional space", J. of Funct. Anal. **92**,No. 1, 155-165

[419] Priouret P., Yor M. (1975), "Processus de diffusion à valeurs dans $\mathbb{R}$ et mesures quasi-invariantes sur $C(\mathbb{R}, \mathbb{R})$", Astérique 22-23, 247-290

[420] Privault N. (1999), "Equivalence of gradients on configuration spaces", Random Oper. Stoch. Eq. **7**, 241-262

[421] Pugachëv O.V. (2001), "On the closability of classical Dirichlet forms in the plane", Dokl. Akad. Nauk **380**, no. 3, 315-318

[422] Ramirez J.A. (2001), "Short time symptotics in Dirichlet spaces", Comm. Appl. Math. **54**, 259-293

[423] Reed M., Simon B. (1980), "Methods of Modern Mathematical Physics. I. Functional Analysis", Academic Press, London

[424] Reed M., Simon B. (1975), "Methods of Modern Mathematical Physics. II. Fourier Analysis, Self-Adjointness", Academic Press, London

[425] Reed M., Simon B. (1979), "Methods of modern mathematical physics. III. Scattering Theory", Academic Press, New York

[426] Reed M., Simon B. (1978), "Methods of modern mathematical physics. IV. Analysis of operators", Academic Press, New York

[427] Rellich F. (1969), "Perturbation Theory of Eigenvalue Problem", Gordon and Breach, New York

[428] Röckle H. (1997), "Banach Space-Valued Ornstein-Uhlenbeck Processes with General Drift Coefficients", Acta Applicandae Mathematicae **47**, 323-349

[429] Röckner M. (1985), "A Dirichlet problem for distributions and specifications for random fields", Memoirs of the American Mathematical Society **324**

[430] Röckner M. (1988), "Traces of harmonic functions and a new path space for the free quantum field", J. Funct. Anal. **79**, 211-249

[431] Röckner M. (1992), "Dirichlet Forms on infinite dimensional state space and applications", pp. 131-186 in H.Korezlioglu, A.S. Üstümel, eds., Birkhäuser, Boston

[432] Röckner M. (1993), "General theory of Dirichlet forms and applications", Lecture Notes in Math. **1563**, 129-193

[433] Röckner M. (1995), "Dirichlet forms on infinite-dimensional 'manifold-like' state spaces: a survey of recent results and some prospects for the future", Springer, New York, Lecture Notes in Statis. **128**, 287-306

[434] Röckner M. (1998), "Stochastic analysis on configuration spaces: Basic ideas and recent results", New Directions in Dirichlet Forms (eds. J. Jost et al.), Studies in Advanced Mathematics, Vol. 8, American Math. Soc., 157-232

[435] Röckner M. (1998), "L^p-Analysis of Finite and Infinite Dimensional Diffusion Operators", Lect. Notes in Math. **1715**, 65-116

[436] Röckner M., Schied A. (1998), "Rademacher's Theorem on Configuration Spaces and Applications", J. Funct. Anal. **169**, 325-356

[437] Röckner M., Schmuland B. (1995), "Quasi-regular Dirichlet forms: examples and counter examples", Canad. J. math. **47**, 165-200

[438] Röckner M., Schmuland B. (1998), "A support property for infinite-dimensional interacting diffusion processes", C.R. Acad. Sci. Paris Sér. I Math. **326**, no. 3, 359-364

[439] Röckner M., Zhang T.S. (1991), "Decomposition of Dirichlet processes on Hilbert space", London Math. Soc. Lecture Note Ser. **167**, 321-332

[440] Röckner M., Zhang T.S. (1992), "Uniqueness of generalized Schrödinger operators and applications", J. Funct. Anal. **105**, no. 1, 187-231

[441] Röckner M., Zhang T.S. (1994), "Uniqueness of generalized Schrödinger operators", J. Funct. Anal. **119**, no. 2, 455-467

[442] Röckner M., Zhang T.S. (1996), "Finite-dimensional approximation of diffusion processes of infinite-dimensional spaces", Stochastics Stochastics Rep. **57**, no. 1-2, 37-55

[443] Röckner M., Zhang T.S. (1997), "Convergence of operators semigroups generated by elliptic operators", Osaka J. Math. **34**, no. 4, 923-932

[444] Röckner M., Zhang T.S. (1998), "On the strong Feller property of the semigroups generated by non-divergence operators with L^p-drift", Progr. Probab. **42**, 401-408, Birkhäuser, Boston

[445] Röckner M., Zhang T.S. (1999), "Probabilistic representations and hyperbound estimates for semigroups", Infin. Dimens. Anal. Quantum Probab. Relat. Top. **2**, no. 3, 337-358

[446] Roelly S., Zessin H. (1993), "Une caractérisation des mesures de Gibbs sur $C((0,1),\mathbb{Z}^d)$ par le calcul des variations stochastiques", Ann. Inst. H. Poincaré Prob. Stat. **29**, 327-338

[447] Roelly S., Zessin H. (1995), "Une caractérization des champs gibbsiens sur un espace de trajectories", C.R. Acad. Sci. Paris Ser I. Math. **321**, 1377-1382

[448] Roman L.J., Zhang X., Weian Z. (2002), "Rate of convergence in homogenization of parabolic PDEs", preprint

[449] Roth J.P. (1976), "Les Operateurs Elliptiques comme Générateurs Infinitésimaux de Semi-Groupes de Feller", Lecture Notes in Math **681**, 234-251

[450] Royer G., Yor M. (1976), "Représentation intégrale de certaines measures quasi-invariantes sur $C(R)$ mesures extrémales et propriété de Markov", Ann. Inst. Fourier (Grenoble) **26**, no. 2, ix, 7-24

[451] Rüdiger B. (2002), "Processes with jumps properly associated to non local quasi-regular (non symmetric) Dirichlet forms obtained by subordination", in preparation

[452] Rüdiger B. (2002), "Stochastic integration w.r.t. compensated Poisson random measures on separable Banach spaces", in preparation

[453] Rüdiger B., Wu J.L. (2000), "Construction by subordination of processes with jumps on infinite-dimensional state spaces and corresponding non local Dirichlet forms", CMS Conf. Proc. **29**, 559-571

[454] Saloff-Coste (2002), "Aspects of Sobolev-Type Inequalities", Cambridge University Press

[455] Schachermayer W. (2002), , St. Flour Lectures 2000, this volume

[456] Scheutzow M., v. Weizsäcker H. (1998), "Which moments of a logarithmic derivative imply quasiinvariance", Doc. Math. **3**, 261-272

[457] Schmuland B. (1990), "An alternative compactification for classical Dirichlet forms on topological vector spaces", Stochastics **33**, 75-90

[458] Schmuland B. (1993), "Non-symmetric Ornstein-Uhlenbeck processes in Banach space via Dirichlet forms", Canad. J. Statis. **45**,no. 6, 1324-1338

[459] Schmuland B. (1994), "A Dirichlet Form Primer", American Mathematical Society, CRM Proceedings and Lecture Notes **5**, 187-197

[460] Schmuland B. (1998), "Some Regularity Results on Infinite Dimensional Diffusions via Dirichlet Forms", Stochastic Analysis and Applications **6**(3), 327-348

[461] Schmuland B. (1999), "Dirichlet forms: some infinite dimensional examples", Canad. J. Statist. **27**, No. 4, 683-700

[462] Schwartz L. (1973), "Random measures on arbitrary topological spaces and cylindrical measures", Oxford University Press

[463] Sharpe M. (1988), "General Theory of Markov Processes", Academic Press, New York

[464] Shigekawa I. (2002), "Square root of a Schrödinger operator and its L^p norms", Stochastic Analysis and Related Topics, RIMS International Project 2002, 25-26

[465] Shigekawa I., Taniguchi S. (1992), "Dirichlet forms on separable metric spaces", Ed. A.N. Shizyaev et al., World Scient. Singapore, Probability Theory and Mathematical Statistics, 324-353

[466] Simon B. (1974), "The $P(\varphi)_2$ Euclidean quantum field theory", Princeton Univ. Press

[467] Simon B. (1979), "Functional Integration", Academic Press, New York

[468] Silverstein M.L. (1974), "Symmetric Markov Processes", Springer Verlag, Lect. Notes in Math. **426**

[469] Silverstein M.L. (1976), "Boundary theory for symmetric Markov processes", Springer Verlag, Lecture Notes in Math. **516**

[470] Smolyanov O.G., von Weizsäcker H. (1999), "Smooth probability measures and associated differential operators", Infin. Dimens. Anal. Quantum Probab. Relat. Top. **2**, no. 1, 51-78

[471] Song S. (1994), "A study on markovian maximality, change of probability and regularity", Potential Anal. **3**, 391-422

[472] Spönemann U. (1997), "An existence and a structural result for Dirichlet forms", Ph. D. thesis, University Bielefeld

[473] Srimurthy V.K. (2000), "On the equivalence of measures on loop space", Prob. Th. Rel. Fields **18**, 522-546

[474] Stannat W. (1999), "The Theory of Generalized Dirichlet Forms and Its Applications in Analysis and Stochastics", Memoirs of the American Mathematical Society **678**

[475] Stannat W. (1999), "(Nonsymmetric) Dirichlet Operators on L^1: Existence, Uniqueness and Associated Markov Processes", Ann. Scuola Norm. Sup. Pisa **28**, 99-140

[476] Stannat W. (2000), "Long-Time Behaviour and Regularity Properties of Transition Semigroups of Fleming-Viot Processes", Probab. Theory Related Fields **122**,no. 3, 431-469

[477] Stannat W. (2000), "On the validity of the log- Sobolev inequality for symmetric Fleming-Viot operators", Ann. Prob. **28**, 667-684

[478] Streit L. (1981), "Energy Forms: Schrödinger Theory, Processes", Physics Reports **77**, no. 3, 363-375
[479] Streit L. (1985), "Quantum theory and stochastic processes - some contact points", Lect. Notes Math. **1203**, 197-213
[480] Sturm K.T. (1995), "On the geometry defined by Dirichlet forms", Birkhäuser Verlag, Seminar on stochastic analysis, random fields and applications (Ascona 1993)
[481] Sturm K.T. (1997), "Monotone approximation of energy functionals for mappings into metric spaces", J. Reine Angew. Math. **468**, 129-151
[482] Sturm K.T. (1998), "Diffusion processes and heat kernels on metric spaces", Ann. Probab. **26**, no. 1, 1-55
[483] Sturm K.T. (1998), "The geometric aspect of Dirichlet forms", AMS/IP Stud. Adv. Math. **8**, 233-277
[484] Sturm K.T. (2002), "Nonlinear martingale theory for processes with values in metric spaces of nonpositive curvature", Annals of Probab.
[485] Sturm K.T. (2002), "Nonlinear Markov operators, discrete heat flow, and harmonic maps between singular spaces", Potential Anal. **16**, no.4, 305-340
[486] Sun W. (1999), "Mosco convergence of quasi-regular Dirichlet forms", Acta Math. Appl. Sinica **15**, no. 3, 225-232
[487] Takeda M. (1992), "The maximum Markovian self-adjoint extensions of generalized Schrödinger operators", J. Math. Soc. Jap. **44**, 113-130
[488] Takeda M. (1994), "Transformations of local Dirichlet forms by supermartingale multiplicative functionals", preprint 94, to appear in Proc. ICDFSP (Z.M. Ma et al., eds.) Walter de Gruyter, Berlin and Hawthorne, NY
[489] Takeda M. (1999), "Topics on Dirichlet forms and symmetric Markov processes", Sugaku Expositions **12**, no. 2, 201-222
[490] Takeda M. (2002), "Conditional gaugeability of generalized Schrödinger operators", Stochastic Analysis and Related Topics, RIMS International Project 2002, 15-16
[491] Tanemura H. (1997), "Uniqueness of Dirichlet forms associated with systems of infinitely many Brownian balls in $\mathbb{R}^d$", Probab. Theory Relat. Fields **109**, 275-299
[492] Trutnau G. (2000), "Stochastic Calculus of Generalized Dirichlet Forms and Applications", Osaka J. Math. **37**, no. 2, 315-343
[493] Uemura T. (1995), "On Weak Convergence of Diffusion Processes Generated by Energy Forms", Osaka J. Math. **32**, 861-868
[494] Üstünel A.S. (1995), "An Introduction to Analysis on Wiener Space", LN Math. **1610**, Springer Berlin
[495] Weidmann J. (1976), "Lineare Operatoren in Hilberträumen", Teubner, Stuttgart
[496] Wu L. (1997), "Uniqueness of Schrödinger Operators Restricted in a Domain", Journal of Functional Analysis **153**, No. 2, 276-319
[497] Wu L. (1999), "Uniqueness of Nelson's diffusions", Probab. Theory Relat. Fields **114**, 549-585
[498] Wu L. (1999), "Uniqueness of Nelson's diffusions II: Infinite Dimensional Setting and Applications", Potential Analysis **13**, 269-301
[499] Wu L. (2001), "L^p-uniqueness of Schrödinger operators and capacitary positive improving property", J. Funct. Anal **182**, 51-80

[500] Wu L. (2001), "Martingale and Markov uniqueness of infinite dimensional Nelson diffusions", Birkhäuser, Boston, Progr. Probab. **50**, 139-150

[501] Wu L., in preparation

[502] Yalovenko I. (1998), "Modellierung des Finanzmarktes und unendlich dimensionale stochastische Prozesse", Diplomarbeit, Bochum

[503] Ying J. (1997), "Dirichlet forms perturbated by additive functionals of extended Kato class", Osaka J. Math. **34**, no. 4, 933-952

[504] Yor M. (1976), "Une remarque sur les formes de Dirichlet et les semimartingales", Lect. Notes. Math. **563**, 283-292

[505] Yoshida M. (1996), "Construction of infinite dimensional interacting diffusion processes through Dirichlet forms", Prob. Theory related Fields **106**, 265-297

[506] K.Yosida (1980), "Functional analysis", Springer, Berlin (VI ed.)

[507] Zambotti L.(2000), "A reflected heat equation as symmetric dynamics with respect to 3-*d* Bessel Bridge", J. Funct. Anal. **180**, no. 1, 195-209

[508] Zhang T.S. (2000), "On the small time asymptotics of diffusion processes on Hilbert spaces", Ann. Prob. **28**, 537-557

[509] Zheng W. (1995), "Conditional propagation of chaos and a class of quasilinear PDE", Ann. Probab. **23**, no. 3, 1389-1413

Index